AF316758

Springer Texts in Statistics

Springer Texts in Statistics

(continued after index)

Jayanta K. Ghosh
Mohan Delampady
Tapas Samanta

An Introduction to Bayesian Analysis
Theory and Methods

With 13 Illustrations

 Springer

Jayanta K. Ghosh
Department of Statistics
Purdue University
150 N. University Street
West Lafayette,
IN 47907-2067
USA
ghosh@stat.purdue.edu
and
Indian Statistical Institute
203 B.T. Road
Kolkata 700108, India
jayanta@isical.ac.in

Mohan Delampady
Indian Statistical Institute,
8th Mile, Mysore Road,
R.V. College Post,
Bangalore 560059, India
mohan@isibang.ac.in

Tapas Samanta
Indian Statistical Institute
203 B.T. Road
Kolkata 700108, India
tapas@isical.ac.in

Library of Congress Control Number: 2006922766

ISBN-10: 0-387-40084-2   e-ISBN: 0-387-35433-6
ISBN-13: 978-0387-40084-6

Printed on acid-free paper.

Printed in the United States of America.   (MVY)

9 8 7 6 5 4 3 2 1

springer.com

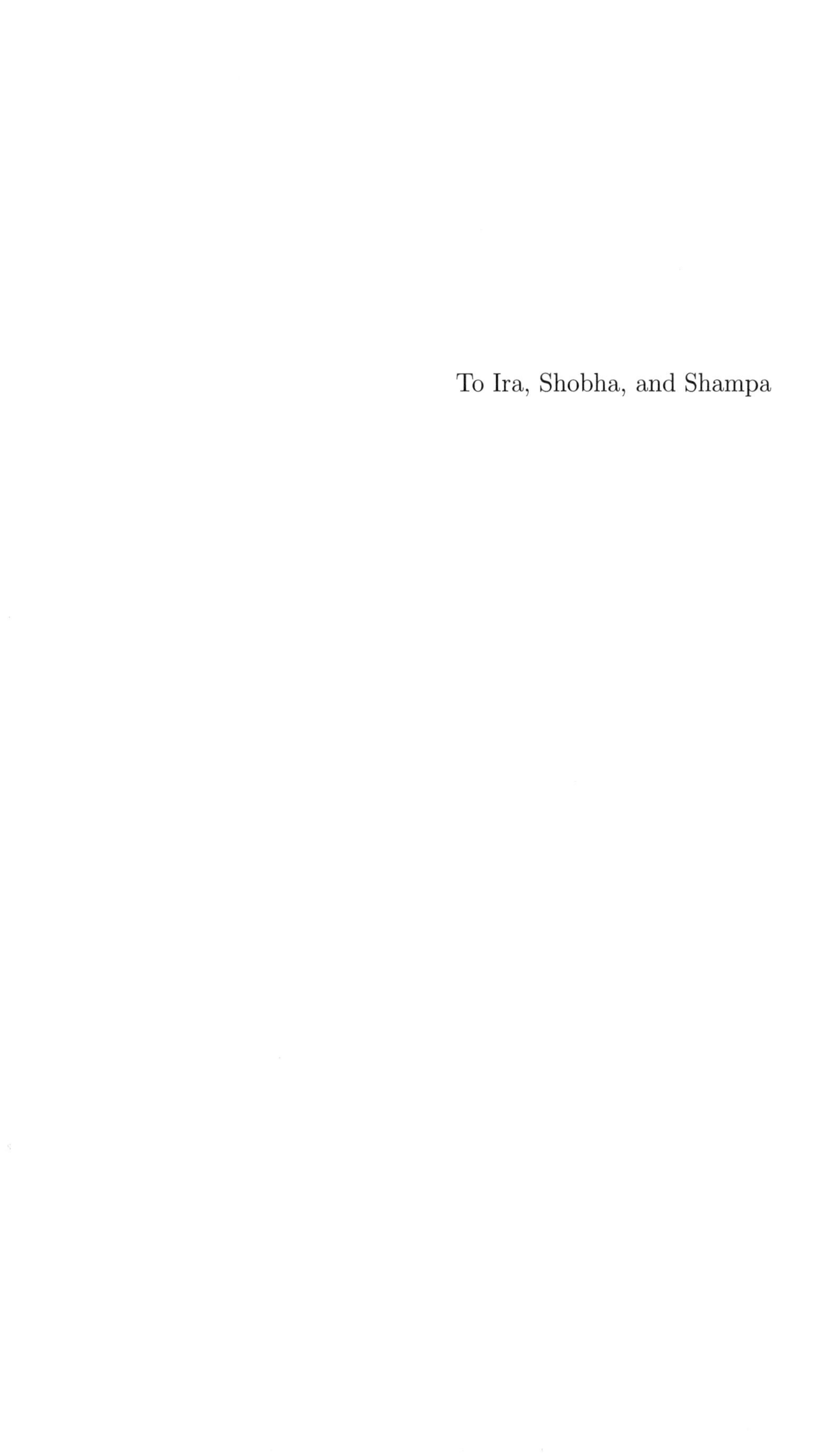

To Ira, Shobha, and Shampa

Preface

Though there are many recent additions to graduate-level introductory books on Bayesian analysis, none has quite our blend of theory, methods, and applications. We believe a beginning graduate student taking a Bayesian course or just trying to find out what it means to be a Bayesian ought to have some familiarity with all three aspects. More specialization can come later.

Each of us has taught a course like this at Indian Statistical Institute or Purdue. In fact, at least partly, the book grew out of those courses. We would also like to refer to the review (Ghosh and Samanta (2002b)) that first made us think of writing a book. The book contains somewhat more material than can be covered in a single semester. We have done this intentionally, so that an instructor has some choice as to what to cover as well as which of the three aspects to emphasize. Such a choice is essential for the instructor. The topics include several results or methods that have not appeared in a graduate text before. In fact, the book can be used also as a second course in Bayesian analysis if the instructor supplies more details.

Chapter 1 provides a quick review of classical statistical inference. Some knowledge of this is assumed when we compare different paradigms. Following this, an introduction to Bayesian inference is given in Chapter 2 emphasizing the need for the Bayesian approach to statistics. Objective priors and objective Bayesian analysis are also introduced here. We use the terms *objective* and *nonsubjective* interchangeably. After briefly reviewing an axiomatic development of utility and prior, a detailed discussion on Bayesian robustness is provided in Chapter 3. Chapter 4 is mainly on convergence of posterior quantities and large sample approximations. In Chapter 5, we discuss Bayesian inference for problems with low-dimensional parameters, specifically objective priors and objective Bayesian analysis for such problems. This covers a whole range of possibilities including uniform priors, Jeffreys' prior, other invariant objective priors, and reference priors. After this, in Chapter 6 we discuss some aspects of testing and model selection, treating these two problems as equivalent. This mostly involves Bayes factors and bounds on these computed over large classes of priors. Comparison with classical P-value is also made whenever appropriate. Bayesian P-value and nonsubjective Bayes factors such as the intrinsic and fractional Bayes factors are also introduced.

Chapter 7 is on Bayesian computations. Analytic approximation and the E-M algorithm are covered here, but most of the emphasis is on Markov chain based Monte Carlo methods including the M-H algorithm and Gibbs sampler, which are currently the most popular techniques. Follwing this, in Chapter 8 we cover the Bayesian approach to some standard problems in statistics. The next chapter covers more complex problems, namely, hierarchical Bayesian (HB) point and interval estimation in high-dimensional problems and parametric empirical Bayes (PEB) methods. Superiority of HB and PEB methods to classical methods and advantages of HB methods over PEB methods are discussed in detail. Akaike information criterion (AIC), Bayes information criterion (BIC), and other generalized Bayesian model selection criteria, high-dimensional testing problems, microarrays, and multiple comparisons are also covered here. The last chapter consists of three major methodological applications along with the required methodology.

We have marked those sections that are either very technical or are very specialized. These may be omitted at first reading, and also they need not be part of a standard one-semester course.

Several problems have been provided at the end of each chapter. More problems and other material will be placed at `http://www.isical.ac.in/~tapas/book`

Many people have helped – our mentors, both friends and critics, from whom we have learnt, our family and students at ISI and Purdue, and the anonymous referees of the book. Special mention must be made of Arijit Chakrabarti for Sections 9.7 and 9.8, Sudipto Banerjee for Section 10.1, Partha P. Majumder for Appendix D, and Kajal Dihidar and Avranil Sarkar for help in several computations. We alone are responsible for our philosophical views, however tentatively held, as well as presentation.

Thanks to John Kimmel, whose encouragement and support, as well as advice, were invaluable.

Indian Statistical Institute and Purdue University   *Jayanta K. Ghosh*
Indian Statistical Institute   *Mohan Delampady*
Indian Statistical Institute   *Tapas Samanta*
February 2006

Contents

1

Statistical Preliminaries

We review briefly some of the background that is common to both classical statistics and Bayesian analysis. More details are available in Casella and Berger (1990), Lehmann and Casella (1998), and Bickel and Doksum (2001). The reader interested in Bayesian analysis can go directly to Chapter 2 after reading Section 1.1.

1.1 Common Models

A statistician, who has been given some data for analysis, begins by providing a probabilistic model of the way his data have been generated. Usually the data can be treated as generated by random sampling or some other random mechanism. Once a model is chosen, the data are treated as a random vector $X = (X_1, X_2, \ldots, X_n)$. The probability distribution of X is specified by $f(x|\theta)$ which stands for a joint density (or a probability mass function), and θ is an unknown constant or a vector of unknown constants called a parameter. The parameter θ may be the unknown mean and variance of a population from which X is a random sample, e.g., the mean life of an electric bulb or the probability of doing something, vide Examples 1.1, 1.2, and 1.3 below. Often the data X are collected to learn about θ, i.e., the modeling precedes collection of data. The set of possible values of θ, called the parameter space, is denoted by Θ, which is usually a p-dimensional Euclidean space $\mathcal{R}^p$ or some subset of it, p being a positive integer. Our usual notation for data vector and parameter vector are X and θ, respectively, but we may use X and θ if there is no fear of confusion.

Example 1.1. (normal distribution). $X_1, X_2, \ldots, X_n$ are heights of n adults (all males or all females) selected at random from some population. A common model is that they are independently, normally distributed with mean μ and variance σ^2, where $-\infty < \mu < \infty$ and $\sigma^2 > 0$, i.e., with $\theta = (\mu, \sigma^2)$,

$$f(\boldsymbol{x}|\boldsymbol{\theta}) = \prod_{i=1}^{n} f(x_i|\boldsymbol{\theta}) = \prod_{i=1}^{n} \left\{ \frac{1}{\sqrt{2\pi}\sigma} \exp\left(-\frac{(x_i - \mu)^2}{2\sigma^2} \right) \right\}.$$

We write this as X_i's are i.i.d. (independently and identically distributed) $N(\mu, \sigma^2)$.

If one samples both genders the model would be much more complicated – X_i's would be i.i.d. but the distribution of each X_i would be a mixture of two normals $N(\mu_F, \sigma_F^2)$ and $N(\mu_M, \sigma_M^2)$ where F and M refer to females and males.

Example 1.2. (exponential distribution). Suppose a factory is producing some electric bulbs or electronic components, say, switches. If the data are a random sample of lifetimes of one kind of items being produced, we may model them as i.i.d. with common exponential density

$$f(x_i|\theta) = \frac{1}{\theta} e^{-x_i/\theta}, \ x_i > 0, \ \theta > 0.$$

Example 1.3. (Bernoulli, binomial distribution). Suppose we have n students in a class with

$$X_i = \begin{cases} 1 & \text{if } i\text{th student has passed a test;} \\ 0 & \text{otherwise.} \end{cases}$$

We model X_i's as i.i.d. with the Bernoulli distribution:

$$f(x_i|\theta) = \begin{cases} \theta & \text{if } x_i = 1; \\ 1 - \theta & \text{if } x_i = 0, \end{cases}$$

which may be written more compactly as $\theta^{x_i}(1 - \theta)^{1-x_i}$. The parameter θ is the probability of passing. The joint probability function of $X_1, X_2, \ldots, X_n$ is

$$f(\boldsymbol{x}|\theta) = \prod_{i=1}^{n} f(x_i|\theta) = \prod_{i=1}^{n} \left\{ \theta^{x_i}(1 - \theta)^{1-x_i} \right\}, \ \theta \in (0, 1).$$

If $Y = \sum_{1}^{n} X_i$, the number of students who pass, then $P(Y = y) = \binom{n}{y}\theta^y(1 - \theta)^{n-y}$, which is a binomial distribution, denoted $B(n, \theta)$.

Example 1.4. (binomial distribution with unknown n's and unknown p). Suppose $Y_1, Y_2, \ldots, Y_k$ are the number of reported burglaries in a place in k years. One may model Y_i's as independent $B(n_i, p)$, where n_i is the number of actual burglaries (some reported, some not) in ith year and p is the probability that a burglary is reported. Here $\boldsymbol{\theta}$ is $(n_1, \ldots, n_k, p)$.

Example 1.5. (Poisson distribution). Let $X_1, X_2, \ldots, X_n$ be the number of accidents on a given street in n years. X_i's are modeled as i.i.d $\mathcal{P}(\lambda)$, i.e., Poisson with mean λ,

$$P(X_i = x_i) = f(x_i|\lambda) = \exp(-\lambda)\frac{\lambda^{x_i}}{x_i!}, \ x_i = 0, 1, 2, \ldots, \qquad \lambda > 0.$$

Example 1.6. (relation between binomial and Poisson). It is known $B(n,p)$ is well approximated by $\mathcal{P}(\lambda)$ if n is large, p is small but $np = \lambda$ is nearly constant, or, more precisely, $n \to \infty, p \to 0$ in such a way that $np \to \lambda$. This is used in modeling distribution of defective items among some particular products, e.g., bulbs or switches or clothes. Suppose a lot size n is large. Then the number of defective items, say X, is assumed to have a Poisson distribution.

Closely related to the binomial are three other distributions, namely, geometric, negative binomial, which includes the geometric distribution, and the multinomial. All three, specially the last, are important.

Example 1.7. (geometric). Consider an experiment or trial with two possible outcomes – success with probability p and failure with probability $1 - p$. For example, one may be trying to hit a bull's eye with a dart. Let X be the number of failures in a sequence of independent trials until the first success is observed. Then

$$P\{X = x\} = (1 - p)^x p, \ x = 0, 1, \ldots$$

This is a discrete analogue of the exponential distribution.

Example 1.8. (Negative binomial). In the same setup as above, let k be given and X be the number of failures until k successes are observed. Then

$$P\{X = x\} = \binom{x + k - 1}{k - 1} p^k (1 - p)^x, \ x = 0, 1, \ldots$$

This is the negative binomial distribution. The geometric distribution is a special case.

Example 1.9. (multinomial). Suppose an urn has N balls of k colors, the number of balls of jth color is $N_j = Np_j$ where $0 \le p_j \le 1, \sum_1^k p_j = 1$. We take a random sample of n balls, one by one and with replacement of the drawn ball before the next draw. Let $X_i = j$ if the ith ball drawn is of jth color and let $n_j =$ frequency of balls of the jth color in the sample. Then the joint probability function of $X_1, X_2, \ldots, X_n$ is

$$f(\boldsymbol{x}|\boldsymbol{p}) = \prod_{j=1}^{k} p_j^{n_j},$$

and the joint probability function of $n_1, \ldots, n_k$ is

$$\frac{n!}{n_1! n_2! \cdots n_k!} \prod_{j=1}^{k} p_j^{n_j}.$$

The latter is called a multinomial distribution. We would also refer to the joint distribution of X's as multinomial.

Instead of considering specific models, we introduce now three families of models that unify many theoretical discussions. In the following $\boldsymbol{X}$ is a k-dimensional random vector unless it is stated otherwise, and f has the same connotation as before.

1.1.1 Exponential Families

Consider a family of probability models specified by $f(\boldsymbol{x}|\boldsymbol{\theta})$, $\boldsymbol{\theta} \in \Theta$. The family is said to be an exponential family if $f(\boldsymbol{x}|\boldsymbol{\theta})$ has the representation

$$f(\boldsymbol{x}|\boldsymbol{\theta}) = \exp\left\{ c(\boldsymbol{\theta}) + \sum_{j=1}^{p} t_j(\boldsymbol{x}) A_j(\boldsymbol{\theta}) \right\} h(\boldsymbol{x}), \tag{1.1}$$

where $c(.)$, $A_j(.)$ depend only on $\boldsymbol{\theta}$ and $t_j(.)$ depends only on $\boldsymbol{x}$. Note that the support of $f(\boldsymbol{x}|\boldsymbol{\theta})$, namely, the set of $\boldsymbol{x}$ where $f(\boldsymbol{x}|\boldsymbol{\theta}) > 0$, is the same as the set where $h(\boldsymbol{x}) > 0$ and hence does not depend on $\boldsymbol{\theta}$. To avoid trivialities, we assume that the support does not reduce to a single point.

Problem 1 invites you to verify that Examples 1.1 through 1.3 and Example 1.5 are exponential families.

It is easy to verify that if $\boldsymbol{X}_i$, $i = 1, \ldots, n$, are i.i.d. with density $f(\boldsymbol{x}|\boldsymbol{\theta})$, then their joint density is also exponential:

$$\prod_{i=1}^{n} f(\boldsymbol{x}_i|\boldsymbol{\theta}) = \exp\left\{ nc(\boldsymbol{\theta}) + \sum_{j=1}^{p} T_j A_j(\boldsymbol{\theta}) \right\} \prod_{i=1}^{n} h(\boldsymbol{x}_i),$$

with $T_j = \sum_{i=1}^{n} t_j(\boldsymbol{x}_i)$.

There are two convenient reparameterizations. Using new parameters we may assume $A_j(\boldsymbol{\theta}) = \theta_j$. Then

$$f(\boldsymbol{x}|\boldsymbol{\theta}) = \exp\left\{ c(\boldsymbol{\theta}) + \sum_{j=1}^{p} t_j(\boldsymbol{x}) \theta_j \right\} h(\boldsymbol{x}). \tag{1.2}$$

The general theory of exponential families, see, e.g., Brown (1986), ensures one can interchange differentiation and integration. Differentiation once under the integral sign leads to

$$0 = E_{\boldsymbol{\theta}}\left(\frac{\partial}{\partial \theta_j} \log f(\boldsymbol{X}|\boldsymbol{\theta}) \right) = \frac{\partial c}{\partial \theta_j} + E_{\boldsymbol{\theta}} t_j(\boldsymbol{X}), \ j = 1, \ldots, p. \tag{1.3}$$

In a similar way,

$$E_{\theta}\left(\frac{\partial^2 \log f}{\partial \theta_j \partial \theta_{j'}} \right) = -E_{\theta}\left(\frac{\partial \log f}{\partial \theta_j} \frac{\partial \log f}{\partial \theta_{j'}} \right). \tag{1.4}$$

In the second parameterization, we set $\eta_j = E_{\boldsymbol{\theta}}(t_j(\boldsymbol{X}))$, i.e.,

$$\eta_j = -\frac{\partial c}{\partial \theta_j}, \quad j = 1, \ldots, p. \tag{1.5}$$

In Problem 3, you are asked to verify for $p = 1$ that η_1 is a one-one function of θ. A similar argument shows $\boldsymbol{\eta} = (\eta_1, \ldots, \eta_p)$ is a one-one function of $\boldsymbol{\theta}$.

The parameters $\boldsymbol{\theta}$ are convenient mathematically, while the usual statistical parameters are closer to $\boldsymbol{\eta}$. You may wish to calculate η's and verify this for Examples 1.1 through 1.3 and Example 1.5.

1.1.2 Location-Scale Families

Definition 1.10. *Let X be a real- valued random variable, with density*

$$f(x|\mu, \sigma) = \frac{1}{\sigma} g\left(\frac{x - \mu}{\sigma}\right),$$

where g is also a density function, $-\infty < \mu < \infty$, $\sigma > 0$. The parameters μ and σ are called location and scale parameters.

With X as above, $Z = (X - \mu)/\sigma$ has density g. The normal $N(\mu, \sigma^2)$ is a location-scale family with Z being the standard normal, $N(0, 1)$. Example 1.2 is a scale family with $\mu = 0$, $\sigma = \theta$. We can make it a location-scale family if we set

$$f(x|\mu, \sigma) = \begin{cases} \frac{1}{\sigma} \exp\left(-\frac{x-\mu}{\sigma}\right) & \text{for } x > \mu; \\ 0 & \text{otherwise.} \end{cases}$$

but then it ceases to be an exponential family for its range depends on μ. The other examples, namely, Bernoulli, binomial, and Poisson are not location-scale families.

Example 1.11. Let X have uniform distribution over (θ_1, θ_2) so that

$$f(x|\boldsymbol{\theta}) = \begin{cases} \frac{1}{\theta_2 - \theta_1} & \text{if } \theta_1 < x < \theta_2; \\ 0 & \text{otherwise.} \end{cases}$$

This is also a location-scale family, with a reparameterization, which is not an exponential family.

Example 1.12. The Cauchy distribution specified by the density

$$f(x|\mu, \sigma) = \frac{1}{\pi} \frac{\sigma}{\sigma^2 + (x - \mu)^2}, \quad -\infty < x < \infty$$

is a location-scale family that is not exponential. It has several interesting properties. As $|x| \to \infty$, it tends to zero but at a much slower rate than the

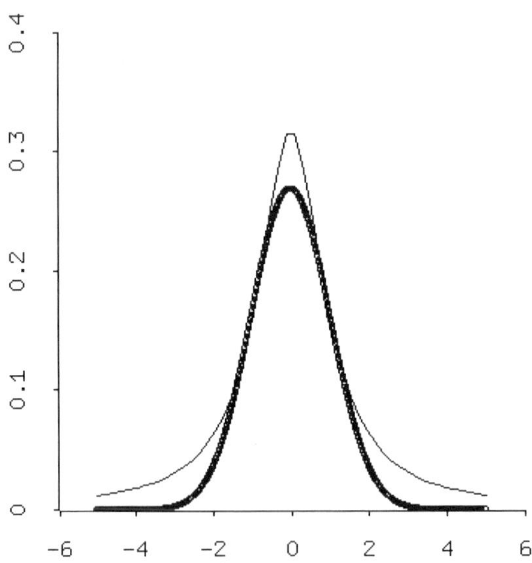

Fig. 1.1. Densities of Cauchy(0, 1) and normal(0, 2.19).

normal. One can verify that $E(|X|^r) = \infty$ for $r = 1, 2, \ldots$ under any μ, σ. So Cauchy has no finite moment. However, Figure 1.1 shows remarkable similarity between the normal and Cauchy, except near the tails. The Cauchy density is much flatter at the tails than the normal, which means x's that deviate quite a bit from μ will appear in data from time to time. Such deviations from μ would be unusual under a normal model and so may be treated as outliers by a data analyst. It provides an important counter-example to the *law of large numbers* or *central limit theorem* when one has infinite moments. It also plays an important role in robustness studies (see, e.g., Section 3.9).

Finally, many of the attractive statistical properties of the normal arise from the fact that it is both an exponential and a location-scale family, thereby inheriting interesting properties of both.

1.1.3 Regular Family

We end this section with a third very general family, defined by what are called mathematical regularity conditions.

Definition 1.13. *A family of densities $f(x|\theta)$ is said to satisfy Cramer-Rao type regularity conditions if the support of $f(x|\theta)$, i.e., the set of x for which $f(x|\theta) > 0$, does not depend on θ, f is k times continuously differentiable with respect to θ (with k usually equal to two or three) and one can differentiate under the integral sign as indicated below for real-valued θ:*

$$E_\theta \left(\frac{d}{d\theta} \log f(\boldsymbol{X}|\theta) \right) = \int_{-\infty}^{\infty} \left\{ \frac{d}{d\theta} \log f(\boldsymbol{x}|\theta) \right\} f(\boldsymbol{x}|\theta) \, d\boldsymbol{x}$$

$$= \int_{-\infty}^{\infty} \frac{d}{d\theta} f(\boldsymbol{x}|\theta) \, d\boldsymbol{x} = \frac{d}{d\theta} \int_{-\infty}^{\infty} f(\boldsymbol{x}|\theta) \, d\boldsymbol{x} = 0, \quad (1.6)$$

and similarly,

$$E_\theta \left(\frac{d^2}{d\theta^2} \log f(\boldsymbol{X}|\theta) \right) = - \int_{-\infty}^{\infty} \left(\frac{d}{d\theta} \log f(\boldsymbol{x}|\theta) \right)^2 f(\boldsymbol{x}|\theta) \, d\boldsymbol{x}. \quad (1.7)$$

The condition that the support of $f(\cdot|\theta)$ is free of θ is required for the last two relations to hold. The results of Chapter 4 require regularity conditions of this kind. The exponential families satisfy these regularity conditions. Location-scale families may or may not satisfy, usually the critical assumption is that relating to the support of f. Thus the Cauchy location-scale family satisfies these conditions but not the uniform or the exponential density

$$f(x|\mu, \sigma) = \frac{1}{\sigma} \exp \left(-\frac{x - \mu}{\sigma} \right), \ x > \mu.$$

1.2 Likelihood Function

A concept of fundamental importance is the *likelihood function*. Informally, for fixed $\boldsymbol{x}$, the joint density or probability mass function (p.m.f.) $f(\boldsymbol{x}|\boldsymbol{\theta})$, regarded as a function of $\boldsymbol{\theta}$, is called the likelihood function. When we think of f as the likelihood function we often suppress $\boldsymbol{x}$ and write f as $L(\boldsymbol{\theta})$. The likelihood function is not unique in that for any $c(\boldsymbol{x}) > 0$ that may depend on $\boldsymbol{x}$ but not on $\boldsymbol{\theta}$, $c(\boldsymbol{x})f(\boldsymbol{x}|\boldsymbol{\theta})$ is also a likelihood function. What is unique are the likelihood ratios $L(\boldsymbol{\theta_2})/L(\boldsymbol{\theta_1})$, which indicate how plausible is $\boldsymbol{\theta_2}$, relative to $\boldsymbol{\theta_1}$, in the light of the given data $\boldsymbol{x}$. In particular, if the ratio is large, we have a lot of confidence in $\boldsymbol{\theta_2}$ relative to $\boldsymbol{\theta_1}$ and the reverse situation holds if the ratio is small. Of course the threshold for what is large or small isn't easy to determine.

It is important to note that the likelihood is a point function. It can provide information on relative plausibility of two points $\boldsymbol{\theta_1}$ and $\boldsymbol{\theta_2}$, but not of two $\boldsymbol{\theta}$-sets, say, two non-degenerate intervals.

If the sample size n is large, usually the likelihood function has a sharp peak as shown in the following figure. Let the value of $\boldsymbol{\theta}$ where the maximum is attained be denoted as the maximum likelihood estimate (MLE) $\hat{\boldsymbol{\theta}}$; we define it formally later. In situations like this, one feels $\hat{\boldsymbol{\theta}}$ is very plausible as an estimate of $\boldsymbol{\theta}$ relative to any other points outside a small interval around $\hat{\boldsymbol{\theta}}$. One would then expect $\hat{\boldsymbol{\theta}}$ to be a good estimate of the unknown $\boldsymbol{\theta}$, at least in the sense of being close to it in some way (e.g., of being consistent, i.e, converging to $\boldsymbol{\theta}$ in probability). We discuss these things more carefully below.

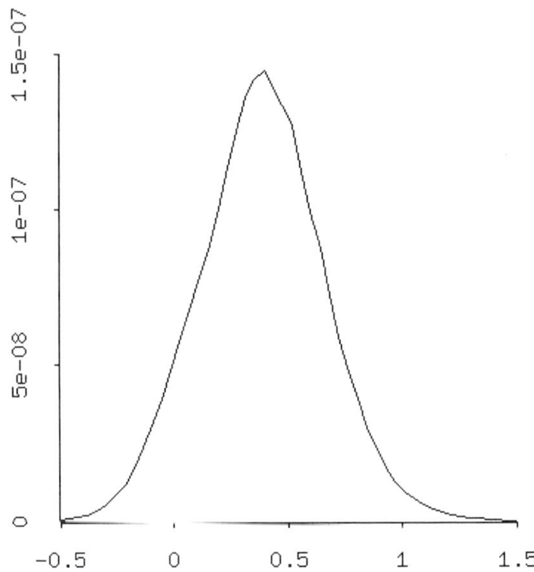

Fig. 1.2. $L(\theta)$ for the double exponential model when data is normal mixture.

Classical statistics also asserts that under regularity conditions and for large n, the maximum likelihood estimate minimizes the variance approximately within certain classes of estimates. Problem 10 provides a counter-example due to Basu (1988) when regularity conditions do not hold.

Definition 1.14. *The maximum likelihood estimate (MLE) $\widehat{\boldsymbol{\theta}}$ is a value of $\boldsymbol{\theta}$ where the likelihood function $L(\boldsymbol{\theta}) = f(\boldsymbol{x}|\boldsymbol{\theta})$ attains its supremum, i.e.,*

$$\sup_{\boldsymbol{\theta}} f(\boldsymbol{x}|\boldsymbol{\theta}) = f(\boldsymbol{x}|\widehat{\boldsymbol{\theta}}).$$

Usually, the MLE can be found by solving the likelihood equation

$$\frac{\partial}{\partial \theta_j} \log f(\boldsymbol{x}|\boldsymbol{\theta}) = 0, \ j = 1, \ldots, p. \tag{1.8}$$

In Problem 4(b), you are asked to show the likelihood function is log-concave, i.e., its logarithm is a concave function. In this case, if (1.8) has a solution, it is unique and provides a global maximum. There are well-known theorems, see, e.g., Rao (1973), which show the existence of a solution of (1.8) which converges in probability to the unknown true $\boldsymbol{\theta}$ if the dimension is fixed and Cramer-Rao type regularity conditions hold. If (1.8) has multiple roots, one has to be careful. A simple solution is to first find a $\sqrt{n}$-consistent estimate T_n, i.e., an estimate T_n such that $\sqrt{n}(T_n - \boldsymbol{\theta})$ is bounded in probability. Then choose a solution that is nearest to T_n.

1.3 Sufficient Statistics and Ancillary Statistics

Given the importance of likelihood function, it is interesting and useful to know what is the smallest set of statistics $T_1(\boldsymbol{x}), \ldots, T_m(\boldsymbol{x})$ in terms of which one can write down the likelihood function. As expected this makes it necessary to introduce *sufficient statistics.*

Definition 1.15. *Let $\boldsymbol{X}$ be distributed with density $f(\boldsymbol{x}|\boldsymbol{\theta})$. Then $\boldsymbol{T} = \boldsymbol{T}(\boldsymbol{X})$ $= (T_1(\boldsymbol{X}), \ldots, T_m(\boldsymbol{X}))$ is sufficient for $\boldsymbol{\theta}$ if the conditional distribution of $\boldsymbol{X}$ given $\boldsymbol{T}$ is free of $\boldsymbol{\theta}$.*

A basic fact for verifying whether $\boldsymbol{T}$ is sufficient is the following factorization theorem: $\boldsymbol{T}$ is sufficient for $\boldsymbol{\theta}$ iff $f(\boldsymbol{x}|\boldsymbol{\theta}) = g(T_1(\boldsymbol{x}), \ldots, T_m(\boldsymbol{x}), \boldsymbol{\theta})h(\boldsymbol{x})$.

Using this, you are invited to prove (Problem 20) that the likelihood function can be written in terms of $\boldsymbol{T}$ iff $\boldsymbol{T}$ is sufficient.

Thus the problem of finding the smallest $\boldsymbol{T}$ in terms of which one can write down the likelihood function reduces to the problem of finding what are called *minimal sufficient statistics.*

Definition 1.16. *A sufficient statistic $\boldsymbol{T}_0$ is minimal sufficient (or smallest among sufficient statistics) if $\boldsymbol{T}_0$ is a function of every sufficient statistic.*

Clearly, a one-one function of a minimal sufficient statistic is also minimal sufficient. In spite of the somewhat abstract definition, minimal sufficient statistics are usually easy to find by inspection. Most examples in this book would be covered by the following fact (Problem 19).

Fact. Suppose $X_i, i = 1, 2, \ldots, n$ are i.i.d. from exponential family. Then $(T_j = \sum_{i=1}^{n} t_j(\boldsymbol{X}_i), j = 1, \ldots, p)$ together form a minimal sufficient statistics and hence is the smallest set of statistics in terms of which we may write down the likelihood function.

Using this, you can prove $(\sum_1^n X_i, \sum_1^n X_i^2)$ is minimal sufficient for μ and σ^2 if $X_1, X_2, \ldots, X_n$ are i.i.d. $N(\mu, \sigma^2)$. This in turn implies $(\bar{X}, s^2 = \frac{1}{n-1}\sum(X_i - \bar{X})^2)$ is also minimal sufficient for (μ, σ^2), being a one-one function of $(\sum_1^n X_i, \sum_1^n X_i^2)$. In the same way, $\bar{X}$ is minimal sufficient for both i.i.d. $B(1, p)$ and $\mathcal{P}(\lambda)$. In Problem 10, one has to show $X_{(1)} = \min(X_1, X_2, \ldots, X_n)$ and $X_{(n)} = \max(X_1, X_2, \ldots, X_n)$ are together minimal sufficient for $U(\theta, 2\theta)$. A bad case is that of i.i.d. Cauchy(μ, σ^2). It is known (see, e.g., Lehmann and Casella (1998)) that the minimal sufficient statistic is the set of all order statistics $(X_{(1)}, X_{(2)}, \ldots, X_{(n)})$ where $X_{(1)}$ and $X_{(n)}$ have been defined earlier and $X_{(r)}$ is the rth X when the X_i's are arranged in ascending order (assuming all X_i's are distinct). This is a bad case because the order statistics together are always sufficient when X_i's are i.i.d., and so if this is the minimal sufficient statistic, it means the density is so complicated that the likelihood cannot be expressed in terms of a smaller set of statistics. The advantage of sufficiency is that we can replace the original data set $\boldsymbol{x}$ by the minimal sufficient statistic. Such reduction works well for i.i.d. random variables with an exponential family of distributions or special examples like $U(\theta_1, \theta_2)$. It doesn't work well in other cases including location-scale families.

There are various results in classical statistics that show a sufficient statistic contains all the information about θ in the data $\boldsymbol{X}$. At the other end is a statistic whose distribution does not depend on θ and so contains no information about θ. Such a statistic is called *ancillary*.

Ancillary statistics are easy to exhibit if $X_1, \ldots, X_n$ are i.i.d. with a location-scale family of densities. In fact, for any four integers a, b, c, and d, the ratio

$$\frac{X_{(a)} - X_{(b)}}{X_{(c)} - X_{(d)}} = \frac{Z_{(a)} - Z_{(b)}}{Z_{(c)} - Z_{(d)}}$$

is ancillary because the right-hand side is expressed in terms of order statistics of Z_i's where $Z_i = (X_i - \mu)/\sigma$, $i = 1, \ldots, n$ are i.i.d. with a distribution free of μ and σ.

There is an interesting technical theorem, due to Basu, which establishes independence of a sufficient statistic and an ancillary statistic. The result is useful in many calculations. Before we state Basu's theorem, we need to introduce the notion of *completeness*.

Definition 1.17. *A statistic T or its distribution is said to be complete if for any real valued function $\psi(T)$,*

$$E_\theta \psi(T(X)) = 0 \; \forall \; \theta \; \text{implies} \; \psi(T(X)) = 0$$

(with probability one under all θ).

Suppose T is discrete. The condition then simply means the family of p.m.f.'s $f^T(t|\theta)$ of T is rich enough that there is no non-zero $\psi(t)$ that is orthogonal to $f^T(t|\theta)$ for all θ in the sense $\sum_t \psi(t) f^T(t|\theta) = 0$ for all θ.

Theorem 1.18. (Basu). *Suppose T is a complete sufficient statistic and U is any ancillary statistic. Then T and U are independent for all θ.*

Proof. Because T is sufficient, the conditional probability of U being in some set B given T is free of θ and may be written as $P_\theta(U \in B|T) = \phi(T)$. Since U is ancillary, $E_\theta(\phi(T)) = P_\theta(U \in B) = c$, where c is a constant. Let $\psi(T) = \phi(T) - c$. Then $E_\theta \psi(T) = 0$ for all θ, implying $\psi(T) = 0$ (with probability one), i.e., $P_\theta(U \in B|T) = P_\theta(U \in B)$. $\square$

It can be shown that a complete sufficient statistic is minimal sufficient. In general, the converse isn't true. For exponential families, the minimal sufficient statistic $(T_1, \ldots, T_p) = (\sum_1^n t_1(\boldsymbol{X}_i), \ldots, \sum_1^n t_p(\boldsymbol{X}_i))$ is complete. For $X_1, X_2, \ldots, X_n$ i.i.d. $U(\theta_1, \theta_2)$, $(X_{(1)}, X_{(n)})$ is a complete sufficient statistic. Here are a couple of applications of Basu's theorem.

Example 1.19. Suppose $X_1, X_2, \ldots, X_n$ are i.i.d. $N(\mu, \sigma^2)$. Then $\bar{X}$ and $s^2 = \frac{1}{n-1}\sum(X_i - \bar{X})^2$ are independent. To prove this, treat σ^2 as fixed to start with and μ as the parameter. Then $\bar{X}$ is complete sufficient and s^2 is ancillary. Hence $\bar{X}$ and s^2 are independent by Basu's theorem.

Example 1.20. Suppose $X_1, X_2, \ldots, X_n$ are i.i.d $U(\theta_1, \theta_2)$. Then for any $1 < r < n$, $Y = (X_{(r)} - X_{(1)})/(X_{(n)} - X_{(1)})$ is independent of $(X_{(1)}, X_{(n)})$. This follows because Y is ancillary.

A somewhat different notion of sufficiency appears in Bayesian analysis. Its usefulness and relation to (classical) sufficiency is discussed in Appendix E.

1.4 Three Basic Problems of Inference in Classical Statistics

For simplicity, we take $p = 1$, so θ is a real-valued parameter. Informally, inference is an attempt to learn about θ. There are three natural things one may wish to do. One may wish to estimate θ by a single number. A classical estimate used in large samples is the MLE $\hat{\theta}$. Secondly, one may wish to choose an interval that covers θ with high probability. Thirdly, one may test hypotheses about θ, e.g., test what is called a null hypothesis $H_0 : \theta = 0$ against a two-sided alternative $H_1 : \theta \neq 0$. More generally, one can test $H_0 : \theta = \theta_0$ against $H_1 : \theta \neq \theta_0$ where θ_0 is a value of some importance. For example, θ is the effect of some new drug on one of the two blood pressures, or θ_0 is the effect of an alternative drug in the market and one is trying to test whether the new drug has different effects. If one wants to test whether the new drug is better then instead of $H_1 : \theta \neq \theta_0$, one may like to consider one-sided alternatives $H_1 : \theta < \theta_0$ or $H_1 : \theta > \theta_0$.

1.4.1 Point Estimates

In principle, any statistic $T(\boldsymbol{X})$ is an estimate though the context usually suggests some special reasonable candidates like sample mean $\bar{X}$ or sample median for a population mean like μ of $N(\mu, \sigma^2)$. To choose a satisfactory or optimal estimate one looks at the properties of its distribution. The two most important quantities associated with a distribution are its mean and variance or mean and the standard deviation, usually called the standard error of the estimate. One would usually report a good estimate and estimate of the standard error. So one judges an estimate T by its mean $E(T|\theta)$ and variance $\mathrm{Var}(T|\theta)$. If we are trying to estimate θ, we calculate the bias $E(T|\theta) - \theta$. One prefers small absolute values of bias, one possibility is to consider only unbiased estimates of θ and so one requires $E(T|\theta) = \theta$. Problem 17 requires you to show both $\bar{X}$ and the sample median are unbiased estimates for μ in

$N(\mu, \sigma^2)$. If the object is to estimate some real-valued function $\tau(\theta)$ of θ, one would require $E(T|\theta) = \tau(\theta)$.

For unbiased estimates of τ, $\mathrm{Var}(T|\theta) = E\{(T - \tau(\theta))^2|\theta\}$ measures how dispersed T is around $\tau(\theta)$. The smaller the variance the better, so one may search for an unbiased estimate that minimizes the variance. Because θ is not known, one would have to try to minimize variance for all θ. This is a very strong condition but there is a good theory that applies to several classical examples. In general, however it would be unrealistic to expect that such an optimal estimate exists. We will see the same difficulty in other problems of classical inference. We now summarize the basic theory in a somewhat informal manner.

Theorem 1.21. Cramer-Rao Inequality (information inequality). *Let T be an unbiased estimate of $\tau(\theta)$. Suppose we can interchange differentiation and integration to get*

$$\frac{d}{d\theta} E(T|\theta) = \int_{-\infty}^{\infty} \cdots \int_{-\infty}^{\infty} T(\boldsymbol{x}) f'(\boldsymbol{x}|\theta) \, d\boldsymbol{x},$$

and

$$0 = \frac{d}{d\theta}\left[\int_{-\infty}^{\infty} \cdots \int_{-\infty}^{\infty} f(\boldsymbol{x}|\theta) \, d\boldsymbol{x}\right] = \int_{-\infty}^{\infty} \cdots \int_{-\infty}^{\infty} f'(\boldsymbol{x}|\theta) \, d\boldsymbol{x}.$$

Then,

$$\mathrm{Var}(T|\theta) \geq \frac{[\tau'(\theta)]^2}{I_n(\theta)},$$

where the $'$ in τ and f indicates a derivative with respect to θ and $I_n(\theta)$ is Fisher information in $\boldsymbol{x}$, namely,

$$I_n(\theta) = E\left\{\left(\frac{d}{d\theta} \log f(\boldsymbol{X}|\theta)\right)^2 \middle| \theta\right\}.$$

Proof. Let $\psi(\boldsymbol{X}, \theta) = \frac{d}{d\theta} \log f(\boldsymbol{X}|\theta)$. The second relation above implies $E(\psi(\boldsymbol{X}, \theta)|\theta) = 0$ and then, $\mathrm{Var}(\psi(\boldsymbol{X}, \theta)|\theta) = I_n(\theta)$. The first relation implies

$$\mathrm{Cov}(T, \psi(\boldsymbol{X}, \theta) \mid \theta) = \tau'(\theta).$$

It then follows that

$$\mathrm{Var}(T|\theta) \geq \frac{[\mathrm{Cov}(T, \psi(\boldsymbol{X}, \theta)|\theta)]^2}{\mathrm{Var}(\psi(\boldsymbol{X}, \theta)|\theta)} \geq \frac{\tau'(\theta)^2}{I_n(\theta)}. \qquad \square$$

If $X_1, \ldots, X_n$ are i.i.d. $f(x|\theta)$, then

$$I_n(\theta) = nI(\theta)$$

where $I(\theta)$ is the Fisher information in a single observation,

$$I(\theta) = E\left\{\left(\frac{d}{d\theta}\log f(X_1|\theta)\right)^2 \Big| \theta\right\}.$$

To get a feeling for $I_n(\theta)$, consider an extreme case where $f(\boldsymbol{x}|\theta)$ is free of θ. Clearly, in this case there can be no information about θ in $\boldsymbol{X}$. On the other hand, if $I_n(\theta)$ is large, then on an average a small change in θ leads to a big change in $\log f(\boldsymbol{x}|\theta)$, i.e., f depends strongly on θ and one expects there is a lot that can be learned about θ and hence $\tau(\theta)$. A large value of $I_n(\theta)$ diminishes the lower bound making it plausible that one may be able to get an unbiased estimate with small variance.

Finally, if the lower bound is attained at all θ by T, then clearly T is a uniformly minimum variance unbiased (UMVUE) estimate. We would call them best unbiased estimates.

A more powerful method of getting best unbiased estimates is via the Rao-Blackwell theorem.

Theorem 1.22. (Rao-Blackwell). *If T is an unbiased estimate of $\tau(\theta)$ and S is a sufficient statistic, the $T' = E(T|S)$ is also unbiased for $\tau(\theta)$ and*

$$Var(T'|\theta) \leq Var(T|\theta)\ \forall\theta.$$

Corollary 1.23. *If T is complete and sufficient, then T' as constructed above is the best unbiased estimate for $\tau(\theta)$.*

Proof. By the property of conditional expectations,

$$E\left(T'|\theta\right) = E\left\{E(T|S)\mid\theta\right\} = E\left(T|\theta\right).$$

(You may want to verify this at least for the discrete case.) Also,

$$\begin{aligned}
Var(T|\theta) &= E\left[\{(T - T') + (T' - \tau(\theta))\}^2 \mid \theta\right] \\
&= E\left\{(T - T')^2 \mid \theta\right\} + E\left\{(T' - \tau(\theta))^2 \mid \theta\right\},
\end{aligned}$$

because

$$\begin{aligned}
\mathrm{Cov}\left\{T - T', T' - \tau(\theta)\mid\theta\right\} &= E\left\{(T - T')(T' - \tau(\theta)\mid\theta\right\} \\
&= E\left[E\left\{(T' - \tau(\theta))(T - T')\mid S\right\}|\theta\right] \\
&= E\left[(T' - \tau(\theta))E(T - T'|S)\mid\theta\right] \\
&= 0.
\end{aligned}$$

The decomposition of $Var(T|\theta)$ above shows that it is greater than or equal to $Var(T'|\theta)$. $\square$

The theorem implies that in our search for the best unbiased estimate, we may confine attention to unbiased estimates of $\tau(\theta)$ based on S. However, under completeness, T' is the only such estimate.

Example 1.24. Consider a random sample from $N(\mu, \sigma^2)$, σ^2 assumed known. Note that by either of the two previous results, $\bar{X}$ is the best unbiased estimate for μ. The best unbiased estimate for μ^2 is $\bar{X}^2 - \sigma^2/n$ by the Rao-Blackwell theorem. You can show it does not attain the Cramer-Rao lower bound. If a T attains the Cramer-Rao lower bound, it has to be a linear function (with $\theta \equiv \mu$),

$$T(\boldsymbol{x}) = a(\theta) + b(\theta)\frac{d}{d\theta} \log f(\boldsymbol{x}|\theta),$$

i.e., must be of the form

$$T(\boldsymbol{x}) = c(\theta) + d(\theta)\bar{x}.$$

But T, being a statistic, this means

$$T(\boldsymbol{x}) = c + d\bar{x},$$

where c, d are constants.

A similar argument holds for any exponential family. Conversely, suppose a parametric model $f(\boldsymbol{x}|\theta)$ allows a statistic T to attain the Cramer-Rao lower bound. Then,

$$T(x) = a(\theta) + b(\theta)\frac{d}{d\theta} \log f(\boldsymbol{x}|\theta),$$

which implies

$$\frac{T(x) - a(\theta)}{b(\theta)} = \frac{d}{d\theta} \log f(\boldsymbol{x}|\theta).$$

Integrating both sides with respect to θ,

$$T(\boldsymbol{x}) \int (b(\theta))^{-1} \, d\theta - \int a(\theta)(b(\theta))^{-1} \, d\theta + d(\boldsymbol{x}) = \log f(\boldsymbol{x}|\theta),$$

where $d(\boldsymbol{x})$ is the constant of integration. If we write $A(\theta) = \int (b(\theta))^{-1} d\theta$, $c(\theta) = \int a(\theta)b(\theta)^{-1} \, d\theta$ and $d(\boldsymbol{x}) = \log h(\boldsymbol{x})$, we get an exponential family.

The Cramer-Rao inequality remains important because it provides information about variance of T. Also, even if a best unbiased estimate can't be found, one may be able to find an unbiased estimate with variance close to the lower bound. A fascinating recent application is Liu and Brown (1992).

An unpleasant feature of the inequality as formulated above is that it involves conditions on T rather than only conditions on $f(\boldsymbol{x}|\theta)$. A considerably more technical version without this drawback may be found in Pitman (1979).

The theory for getting best unbiased estimates breaks down when there is no complete sufficient statistic. Except for the examples we have already seen, complete sufficient statistics rarely exist. Even when a complete sufficient statistic exists, one has to find an unbiased estimate based on the complete sufficient statistic S. This can be hard. Two heuristic methods work sometimes. One is the method of indicator functions, illustrated in Problem 5.

The other is to start with a plausible estimate and then make a suitable adjustment to make it unbiased. Thus to get an unbiased estimate for μ^2 for $N(\mu, \sigma^2)$, one would start with $\bar{X}^2$. We know for sure $\bar{X}^2$ can't be unbiased since $E(\bar{X}^2|\mu, \sigma^2) = \mu^2 + \sigma^2/n$. So if σ^2 is known, we can use $\bar{X}^2 - \sigma^2/n$. If σ^2 is unknown, we can use $\bar{X}^2 - s^2/n$, where $s^2 = \sum(X_i - \bar{X})^2/(n-1)$ is an unbiased estimate of σ^2. Note that $\bar{X}^2 - s^2/n$ is a function of the complete, sufficient statistic $(\bar{X}, s^2)$ but may be negative even though μ^2 is a positive quantity.

For all these reasons, unbiasedness isn't important in classical statistics as it used to be. Exceptions are in unbiased estimation of risk (see Berger and Robert (1990), Lu and Berger (1989a, b)) with various applications and occasionally in variance estimation, specially in high-dimensional problems. See Chapter 9 for an application.

We note finally that for relatively small values of p and relatively large values of n, it is easy to find estimates that are approximately unbiased and approximately attain the Cramer-Rao lower bound in a somewhat weak sense. An informal introduction to such results appears below.

Under regularity conditions, it can be shown that

$$\sqrt{n}\left(\hat{\theta} - \theta\right) - \frac{1}{\sqrt{nI(\theta)}}\left(\sum \frac{d}{d\theta} \log f(X_i|\theta)\right) \xrightarrow{P} 0.$$

This implies $\hat{\theta}$ is approximately normal with mean θ and variance $(nI(\theta))^{-1}$, which is the Cramer-Rao lower bound when we are estimating θ. Thus $\hat{\theta}$ is approximately normal with expectation equal to θ and variance equal to the Cramer-Rao lower bound for $\tau(\theta) = \theta$. For a general differentiable $\tau(\theta)$, we show $\tau(\hat{\theta})$ has similar properties. Observe that $\tau(\hat{\theta}) = \tau(\theta) + (\hat{\theta} - \theta)\tau'(\theta) +$ *smaller terms*, which exhibits $\tau(\hat{\theta})$ as an approximately linear function of $\hat{\theta}$. Hence $\tau(\hat{\theta})$ is also approximately normal with

$$\text{mean} = \tau(\theta) + \text{(approximate) mean of } \left(\hat{\theta} - \theta\right)\tau'(\theta) = \tau(\theta), \text{ and}$$

$$\text{variance} = \left(\tau'(\theta)\right)^2 \times \text{approximate variance of } \left(\hat{\theta} - \theta\right) = (\tau'(\theta))^2 \frac{1}{nI(\theta)}.$$

The last expression is the Cramer-Rao lower bound for $\tau(\theta)$. The method of approximating $\tau(\hat{\theta})$ by a linear function based on Taylor expansion is called the *delta method*.

For $N(\mu, \sigma^2)$ and fixed x, let $\tau(\theta) = \tau(\theta, x) = P\{X \leq x|\mu, \sigma\}$. An approximately best unbiased estimate is $P\{X \leq x|\hat{\mu}, \hat{\sigma}\} = \Phi(\frac{x-\bar{X}}{s'})$ where $s' = \sqrt{\frac{1}{n}\sum(X_i - \bar{X})^2}$ and $\Phi(.)$ is the standard normal distribution function. The exact best unbiased estimate can be obtained by the method of indicator functions. Let

$$I(X_1) = \begin{cases} 1 & \text{if } X_1 \leq x; \\ 0 & \text{otherwise .} \end{cases}$$

Then I is an unbiased estimate of $\tau(\theta)$, so the best unbiased estimate is $E(I|\bar{X}, s') = P\{X_1 \leq x|\bar{X}, s'\}$. The explicit form is given in Problem 5.

1.4.2 Testing Hypotheses

We consider only the case of real-valued θ_0, the null hypothesis $H_0 : \theta = \theta_0$ and the two-sided alternative $H_1 : \theta \neq \theta_0$ or, one-sided null and one-sided alternatives, e.g., $H_0 : \theta \leq \theta_0$ and $H_1 : \theta > \theta_0$. In this formulation, the null hypothesis represents *status quo* as in the drug example. It could also mean an accepted scientific hypothesis, e.g., on the value of the gravitational constant or velocity of light in some medium. This suggests that one should not reject the null hypothesis unless there is compelling evidence in the data in favor of H_1. This fact will be used below.

A test is a rule that tells us for each possible data set (under our model $f(\boldsymbol{x}|\theta)$) whether to accept or reject H_0. Let W be the set of $\boldsymbol{x}$'s for which a given test rejects H_0 and W^c be the set where the test accepts H_0. The region W, called a critical region or rejection region, completely specifies the test. Sometimes one works with the indicator of W rather than W itself. The collection of all subsets W in $\mathcal{R}^n$ or their indicators correspond to all possible tests. How does one evaluate them in principle or choose one in some optimal manner? The error committed by rejecting H_0 when H_0 is true is called the error of first kind. Avoiding this is considered to be more important than the so called second kind of error committed when H_0 is accepted even though H_1 is true. For any given W,

$$\text{Probability of error of first kind} \ = P_{\theta_0}\left(\boldsymbol{X} \in W\right) = E_{\theta_0}\left(I(\boldsymbol{X})\right),$$

where $I(\boldsymbol{x})$ is the indicator of W,

$$I(\boldsymbol{x}) = \begin{cases} 1 & \text{if } \boldsymbol{x} \in W; \\ 0 & \text{if } \boldsymbol{x} \in W^c. \end{cases}$$

Probability of error of second kind $= P_\theta(\boldsymbol{X} \in W^c) = 1 - E_\theta(I(\boldsymbol{X}))$, for θ as in H_1. One also defines the power of detecting H_1 as $1 - P_\theta(\boldsymbol{X} \in W^c) = E_\theta(I(\boldsymbol{X}))$ for θ as in H_1.

It turns out that in general if one tries to reduce one error probability the other error probability goes up, so one cannot reduce both simultaneously. Because probability of error of first kind is more important, one first makes it small,

$$E_{\theta_0}\left(I(\boldsymbol{X})\right) \leq \alpha, \tag{1.9}$$

where α, conventionally .05, .01, etc., is taken to be a small number. Among all tests satisfying this, one then tries to minimize the probability of committing error of second kind or equivalently, to maximize the power uniformly for all θ as in H_1. You can see the similarity of (1.9) with restriction to unbiased estimates and the optimization problem subject to (1.9) as the problem of minimizing variance among unbiased estimates. The best test is called uniformly most powerful (UMP) after Neyman and Pearson who developed this theory.

It turns out that for exponential families and some special cases like $U(0, \theta)$ or $U(\theta - \frac{1}{2}, \theta + \frac{1}{2})$, one can find UMP tests for one-sided alternatives. The basic tool is the following result about a simple alternative $H_1 : \theta = \theta_1$.

Lemma 1.25. (Neyman-Pearson). *Consider $H_0 : \theta = \theta_0$ versus $H_1 : \theta = \theta_1$. Fix $0 \leq \alpha \leq 1$.*
A. *Suppose there exists a non-negative k and a test given by the indicator function I_0 such that*

$$I_0(\boldsymbol{x}) = \begin{cases} 1 & \text{if } f(\boldsymbol{x}|\theta_1) > kf(\boldsymbol{x}|\theta_0); \\ 0 & \text{if } f(\boldsymbol{x}|\theta_1) < kf(\boldsymbol{x}|\theta_0), \end{cases}$$

with no restriction on I_0 if $f(\boldsymbol{x}|\theta_1) = kf(\boldsymbol{x}|\theta_0))$, such that $E_{\theta_0}(I_0(\boldsymbol{X})) = \alpha$. Then

$$E_{\theta_1}(I_0(\boldsymbol{X})) \geq E_{\theta_1}(I_1(\boldsymbol{X}))$$

for all indicators I_1 satisfying

$$E_{\theta_0}(I_1(\boldsymbol{X})) \leq \alpha.$$

i.e., the test given by I_0 is MP among all tests satisfying the previous inequality.
B. *Suppose g is a given integrable function and we want all tests to satisfy*

$$E_{\theta_0}(I(\boldsymbol{X})) = \alpha \quad \text{and} \quad \int \cdots \int I(\boldsymbol{x})g(\boldsymbol{x})\,d\boldsymbol{x} = c \text{ (same for all } I\text{).} \quad (1.10)$$

Then among all such I, $E_{\theta_1}(I(\boldsymbol{X}))$ is maximum at

$$I_0(\boldsymbol{x}) = \begin{cases} 1 & \text{if } f(\boldsymbol{x}|\theta_1) > k_1 f(\boldsymbol{x}|\theta_0) + k_2 g(\boldsymbol{x}); \\ 0 & \text{if } f(\boldsymbol{x}|\theta_1) < k_1 f(\boldsymbol{x}|\theta_0) + k_2 g(\boldsymbol{x}), \end{cases}$$

where k_1 and k_2 are two constants such that I_0 satisfies the two constraints given in (1.10).
C. *If I_0 exists in **A** or **B** and I_1 is an indicator having the same maximizing property as I_0 under the same constraints, then $I_0(\boldsymbol{x})$ and $I_1(\boldsymbol{x})$ are same if $f(\boldsymbol{x}|\theta_1) - kf(\boldsymbol{x}|\theta_0) \neq 0$, in case of **A** and $f(\boldsymbol{x}|\theta_1) - k_1 f(\boldsymbol{x}|\theta_0) - k_2 g(\boldsymbol{x}) \neq 0$, in case of **B**.*

Proof. **A.** By definition of I_0 and the fact that $0 \leq I_1(\boldsymbol{x}) \leq 1$ for all I_1, we have that

$$\int_{\mathcal{X}} \{(I_0(\boldsymbol{x}) - I_1(\boldsymbol{x})) \, (f(\boldsymbol{x}|\theta_1) - kf(\boldsymbol{x}|\theta_0))\} \, d\boldsymbol{x} \geq 0, \qquad (1.11)$$

which implies

$$\int_{\mathcal{X}} \{I_0(\boldsymbol{x}) - I_1(\boldsymbol{x})\}\, f(\boldsymbol{x}|\theta_1)\, d\boldsymbol{x} \geq k \int_{\mathcal{X}} I_0(\boldsymbol{x}) f(\boldsymbol{x}|\theta_0)\, d\boldsymbol{x} - k \int_{\mathcal{X}} I_1(\boldsymbol{x}) f(\boldsymbol{x}|\theta_0)\, d\boldsymbol{x}$$

$$\geq k\alpha - k\alpha = 0.$$

B. The proof is similar to that of **A** except that one starts with

$$\int_{\mathcal{X}} (I_0 - I)\, \{f(\boldsymbol{x}|\theta_1) - k_1 f(\boldsymbol{x}|\theta_0) - k_2 g(\boldsymbol{x})\}\, d\boldsymbol{x} \geq 0.$$

C. Suppose I_0 is as in **A** and I_1 maximizes $\int_{\mathcal{X}} I f(\boldsymbol{x}|\theta_1)\, d\boldsymbol{x}$. i.e.,

$$\int_{\mathcal{X}} I_0 f(\boldsymbol{x}|\theta_1)\, d\boldsymbol{x} = \int_{\mathcal{X}} I_1 f(\boldsymbol{x}|\theta_1)\, d\boldsymbol{x}$$

subjected to

$$\int_{\mathcal{X}} I_0 f(\boldsymbol{x}|\theta_0)\, d\boldsymbol{x} = \alpha, \ \text{and} \ \int_{\mathcal{X}} I_1 f(\boldsymbol{x}|\theta_0)\, d\boldsymbol{x} = \alpha.$$

Then,

$$\int_{\mathcal{X}} \{I_0 - I_1\}\{f(\boldsymbol{x}|\theta_1) - k f(\boldsymbol{x}|\theta_0)\} d\boldsymbol{x} = 0.$$

But the integrand $\{I_0(\boldsymbol{x}) - I_1(\boldsymbol{x})\}\{f(\boldsymbol{x}|\theta_1) - k f(\boldsymbol{x}|\theta_0\boldsymbol{x})\}$ is non-negative for all $\boldsymbol{x}$. Hence

$$I_0(\boldsymbol{x}) = I_1(\boldsymbol{x}) \ \text{if} \ f(\boldsymbol{x}|\theta_1) - k f(\boldsymbol{x}|\theta_0) \neq 0.$$

This completes the proof. $\square$

Remark 1.26. Part **A** is called the sufficiency part of the lemma. Part **B** is a generalization of **A**. Part **C** is a kind of necessary condition for I_1 to be MP provided I_0 as specified in **A** or **B** exists.

If X_i's are i.i.d. $N(\mu, \sigma^2)$, then $\{\boldsymbol{x} : f(\boldsymbol{x}|\theta_1) = k f(\boldsymbol{x}|\theta_0)\}$ has probability zero. This is usually the case for continuous random variables. Then the MP test, if it exists, is unique. It fails for some continuous random variables like X_i's that are i.i.d. $U(0, \theta)$ and for discrete random variables. In such cases the MP test need not be unique.

Using **A** of the lemma we show that for $N(\mu, \sigma^2)$, σ^2 known, the UMP test of $H_0 : \mu = \mu_0$ for a one-sided alternative, say, $H_1 : \mu > \mu_0$ is given by

$$I_0 = \begin{cases} 1 & \text{if } \bar{x} > \mu_0 + z_\alpha \frac{\sigma}{\sqrt{n}}; \\ 0 & \text{if } \bar{x} < \mu_0 + z_\alpha \frac{\sigma}{\sqrt{n}}, \end{cases}$$

where z_α is such that $P\{Z > z_\alpha\} = \alpha$ with $Z \sim N(0, 1)$.

Fix $\mu_1 > \mu_0$. Note that $f(\boldsymbol{x}|\mu_1)/f(\boldsymbol{x}|\mu_0)$ is an increasing function of $\bar{x}$. Hence for any k in **A**, there is a constant c such that

$$f(\boldsymbol{x}|\mu_1) > k f(\boldsymbol{x}|\mu_0) \quad \text{if and only if} \quad \bar{x} > c.$$

So the MP test is given by the indicator

$$I_0 = \begin{cases} 1 & \text{if } \bar{x} > c; \\ 0 & \text{if } \bar{x} < c, \end{cases}$$

where c is such that $E\mu_0(I_0) = \alpha$. It is easy to verify $c = \mu_0 + z_\alpha \sigma/\sqrt{n}$ does have this property. Because this test does not depend on the value of μ_1, it is MP for all $\mu_1 > \mu_0$ and hence it is UMP for $H_1 : \mu > \mu_0$.

In the same way, one can find the UMP test of $H_1 : \mu < \mu_0$ and verify that the test now rejects H_0 if $\bar{x} < \mu_0 - z_\alpha \sigma/\sqrt{n}$. How about $H_0 : \mu \leq \mu_0$ versus $H_1 : \mu > \mu_0$? Here we consider all tests with the property

$$P_{\mu,\sigma^2}(H_0 \text{ is rejected}) \leq \alpha \quad \text{for all } \mu \leq \mu_0.$$

Using Problem 6 (or 7), it is easy to verify that the UMP test of $H_0 : \mu = \mu_0$ versus $H_1 : \mu > \mu_0$ is also UMP when the null is changed to $H_0 : \mu \leq \mu_0$.

One consequence of these calculations and the uniqueness of MP tests (Part **C**) is that there is no UMP test against two-sided alternatives. Each of the two UMP tests does well for its H_1 but very badly at other θ_1, e.g., the UMP test I_0 for $H_0 : \mu = \mu_0$ versus $H_1 : \mu > \mu_0$ obtained above has $E\mu_1(I_0) \to 0$ as $\mu_1 \to -\infty$. To avoid such poor behavior at some θ's, one may require that the power cannot be smaller than α. Then $E_{\theta_0}(I) \leq \alpha$, and $E_\theta(I) \geq \alpha$, $\theta \neq \theta_0$ imply $E_{\theta_0}(I) = \alpha$ and $E_\theta(I)$ has a global and hence a local minimum at $\theta = \theta_0$. Tests of this kind were first considered by Neyman and Pearson who called them unbiased. There is a similarity with unbiased estimates that was later pointed out by Lehmann (1986) (see Chapter 1 there). Because every unbiased I satisfies conditions of Part **B** with $g = f'(\boldsymbol{x}|\theta_0)$, one can show that the MP test for any $\theta_1 \neq \theta_0$ satisfies conditions for I_0. With a little more effort, it can be shown that the MP test is in fact

$$I_0 = \begin{cases} 1 & \text{if } \bar{x} > c_2 \text{ or } \bar{x} < c_1; \\ 0 & \text{if } c_1 < \bar{x} < c_2, \end{cases}$$

for suitable c_1 and c_2. The given constraints can be satisfied if

$$c_1 = \mu_0 - z_{\alpha/2}\frac{\sigma}{\sqrt{n}} \quad \text{and} \quad c_2 = \mu_0 + z_{\alpha/2}\frac{\sigma}{\sqrt{n}}.$$

This is the UMP unbiased test.

We have so far discussed how to control α, the probability of error of first kind and then, subject to this and other constraints, minimize $\beta(\theta)$, the probability of error of second kind. But how do we bring $\beta(\theta)$ to a level that is desired? This is usually done by choosing an appropriate sample size n, see Problem 8.

The general theory for exponential families is similar with $T = \sum_1^n t(x_i)$ or T/n taking on the role of $\bar{x}$. However, the distribution of T may be discrete, as in the case of binomial or Poisson. Then it may not be possible to find the constants c or c_1, c_2. Try, for example, the case of $B(5,p)$, $H_0 : p = \frac{1}{2}$ versus $H_1 : p > \frac{1}{2}$, $\alpha = .05$. In practice one chooses an $\alpha' < \alpha$ and as close to α as possible for which the constants can be found and the lemma applied with α' instead of α.

A different option of some theoretical interest only is to extend the class of tests to what are called randomized tests. A randomized test is given by a function $0 \le \phi(\boldsymbol{x}) \le 1$, which we interpret as the probability of rejecting H_0 given $\boldsymbol{x}$. By setting ϕ equal to an indicator we get back the non-randomized tests. With this extension, one can find a UMP test for binomial or Poisson of the form

$$\phi_0 = \begin{cases} 1 & \text{if } T > c; \\ 0 & \text{if } T < c; \\ \gamma & \text{if } T = c, \end{cases}$$

where $0 \le \gamma \le 1$ is chosen along with c so that $E_{\theta_0}(\phi_0) = \alpha$. Such use of randomization has some other theoretical advantages. Randomization is sometimes needed to get a minimax test (i.e., a test that minimizes maximum probability or error of either kind), vide, Problem 14. Most important of all, randomization leads to the convexity of the collection of all tests in the sense that if $\phi_1(\boldsymbol{x})$ and $\phi_2(\boldsymbol{x})$ are two randomized or non-randomized tests, the convex combination $\lambda\phi_1 + (1 - \lambda)\phi_2$, $0 < \lambda < 1$, is again a function $\phi(x)$ lying between 0 and 1 and so it is a randomized test. This leads to convexity of risk set (Problem 15).

Except for exponential families and a few special examples, UMP tests don't exist. However, just as in the case of estimation theory, there are approximately optimum tests based directly on maximum likelihood estimates of θ or the likelihood ratio statistic

$$\lambda = \frac{f(\boldsymbol{x}|\theta_0)}{\sup_{\theta \in \Theta_1} f(\boldsymbol{x}|\theta)},$$

where Θ_1 is the set specified by H_1.

1.4.3 Interval Estimation

A commonly used so called confidence interval for μ in $N(\mu, \sigma^2)$ with σ^2 known is $\bar{X} \pm z_{\alpha/2}\sigma/\sqrt{n}$. This means

$$P_{\mu,\sigma}\{\mu \in \text{ confidence interval }\} = P_{\mu,\sigma}\left\{\bar{X} - z_{\alpha/2}\frac{\sigma}{\sqrt{n}} \le \mu \le \bar{X} + z_{\alpha/2}\frac{\sigma}{\sqrt{n}}\right\}$$

$$= 1 - \alpha.$$

In this statement, as in all other areas of classical statistics, μ is a constant, the probability statement is about $\bar{X}$. So $(1 - \alpha)$ is the proportion of times the interval covers μ over repetitions of the experiment and data sets. If one has a data set with $\bar{X} = 3$, and asks for the probability that μ lies in $3 \pm z_{\alpha/2}\sigma/\sqrt{n}$, the answer isn't $1 - \alpha$ but trivially zero or one depending on the value of μ. Though the idea of such intervals is quite old, it was Neyman who formalized them.

For $\boldsymbol{X} \sim f(\boldsymbol{x}|\theta)$, $\theta \in \mathcal{R}$, one calls $(\underline{\theta}(\boldsymbol{X}), \bar{\theta}(\boldsymbol{X}))$ a confidence interval with confidence coefficient $1 - \alpha$, if $P_\theta \left\{ \underline{\theta}(\boldsymbol{X}) \leq \theta \leq \bar{\theta}(\boldsymbol{X}) \right\} = 1 - \alpha$.

The simplest way to generate them is to find what Fisher called a pivotal quantity, namely, a real valued function $T(\boldsymbol{X}, \theta)$ of both $\boldsymbol{X}$ and θ such that the distribution of $T(\boldsymbol{X}, \theta)$ does not depend on θ. Suppose then we choose two numbers t_1 and t_2 such that $P_\theta \left\{ t_1 \leq T(\boldsymbol{X}, \theta) \leq t_2 \right\} = 1 - \alpha$. If for each $\boldsymbol{X}$, $T(\boldsymbol{X}, \theta)$ is monotone in θ, say, an increasing function of θ, then we can find $\underline{\theta}(\boldsymbol{X})$ and $\bar{\theta}(\boldsymbol{X})$ such that $T(\boldsymbol{X}, \bar{\theta}(\boldsymbol{X})) = t_2$ and $T(\boldsymbol{X}, \underline{\theta}(\boldsymbol{X})) = t_1$. Clearly $(\underline{\theta} \leq \theta \leq \bar{\theta})$ iff $t_1 \leq T \leq t_2$ and hence $\underline{\theta} \leq \theta \leq \bar{\theta}$ with probability $1 - \alpha$.

In the normal example, $T(\boldsymbol{X}, \mu) = \bar{X} - \mu$, the distribution of which is $N(0, \sigma^2/n)$.

Neyman showed one can also derive confidence intervals from tests. We illustrate this with the normal. For each μ_0, consider the UMPU test

$$
I_0 = \begin{cases} 0 & \text{if } \mu_0 - z_{\alpha/2}\frac{\sigma}{\sqrt{n}} \leq \bar{X} \leq \mu_0 + z_{\alpha/2}\frac{\sigma}{\sqrt{n}}; \\ 1 & \text{otherwise.} \end{cases}
$$

(We have taken $I_0 = 0$ at the two boundaries, which have zero probability anyway.)

We now define a confidence set, say, $A(\boldsymbol{X}) \subset \mathcal{R}$ by,

$$A(\boldsymbol{X}) = \left\{ \mu_0 \text{ such that } H_0 : \mu = \mu_0 \text{ is accepted by its UMPU test} \right\}.$$

Then $P_{\mu_0}\left\{ A(\boldsymbol{X}) \text{ covers } \mu_0 \right\} = P_{\mu_0}\left\{ \mu_0 \text{ is accepted by its UMPU test } \right\} = 1 - \alpha$.

Also $A(\boldsymbol{X})$ is nothing but the interval $\bar{X} \pm z_{\alpha/2}\sigma/\sqrt{n}$. We have just gotten the same interval by a different route.

This approach helps in showing many common intervals have the property of being shortest, i.e., having smallest expected length of all confidence intervals obtainable from a family of unbiased tests. This follows from an application of a simple but somewhat technical result (vide Ghosh-Pratt identity in Encyclopedia of Statistics).

1.5 Inference as a Statistical Decision Problem

The three apparently very different inference problems discussed in Section 1.4 can be unified by formulating them as statistical decision problems. This approach is due to Wald, who not only unified classical inference but proved basic theorems applying to all inference. A couple of his results are mentioned below and in Section 2.3. One gains a certain conceptual clarity as well as a certain broader outlook. However certain special features of each problem, either relating to historical context or relating to such consideration as intuitive appeal or reasonableness, are lost.

A statistical decision problem has a model $f(\boldsymbol{x}|\boldsymbol{\theta})$ and a space $\mathcal{A}$ of actions or decisions "a". A decision rule or decision function is a function $\delta(\boldsymbol{x})$ from the sample space of the data to the action space $\mathcal{A}$, i.e., $\delta(\boldsymbol{x})$ is an action, for each $\boldsymbol{x}$. To implement this rule, one simply takes the action $\delta(\boldsymbol{x})$ if data are $\boldsymbol{x}$.

In estimation $\mathcal{A} = \mathcal{R}$ and $\delta(\boldsymbol{x}) = T(\boldsymbol{x}) \in \mathcal{R}$ is nothing but an estimate of θ. In testing, the action or decision consists of two elements {"accept H_0", "accept H_1"}. We may denote these elements as a_0 and a_1. A decision function has a one-one correspondence with indicator function as follows

$$I(\boldsymbol{x}) = \begin{cases} 1 & \text{iff } \delta(\boldsymbol{x}) = a_1; \\ 0 & \text{iff } \delta(\boldsymbol{x}) = a_0. \end{cases}$$

In interval estimation, action space would be the collection of all intervals $[a, b]$. Each confidence interval is a decision function.

One of the advantages of the new approach is that it liberated classical statistics from some historical legacies like unbiasedness and in this way broadened it. We will discuss this particular point again in the chapter on hierarchical Bayes analysis.

One more concept is needed to evaluate the performance of a decision function. Let the loss $L(\theta, a)$ be a measure of how good the action a is when θ is the value of the parameter: the smaller the loss better the action a relative to θ.

In estimation, a commonly used $L(\theta, a)$ is the squared error loss function. In testing $H_0 : \theta = \theta_0$ versus $H_1 : \theta \neq \theta_0$, a commonly used loss is the 0-1 loss, namely,

$$L(\theta, a) = \begin{cases} 0 & \text{if } \theta = \theta_0 \text{ and } a = a_0 \text{ or } \theta \neq \theta_0 \text{ and } a = a_1; \\ 1 & \text{otherwise.} \end{cases}$$

In interval estimation there is no commonly used loss function. One choice would be a suitable penalty for length and failure to cover θ by a chosen interval $[a, b]$, e.g.,

$$L(\theta, [a, b]) = c_1 L_1(\theta, [a, b]) + c_2(b - a),$$

where

$$L_1(\theta, [a, b]) = \begin{cases} 1 & \text{if } \theta \in [a, b]; \\ 0 & \text{otherwise.} \end{cases}$$

To evaluate a decision function $\delta(\boldsymbol{x})$, one calculates the average loss

$$E_\theta \left(L(\theta, \delta(\boldsymbol{X})) \right) \stackrel{\text{def}}{=} R(\theta, \delta).$$

This is a function of θ.

How would one define an optimal decision function? In estimation, if one confines attention to unbiased decision functions, then $R(\theta, \delta) = \text{Var}(\delta(\boldsymbol{X})|\theta)$. Sometimes one can get a single δ_0 minimizing $R(\theta, \delta)$ for all θ among all unbiased δ. Without the restriction to unbiasedness, this is no longer the case. Similar questions arise in testing and other problems also. Clearly, new principles are called for. We can introduce a weight function $\pi(\theta)$ and minimize the weighted risk

$$R(\pi, \delta) = \int R(\theta, \delta)\pi(\theta)d\theta.$$

A δ_0 minimizing this is called a Bayes rule. This is a problem that we discuss in Chapter 2. There $\pi(\theta)$ is interpreted as a quantification of prior belief and is called a prior distribution of θ. We say δ_0 is a *Bayes rule in the limit* (or *Bayes in the wide sense*, (Wald (1950))), if for a sequence of priors π_i,

$$\lim_{i \to \infty} [R(\pi_i, \delta_0) - \inf_\delta R(\pi_i, \delta)] = 0.$$

A somewhat conservative optimization principle is to minimize

$$\sup_\theta R(\theta, \delta).$$

A decision rule δ_0 is said to be minimax if

$$\sup_\theta R(\theta, \delta_0) = \inf_\delta \sup_\theta R(\theta, \delta).$$

A sufficient condition for a rule δ_0 to be minimax is that δ_0 minimizes $R(\pi, \delta)$ for some π and has constant risk $R(\theta, \delta_0) = c$. Then

$$\sup_\theta R(\theta, \delta_0) = c = R(\pi, \delta_0)$$
$$\leq R(\pi, \delta)$$
$$\leq \sup_\theta R(\theta, \delta).$$

This argument is due to Wald (1950). In Problem 16, you are asked to prove that if a rule δ_0 is Bayes in the limit and has constant risk, then it is minimax.

1.6 The Changing Face of Classical Inference

Because the exact theories of optimal estimates are difficult to apply, attention has shifted to approximate algorithmic methods, like the EM algorithm, simulation, and asymptotics. Along with this, there has been much interest in robust methods that do well under a broad spectrum of models. As an example, we discuss the method of *Bootstrap* due to Efron (see Efron (1982)).

We illustrate the method of Bootstrap by showing how to calculate, say, the variance of $\tau(\hat{\theta})$ for a given τ. The original sample is $(x_1, x_2, \ldots, x_n)$. We sample from this data set n times at random and replacing each chosen item before the next draw. This produces a *pseudo* data set that we denote by $(x_1^*, x_2^*, \ldots, x_n^*)$. We calculate $\hat{\theta}^*$ from this pseudo data and then $\tau(\hat{\theta}^*)$. We repeat this N times (where N is much larger than n) to generate N pseudo data sets and N pseudo values of $\tau(\hat{\theta})$, which we denote as $\tau_1^*, \tau_2^*, \ldots, \tau_N^*$. The estimate for $E(\tau(\hat{\theta})|\theta)$ is $\bar{\tau}^* = \frac{1}{N} \sum_1^N \tau_i^*$ and an estimate for $\mathrm{Var}(\tau(\hat{\theta})|\theta)$ is $\frac{1}{N} \sum (\tau_i^* - \bar{\tau}^*)^2$. There is considerable numerical and theoretical evidence that show the Bootstrap estimates are superior to earlier methods like the delta method discussed in Section 1.4.

Finally, classical statistics has come up with many new methods for dealing with high-dimensional problems. A couple of them will be discussed in Chapter 9.

1.7 Exercises

1. Verify that $N(\mu, \sigma^2)$, exponential with $f(x|\theta) = \frac{1}{\theta} e^{-x/\theta}$, Bernoulli($p$), binomial B($n, p$), and Poisson $\mathcal{P}(\lambda)$, each constitutes an exponential family.
2. Verify (1.4).
3. Assuming $p = 1$ in (1.4), show that $\frac{d\eta}{d\theta} > 0$.
4. (a) Generate data by drawing a sample of size $n = 30$ from $N(\mu, 1)$ with $\mu = 2$. For your data, plot the likelihood function and comment on its shape and how informative it is about μ.
   (b) For an exponential family, show that the likelihood function is log concave, i.e., the matrix with (i, j)th element $\frac{\partial^2 \log L}{\partial\theta_i \partial\theta_j}$ is negative definite. (Hint. The proof is similar to that for Problem 3. By direct calculation

$$\frac{\partial^2 \log L}{\partial\theta_i \partial\theta_j} = \frac{\partial^2 c}{\partial\theta_i \partial\theta_j} = E_{\boldsymbol{\theta}}\left(\frac{\partial^2 \log L}{\partial\theta_i \partial\theta_j}\right) = -E_{\boldsymbol{\theta}}\left(\frac{\partial \log L}{\partial\theta_i}\frac{\partial \log L}{\partial\theta_j}\right).$$

   Now use the fact that a variance-covariance matrix is positive definite, unless the distribution is degenerate).
   (c) Let $X_1, \ldots, X_n$ be i.i.d. with density $f(x|\theta)$, $p = 1$, in an exponential family. Show that MLE of η is $(1/n) \sum_{i=1}^n t(X_i)$ and hence the MLE $\hat{\theta} \xrightarrow{P} \theta$ as $n \to \infty$.
5. Let $X_1, X_2, \ldots, X_n$ be i.i.d $N(\mu, \sigma^2)$, with μ, σ^2 unknown. Let $\tau(\mu, \sigma^2) = P\{X_1 \le 0 | \mu, \sigma^2\}$.
   (a) Calculate $\tau(\hat{\mu}, \hat{\sigma}^2)$, where $\hat{\mu}$, and $\hat{\sigma}^2$ are the MLE of μ and σ^2.
   (b) Show that the best unbiased estimate of $\tau(\mu, \sigma^2)$ is

$$W(\boldsymbol{X}) = E\left(I\{X_1 \le 0\}|\bar{X}, S^2\right) = F(-\bar{X}/S)$$

   where S^2 is the sample variance and F is the distribution function of $(X_1 - \bar{X})/S$.
   (c) For $\mu = 0, \sigma^2 = 1, n = 36$ find the mean squared errors

$E\{(\tau(\hat{\mu}, \hat{\sigma}^2) - \tau(0, 1))^2 | 0, 1\}$ and $E\{(W(\boldsymbol{X}) - \tau(0, 1))^2 | 0, 1\}$ approximately by simulations.

(d) Estimate the mean, variance and the mean squared error of $\tau(\hat{\mu}, \hat{\sigma}^2)$ by (i) delta method, (ii) Bootstrap, and compare with (c).

6. Let $X_1, X_2, \ldots, X_n$ be i.i.d. with density $(1/\sigma)f((x - \mu)/\sigma)$. Show that for fixed σ, $P_{\mu,\sigma}\{\sum_1^n X_i > c\}$ is an increasing function of μ.

7. $X_1, X_2, \ldots, X_n$ are said to have a family of densities $f(\boldsymbol{x}|\theta)$ with monotone likelihood ratio (MLR) in $T(\boldsymbol{x})$ if there exists a sufficient statistic $T(\boldsymbol{x})$ such that $f(\boldsymbol{x}|\theta_2)/f(\boldsymbol{x}|\theta_1)$ is a non-decreasing function of $T(\boldsymbol{x})$ if $\theta_2 > \theta_1$.

   (a) Verity that exponential families have this property.

   (b) If $f(\boldsymbol{x}|\theta)$ has MLR in T, show that $P_\theta\{T(\boldsymbol{X}) > c\}$ is non-decreasing in θ.

8. Let $X_1, X_2, \ldots, X_n$ be i.i.d. $N(\mu, 1)$. Let $0 < \alpha, \beta < 1$, $\Delta > 0$.

   (a) For $H_0 : \mu = \mu_0$ versus $H_1 : \mu > \mu_0$, show the smallest sample size n for which the UMP test has probability of error of first kind equal to α and probability of error of second kind $\leq \beta$ for $\mu \geq \mu_0 + \Delta$ is (approximately) the integer part of $((z_\alpha + z_\beta)/\Delta)^2 + 1$.
   Evaluate n numerically when $\Delta = .5$, $\alpha = .01$, $\beta = .05$.

   (b) For $H_0 : \mu = \mu_0$ versus $H_1 : \mu \neq \mu_0$, show the smallest n such that UMPU test has probability of error of first kind equal to α and probability of error of second kind $\leq \beta$ for $|\mu - \mu_0| \geq \Delta$ is (approximately) the solution of

$$\Phi\left(z_{\alpha/2} - \sqrt{n}\Delta\right) + \Phi\left(z_{\alpha/2} + \sqrt{n}\Delta\right) = 1 + \beta.$$

   Evaluate n numerically when $\Delta = 0.5$, $\alpha = .01$ and $\beta = .05$.

9. Let $X_1, X_2, \ldots, X_n$ be i.i.d $U(0, \theta)$, $\theta > 0$. Find the smallest n such that the UMP test of $H_0 : \theta = \theta_0$ against $H_1 : \theta > \theta_0$ has probability of error of first kind equal to α and probability of error of second kind $\leq \beta$ for $\theta \geq \theta_1$, with $\theta_1 > \theta_0$.

10. (Basu (1988, p.1)) Let $X_1, X_2, \ldots, X_n$ be i.i.d $U(\theta, 2\theta)$, $\theta > 0$.

    (a) What is the likelihood function of θ?

    (b) What is the minimal sufficient statistic in this problem?

    (c) Find $\hat{\theta}$, the MLE of θ.

    (d) Let $X_{(1)} = \min(X_1, \ldots, X_n)$ and $T = (4\hat{\theta} + X_{(1)})/5$. Show that $E\left((T - \theta)^2\right)/E\left((\hat{\theta} - \theta)^2\right)$ is always less than 1, and further,

$$\frac{E\left((T - \theta)^2\right)}{E\left((\hat{\theta} - \theta)^2\right)} \longrightarrow \frac{12}{25} \text{ as } n \to \infty.$$

11. Suppose $X_1, X_2, \ldots, X_n$ are i.i.d $N(\mu, 1)$. A statistician has to test $H_0 : \mu = 0$; he selects his alternative depending on data. If $\bar{X} < 0$, he tests against $H_1 : \mu < 0$. If $\bar{X} > 0$, his alternative is $H_1 : \mu > 0$.

    (a) If the statistician has taken $\alpha = .05$, what is his real α?

    (b) Calculate his power at $\mu = \pm 1$ when his nominal $\alpha = .05$, $n = 25$. Will this power be smaller than the power of the UMPU test with $\alpha = .05$?

12. Consider n patients who have received a new drug that has reduced their blood pressure by amounts $X_1, X_2, \ldots, X_n$. It may be assumed that $X_1, X_2, \ldots, X_n$ are i.i.d. $N(\mu, \sigma^2)$ where σ^2 is assumed known for simplicity. On the other hand, for a standard drug in the market it is known that the average reduction in blood pressure is μ_0. The company producing the new drug claims $\mu = \mu_0$, i.e., it does what the old drug does (and probably costing much less). Discuss what should he H_0 and H_1 here. (This is a problem of bio-equivalence.)

13. (P-values) The error probabilities of a test do not provide a measure of the strength of evidence against H_0 in a particular data set. The P-values defined below try to capture that.

    Suppose $H_0 : \theta = \theta_0$ and your test is to reject H_0 for large values of a test statistic $W(\boldsymbol{X})$, say, you reject H_0 if $W > W_\alpha$. Then, when $\boldsymbol{X} = \boldsymbol{x}$ is observed, the P-value is defined as

$$P(\boldsymbol{x}) = 1 - F_{\theta_0}^W \left(W(\boldsymbol{x}) \right),$$

   where $F_{\theta_0}^W$ = distribution function of W under θ_0.

   (a) Show that if $F_{\theta_0}^W$ is continuous then $P(\boldsymbol{X})$ has uniform distribution on $(0, 1)$.

   (b) Suppose you are not given the value of W but you know P. How will you decide whether to accept or reject H_0 ?

   (c) Let $X_1, X_2, \ldots, X_n$ be i.i.d $N(\mu, 1)$. You are testing $H_0 : \mu = \mu_0$ versus $H_1 : \mu \neq \mu_0$. Define P-value for the UMPU test. Calculate $E_{\mu_0}(P)$ and $E_{\mu_0}(P | P \leq \alpha)$.

14. (a) Let $f(\boldsymbol{x}|\theta_0)$, $f(\boldsymbol{x}|\theta_1)$ and I_0 be as in Part $\mathbf{A}$ of the Neyman-Pearson Lemma. The constant k is chosen not from given α but such that

$$E_{\theta_0}(I_0) = 1 - E_{\theta_1}(I_0).$$

   Then show that I_0 is minimax, i.e., I_0 minimizes the maximum error probability,

$$\max(E_{\theta_0}(I_0), 1 - E_{\theta_1}(I_0)) \leq \max(E_{\theta_0}(I), 1 - E_{\theta_1}(I)).$$

   (b) Let $X_1, X_2, \ldots, X_n$ be i.i.d. $N(\mu, 1)$. Using (a) find the minimax test of $H_0 : \mu = -1$ versus $H_1 : \mu = +1$.

15. (a) Let $\boldsymbol{X}$ have density $f(\boldsymbol{x}|\theta)$ and $\Theta = \{\theta_0, \theta_1\}$. The null hypothesis is $H_0 : \theta = \theta_0$, the alternative is $H_1 : \theta = \theta_1$. Suppose the error probabilities of each randomized test ϕ is denoted by $(\alpha_\phi, \beta_\phi)$ and $S=$ the collection of all points $(\alpha_\phi, \beta_\phi)$. S is called the risk set. Show that S is convex.

   (b) Let X be $B(2, p)$, $p = \frac{1}{2}$ (corresponding with H_0) or $\frac{1}{4}$ (corresponding with H_1). Plot the risk set S as a subset of the unit square.

   (Hint. Identify the lower boundary of S as a polygon with vertices corresponding with non-randomized most powerful tests. The upper boundary connects vertices corresponding with least powerful tests that are similar to I_0 in the N-P lemma but with reverse inequalities.)

16. (a) Suppose δ_0 is a decision rule that has constant risk and is Bayes in the limit (as defined in Section 1.5). Show that δ_0 is minimax.

    (b) Consider i.i.d. observations $X_1, \ldots, X_n$ from $N(\mu, 1)$. Using a normal prior distribution for μ, show that $\bar{X}$ is a minimax estimate for μ under squared error loss.

17. Let $X_1, X_2, \ldots, X_n$ be i.i.d. $N(\mu, \sigma^2)$. Consider estimating μ.

    (a) Show that both $\bar{X}$ and the sample median M are unbiased estimators of μ.

    (b) Further, show that both of them are consistent and asymptotically normal.

    (c) Discuss why you would prefer one over the other.

18. Let $X_1, X_2, \ldots, X_n$ be i.i.d. $N(\mu, \sigma^2)$, $Y_1, Y_2, \ldots, Y_m$ be i.i.d. $N(\eta, \tau^2)$ and let these two samples be independent also. Find the set of minimal sufficient statistics when

    (a) $-\infty < \mu, \eta < \infty$, $\sigma^2 > 0$ and $\tau^2 > 0$.

    (b) $\mu = \eta$, $-\infty < \mu < \infty$, $\sigma^2 > 0$ and $\tau^2 > 0$.

    (c) $-\infty < \mu, \eta < \infty$, $\sigma^2 = \tau^2$, and $\sigma^2 > 0$.

    (d) $\mu = \eta$, $\sigma^2 = \tau^2$, $-\infty < \mu < \infty$, and $\sigma^2 > 0$.

19. Suppose $\mathbf{X}_i, i = 1, 2, \ldots, n$ are i.i.d. from the exponential family with density (1.2) having full rank, i.e., the parameter space contains a p-dimensional open rectangle. Then show that $(T_j = \sum_{i=1}^{n} t_j(\mathbf{X}_i), j = 1, \ldots, p)$ together form a minimal sufficient statistic.

20. Refer to the 'factorization theorem' in Section 1.3. Show that a statistic U is sufficient if and only if for every pair θ_1, θ_2, the ratio $f(x|\theta_2)/f(x|\theta_1)$ is a function of $U(x)$.

2

Bayesian Inference and Decision Theory

This chapter is an introduction to basic concepts and implementation of Bayesian analysis. We begin with subjective probability as distinct from classical or objective probability of an uncertain event based on the long run relative frequency of its occurrence. Subjective probability, along with utility or loss function, leads to Bayesian inference and decision theory, e.g., estimation, testing, prediction, etc.

Elicitation of subjective probability is relatively easy when the observations are exchangeable. We discuss exchangeability, its role in Bayesian analysis, and its importance for science as a whole.

In most cases in practice, quantification of subjective belief or judgment is not easily available. It is then common to choose from among conventional priors on the basis of some relatively simple subjective judgments about the problem and the conventional probability model for the data. Such priors are called objective or noninformative. These priors have been criticized for various reasons. For example, they depend on the form of the likelihood function and usually are improper, i.e., the total probability of the parameter space is infinity. Here in Chapter 2, we discuss how they are applied; some answers to the criticisms are given in Chapter 5.

In Section 2.3 of this chapter, there is a brief discussion of the many advantages of being a Bayesian.

2.1 Subjective and Frequentist Probability

Probability has various connotations. Historically, it has been connected with both personal evaluation of uncertainty, as in gambling or other decision making under uncertainty, and predictions about proportion of occurrence of some uncertain event. Thus when a person says the probability is half that this particular coin will turn up a head, then it will usually mean that in many tosses about half the time it will be a head (a version of the law of large numbers). But it can also mean that if someone puts this bet on head – if head he wins a dollar, if not he loses a dollar – the gamble is fair. The first interpretation is frequentist, the second subjective. Similarly one can have both interpretations in mind when a weather forecast says there is a probability of 60% of rain, but the subjective interpretation matters more. It helps you decide if you will take an umbrella. Finally, one can think up situations, e.g., election of a particular candidate or success of a particular student in a particular test, where only the subjective interpretation is valid.

Some scientists and philosophers, notably Jeffreys and Carnap, have argued that there may be a third kind of probability that applies to scientific hypotheses. It may be called objective or conventional or non-subjective in the sense that it represents a shared belief or shared convention rather than an expression of one person's subjective uncertainty.

Fortunately, the probability calculus remains the same, no matter which kind of probability one uses. A Bayesian takes the view that all unknown quantities, namely the unknown parameter and the data before observation, have a probability distribution. For the data, the distribution, given $\boldsymbol{\theta}$, comes from a model that arises from past experience in handling similar data as well as subjective judgment. The distribution of $\boldsymbol{\theta}$ arises as a quantification of the Bayesian's knowledge and belief. If her knowledge and belief are weak, she may fall back on a common objective distribution in such situations.

Excellent expositions of subjective and objective Bayes approaches are Savage (1954, 1972), Jeffreys (1961), DeGroot (1970), Box and Tiao (1973), and Berger (1985a). Important relatively recent additions to the literature are Bernardo and Smith (1994), O'Hagan (1994), Gelman et al. (1995), Carlin and Louis (1996), Leonard and Hsu (1999), Robert (2001), and Congdon (2001).

2.2 Bayesian Inference

Informally, to make inference about $\boldsymbol{\theta}$ is to learn about the unknown $\boldsymbol{\theta}$ from data $\boldsymbol{X}$, i.e., based on the data, explore which values of $\boldsymbol{\theta}$ are probable, what might be plausible numbers as estimates of different components of $\boldsymbol{\theta}$ and the extent of uncertainty associated with such estimates. In addition to having a model $f(\boldsymbol{x}|\boldsymbol{\theta})$ and a likelihood function, the Bayesian needs a distribution for $\boldsymbol{\theta}$. The distribution is called a prior distribution or simply a prior because it quantifies her uncertainty about $\boldsymbol{\theta}$ prior to seeing data. The prior may represent a blending of her subjective belief and knowledge, in which case it would be a subjective prior. Alternatively, it could be a conventional prior supposed to represent small or no information. Such a prior is called an objective prior. We discuss construction of objective priors in Chapter 5 (and in Section 6.7.3 to some extent). An example of elicitation of subjective prior is given in Section 5.4.

Given all the above ingredients, the Bayesian calculates the conditional probability density of $\boldsymbol{\theta}$ given $\boldsymbol{X} = \boldsymbol{x}$ by Bayes formula

$$\pi(\boldsymbol{\theta}|\boldsymbol{x}) = \frac{\pi(\boldsymbol{\theta})f(\boldsymbol{x}|\boldsymbol{\theta})}{\int_{\Theta} \pi(\boldsymbol{\theta}')f(\boldsymbol{x}|\boldsymbol{\theta}')d\boldsymbol{\theta}'} \tag{2.1}$$

where $\pi(\boldsymbol{\theta})$ is the prior density function and $f(\boldsymbol{x}|\boldsymbol{\theta})$ is the density of $\boldsymbol{X}$, interpreted as the conditional density of $\boldsymbol{X}$ given $\boldsymbol{\theta}$. The numerator is the joint density of $\boldsymbol{\theta}$ and $\boldsymbol{X}$ and the denominator is the marginal density of $\boldsymbol{X}$. The symbol $\boldsymbol{\theta}$ now represents both a random variable and its value. When the parameter $\boldsymbol{\theta}$ is discrete, the integral in the denominator of (2.1) is replaced by a sum.

The conditional density $\pi(\boldsymbol{\theta}|\boldsymbol{x})$ of $\boldsymbol{\theta}$ given $\boldsymbol{X} = \boldsymbol{x}$ is called the posterior density, a quantification of our uncertainty about $\boldsymbol{\theta}$ in the light of data. The transition from $\pi(\boldsymbol{\theta})$ to $\pi(\boldsymbol{\theta}|\boldsymbol{x})$ is what we have learnt from the data.

A Bayesian can simply report her posterior distribution, or she could report summary descriptive measures associated with her posterior distribution. For example, for a real valued parameter $\boldsymbol{\theta}$, she could report the posterior mean

$$E(\boldsymbol{\theta}|\boldsymbol{x}) = \int_{-\infty}^{\infty} \boldsymbol{\theta}\pi(\boldsymbol{\theta}|\boldsymbol{x})d\boldsymbol{\theta}$$

and the posterior variance

$$\mathrm{Var}\,(\boldsymbol{\theta}|\boldsymbol{x}) = E\{(\boldsymbol{\theta} - E(\boldsymbol{\theta}|\boldsymbol{x}))^2|\boldsymbol{x}\}$$
$$= \int_{-\infty}^{\infty} (\boldsymbol{\theta} - E(\boldsymbol{\theta}|\boldsymbol{x}))^2\pi(\boldsymbol{\theta}|\boldsymbol{x})d\boldsymbol{\theta}$$

or the posterior standard deviation. Finally, she could use the posterior distribution to answer more structured problems like estimation and testing. In the case of estimation of θ, one would report the above summary measures. In the case of testing one would report the posterior odds of the relevant hypotheses.

Example 2.1. We illustrate these ideas with an example of inference about μ for normally distributed data $(N(\mu, \sigma^2))$ with mean μ and variance σ^2. The data consist of i.i.d. observations $X_1, X_2, \cdots, X_n$ from this distribution. To keep the example simple we assume $n = 10$ and σ^2 is known. A mathematically convenient and reasonably flexible prior distribution for μ is a normal distribution with suitable prior mean and variance, which we denote by η and τ^2. To fix ideas we take $\eta = 100$. The prior variance τ^2 is a measure of the strength of our belief in the prior mean $\eta = 100$ in the sense that the larger the value of τ^2, the less sure we are about our prior guess about η. Jeffreys (1961) has suggested we can calibrate τ^2 by comparing with σ^2. For example, setting $\tau^2 = \sigma^2/m$ would amount to saying information about η is about as strong as the information in m observations in data. Some support for this interpretation is provided in Chapter 5. By way of illustration, we take $m = 1$. With a little algebra (vide Problem 2), the posterior distribution can be shown to be normal with posterior mean

$$E(\mu|\boldsymbol{X}) = (\frac{1}{\tau^2}\eta + \frac{n}{\sigma^2}\bar{X})/(\frac{1}{\tau^2} + \frac{n}{\sigma^2}) = (\eta + 10\bar{X})/11 \tag{2.2}$$

and posterior variance

$$(\frac{\sigma^2}{n}\tau^2)/(\frac{\sigma^2}{n} + \tau^2) = \sigma^2/11 \tag{2.3}$$

i.e., in the light of the data, μ shifts from prior guess η towards a weighted average of the prior guess about μ and $\bar{X}$, while the variability reduces from σ^2 to $\sigma^2/11$. If the prior information is small, implying large τ^2 or there are lots of data, i.e., n is large, the posterior mean is close to the MLE $\bar{X}$.

We will see later that we can quantify how much we have learnt from the data by comparing $\pi(\mu)$ and $\pi(\mu|\boldsymbol{X})$. The posterior depends on both the prior and the data. As data increase the influence of data tends to wash away the prior. Our second example goes back in principle to Bayes, Laplace, and Karl Pearson (*The Grammar of Science*, 1892).

Example 2.2. Consider an urn with Np red and $N(1 - p)$ black balls, p is unknown but N is a known large number. Balls are drawn at random one by one and with replacement, selection is stopped after n draws. For $i = 1, 2, \ldots, n$, let

$$X_i = \begin{cases} 1 & \text{if the } i\text{th ball drawn is red;} \\ 0 & \text{otherwise.} \end{cases}$$

Then X_i's are i.i.d $B(1, p)$, i.e., *Bernoulli* with probability of success p. Let p have a prior distribution $\pi(p)$. We will consider a family of priors for p that simplifies the calculation of posterior and then consider some commonly used priors from this family. Let

$$\pi(p) = \frac{\Gamma(\alpha + \beta)}{\Gamma(\alpha)\Gamma(\beta)}p^{\alpha-1}(1 - p)^{\beta-1}, \quad 0 \leq p \leq 1; \alpha > 0, \beta > 0. \tag{2.4}$$

This is called a Beta distribution. (Note that for convenience we take p to assume all values between 0 and 1, rather than only $0, 1/N, 2/N$, etc.) The prior mean and variance are $\alpha/(\alpha + \beta)$ and $\alpha\beta/\{(\alpha + \beta)^2(\alpha + \beta + 1)\}$, respectively.

By Bayes formula, the posterior density can be written as

$$\pi(p|\boldsymbol{X} = \boldsymbol{x}) = C(\boldsymbol{x})p^{\alpha+r-1}(1 - p)^{\beta+(n-r)-1} \tag{2.5}$$

where $r = \sum_{i=1}^{n} x_i =$ number of red balls, and $(C(\boldsymbol{x}))^{-1}$ is the denominator in the Bayes formula. A comparison with (2.4) shows the posterior is also a Beta density with $\alpha + r$ in place of α and $\beta + (n - r)$ for β and

$$C(\boldsymbol{x}) = \Gamma(\alpha + \beta + n)/\{\Gamma(\alpha + r)\Gamma(\beta + n - r)\}.$$

The posterior mean and variance are

$$E(p|\boldsymbol{x}) = (\alpha + r)/(\alpha + \beta + n),$$

$$\text{Var}\,(p|\boldsymbol{x}) = \frac{(\alpha + r)(\beta + n - r)}{(\alpha + \beta + n)^2(\alpha + \beta + n + 1)}. \tag{2.6}$$

As indicated earlier, a Bayesian analyst may just report the posterior (2.5), and the posterior mean and variance, which provide an idea of the center and dispersion of the posterior distribution. It will not escape one's attention that if n is large then the posterior mean is approximately equal to the MLE, $\hat{p} = r/n$ and the posterior variance is quite small, so the posterior is concentrated around $\hat{p}$ for large n. We can interpret this as an illustration of a fact mentioned before when we have lots of data, the data tend to wash away the influence of the prior.

The posterior mean can be rewritten as a weighted average of the prior mean and MLE.

$$\frac{(\alpha + \beta)}{(\alpha + \beta + n)} \frac{\alpha}{(\alpha + \beta)} + \frac{n}{(\alpha + \beta + n)} \frac{r}{n}.$$

Once again, the importance of both the prior and the data comes out, the relative importance of the prior and the data being measured by $(\alpha + \beta)$ and n.

Suppose we want to predict the probability of getting a red ball in a new $(n + 1)$-st draw given the above data. This has been called a fundamental problem of science. It would be natural to use $E(p|\boldsymbol{x})$, the same estimate as above. We list below a number of commonly used priors and the corresponding value of $E(p|X_1, X_2, \cdots, X_n)$.

The uniform prior corresponds with $\alpha = \beta = 1$, with posterior mean equal to $(\sum_1^n X_i + 1)/(n + 2)$. This was a favorite of Laplace and Bayes but not so popular anymore. If $\alpha = \beta = \frac{1}{2}$, we have the Jeffreys prior with posterior mean $(\sum_1^n X_i + \frac{1}{2})/(n+1)$. This prior is very popular in the case of one-dimensional θ as here. It is also a reference prior due to Bernardo (1979). Reference priors are very popular. If we take a Beta density with $\alpha = 0$, $\beta = 0$, it integrates to infinity. Such a prior is called improper. If we still use the Bayes formula to produce a posterior density, the posterior is proper unless $r = 0$ or n. The posterior mean is exactly equal to the MLE.

Objective priors are usually improper. To be usable they must have proper posteriors. It is argued in Chapter 5 that improper priors are best understood through the posteriors they produce. One might examine whether the posterior seems reasonable.

Suppose we think of the problem as a representation of production of defective and non-defective items in a factory producing switches, we would take red to mean defective and black to mean a good switch. In this context, there would be some prior information available from the engineers. They may be able to pinpoint the likely value of p, which may be set equal to the prior mean $\alpha/(\alpha + \beta)$. If one has some knowledge of prior variability also, one would have two equations from which to determine α and β. In this particular context, the Jeffreys prior with a lot of mass at the two end points might be adequate if the process maintains a high level of quality (small p) except when it is out of control and has high values of p. The peak of the prior near $p = 1$

Table 2.1. An Epidemiological Study

	Food Eaten			
	Crabmeat		No Crabmeat	
	Potato Salad	No Potato Salad	Potato Salad	No Potato Salad
Ill	120	4	22	0
Not Ill	80	31	24	23

could reflect frequent occurrence of lack of control or a pessimistic prior belief to cope with disasters.

It is worth noting that the uniform, Jeffreys prior, and reference priors are examples of objective priors and that all of them produce a posterior mean that is very close to the MLE even for small n. Also all of them make better sense than the MLE in the extreme case when $\hat{p} = 0$. In most contexts the estimate $\hat{p} = 0$ is absurd, the objective Bayes estimates move it a little towards $p = \frac{1}{2}$, which corresponds with total ignorance in some sense. Such a movement is called a *shrinkage*. Agresti and Caffo (2000) and Brown et al. (2003) have shown that such estimates lead to confidence intervals with closer agreement between nominal and true coverage probability than the usual confidence intervals based on normal approximation to $\hat{p}$ or inversion of tests. In other words, the Bayesian approach seems to lead to a more reasonable point estimate as well as a more reliable confidence interval than the common classical answers based on MLE.

Example 2.3. This example illustrates the advantages of a Bayesian interpretation of probability of making a wrong inference for given data as opposed to classical error probabilities over repetitions. In this epidemiological study repetitions don't make sense.

The data in Table 2.1 on food poisoning at an outing are taken from Bishop et al. (1975) who provide the original source of the study. Altogether 320 people attended the outing, 304 responded to questionnaires.

There was other food also but only two items, potato salad and crabmeat, attracted suspicion. We focus on the main suspect, namely, potato salad. A partial Bayesian analysis of this example will be presented later in Chapter 4.

Example 2.4. Let $X_1, X_2, \ldots, X_n$ be i.i.d $N(\mu, \sigma^2)$ and assume for simplicity σ^2 is known. As in Chapter 1, μ may be the expected reduction of blood pressure due to a new drug. You want to test $H_0 : \mu \leq \mu_0$ versus $H_1 : \mu > \mu_0$, where μ_0 corresponds with a standard drug already in the market.

Let $\pi(\mu)$ be the prior. First calculate the posterior density $\pi(\mu|\boldsymbol{X})$. Then calculate

$$\int_{-\infty}^{\mu_0} \pi(\mu|\boldsymbol{X})d\mu = P\{H_0|\boldsymbol{X}\},$$

and

$$\int_{\mu_0}^{\infty} \pi(\mu|\boldsymbol{X})d\mu = 1 - P\{H_0|\boldsymbol{X}\} = P\{H_1|\boldsymbol{X}\}.$$

One may simply report these numbers or choose one of the two hypotheses if one of the two probabilities is substantially bigger.

We provide some calculations when the prior for μ is $N(\eta, \tau^2)$. We recall from Example 2.1 that the posterior for μ is normal with mean and variance given by equations (2.2) and (2.3). If follows that

$$\pi(\mu \le \mu_0|\boldsymbol{X}) = \Phi(z) \text{ and } \pi(\mu > \mu_0|\boldsymbol{X}) = 1 - \Phi(z)$$

where Φ is the standard normal distribution function and

$$z = \frac{\mu_0 - (\frac{1}{\tau^2}\eta + \frac{n}{\sigma^2}\bar{X})/(\frac{1}{\tau^2} + \frac{n}{\sigma^2})}{\{\frac{\sigma^2}{n}\tau^2/(\frac{\sigma^2}{n} + \tau^2)\}^{\frac{1}{2}}}.$$

A conventional choice is to make $\tau^2 \to \infty$ above, which would give the same result as assuming an improper uniform prior

$$\pi(\mu) = c, \qquad -\infty < \mu < \infty.$$

Any of these would lead to

$$z = (\mu_0 - \bar{X})\frac{\sqrt{n}}{\sigma}.$$

Suppose we wish to reject if the posterior odds against H_0 are 19:1 or more i.e., if posterior probability of H_0 is $\le .05$. Then we reject H_0 if

$$\mu_0 - \bar{X} \le (-1.64)\frac{\sigma}{\sqrt{n}}$$

$$\text{or } \bar{X} \ge \mu_0 + 1.64\frac{\sigma}{\sqrt{n}},$$

which is exactly the same as the classical test for this problem with $\alpha = .05$.

However if we had wished to test the sharp null hypothesis $H_0 : \mu = \mu_0$ against $H_1 : \mu \ne \mu_0$ or $H_1 : \mu > \mu_0$, we have to choose the prior in a different way since the prior we chose would assign zero probability to H_0. Moreover, the answers tend to be very different from classical answers as we shall see in Chapter 6.

2.3 Advantages of Being a Bayesian

The Bayesian approach provides a fairly explicit solution to common problems of statistical inference (Chapters 2 and 8), new problems of high-dimensional data analysis that are coming up because of emergence of high-dimensional data sets (Chapters 9 and 10), as well as complex decision problems of real life (Chapter 10). It can handle presence of prior knowledge or partial prior knowledge, specially constraints like $a \leq \theta \leq b$ relatively easily. In some cases, a subjective prior can be elicited (Chapter 5), and in most other cases one can choose objective priors. Of course, in all cases one would wish to study to some extent the robustness of various aspects of the posterior with respect to modest variation in prior as illustrated in Chapter 3.

In classical decision theory, there are theorems due to Wald that imply that Bayes rules and their limits together form a complete class, i.e., any decision rule that is not of this form can be improved by a rule of this form. In a similar vein and as a sort of converse, Wald (1950) also proved that if a decision rule is admissible then it must be Bayes or limit of Bayes rules. There are various senses in which a decision rule δ can be a Bayes rule in the limit.

In this book, we stress objective priors, because it still seems difficult to elicit fully subjective priors, at least in most problems in practice. If a fully subjective prior is available we would indeed use it. In particular, whatever subjective input is available ought to be used, specially in high-dimensional problems.

The Bayesian approach can be deduced from several sets of axioms. One such set is discussed in Section 3.3. Moreover, the subjective Bayesian approach is free from certain paradoxes or violation of principles that are associated with classical statistics. These unpleasant properties are due to the fact that classical statistics provides either data dependent measures like P-values which are not easy to interpret or evaluations like risk functions or confidence coefficients that are obtained by integrating over the whole sample space and so may be absurd when a particular data set is in hand. The paradoxes can be quite dramatic. The objective Bayesian approach is not completely free from violation of some of these principles. We discuss some of these issues in Section 5.2.

Bayesians usually accept as a principle that some validation in the real world is good whenever possible. Occasionally, a proxy for the real world may be found in conceptual frequentist constructions of possible real world scenarios and a Bayesian may seek some sort of validation in such cases. By validation in the real world we mean predictive ability. One may use a baseball or cricket or soccer player's performance in the first half of the season to predict his performance in the second half. For a successful application of (parametric empirical) Bayes methodology, relative to classical methods, see Morris (1983) and Ghosh and Meeden (1997). By cross validation, one means that a part of data is used to make an inference and the other part to validate it, even if these two parts do not have a connotation of present and future as in the baseball example of Morris (1983). A validation of Bayesian approach to model selection is given in Hoeting et al. (1999). Most Bayesian papers on new methods offer some validation.

It turns out that in objective Bayesian analysis one often has such frequentist validation; see, for example, the concept of probability matching priors (Subsection 5.1.4). Although this provides some reconciliation between the two approaches as far as the decision that is made, only the objective Bayesian approach has a posterior and hence a data dependent method of evaluating the performance of the decision.

Finally, basic Bayesian ideas and measures are easy to interpret and hence easy to communicate.

One may well ask why in spite of all these advantages, an explosive growth and spread of the Bayesian approach has occurred only recently, in the past fifteen or so years. A major factor has been the arrival of MCMC (Markov chain Monte Carlo) in a big way and consequent advances in computation of posteriors for high-dimensional Θ and many real-life applications. A classic paper that ushered in these changes is Gelfand and Smith (1990).

2.4 Paradoxes in Classical Statistics

The evaluation of performance of an inference procedure in classical statistics is based on expected quantities like bias or variance of an estimate, error probabilities for a test, and confidence coefficients of a confidence interval. Such measures are obtained by integrating or summing over the sample space of all possible data. Hence they do not answer how good the inference is for a particular data set. The following two examples show how irrelevant the classical answers can be once the data are in hand.

Example 2.5. (Cox (1958)) To estimate μ in $N(\mu, \sigma^2)$, toss a fair coin. Have a sample of size $n = 2$ if it is a head and take $n = 1000$ if it is a tail. An unbiased estimate of μ is $\bar{X}_n = \sum_{i=1}^{n} X_i/n$ with variance $= \frac{1}{2}\{\frac{\sigma^2}{2} + \frac{\sigma^2}{100}\} \sim \frac{\sigma^2}{4}$. Suppose it was a tail. Would you believe $\sigma^2/4$ is a measure of accuracy of the estimate?

Example 2.6. (Welch (1939)) Let X_1, X_2 be i.i.d. $U(\theta - \frac{1}{2}, \theta + \frac{1}{2})$. Let $\bar{X} \pm C$ be a 95% confidence interval, $C > 0$ being suitably chosen. Suppose $X_1 = 2$ and $X_2 = 1$. Then we know for sure $\theta = (X_1 + X_2)/2$ and hence $\theta \in (\bar{X} - C, \bar{X} + C)$. Should we still claim we have only 95% confidence that the confidence interval covers θ?

One of us (Ghosh) learned of this example from a seminar of D. Basu at the University of Illinois, Urbana-Champaign, in 1965. Basu pointed out how paradoxical is the confidence coefficient in this example. This perspective doesn't seem to be stressed in Welch (1939). The example has been discussed many times, see Lehmann (1986, Chapter 10, Problems 27 and 28), Pratt (1961), Kiefer (1977), Berger and Wolpert (1988), and Chatterjee and Chattopadhyay (1994).

Fisher was aware of this phenomenon and suggested we could make inference conditional on a suitable ancillary statistic. In Cox's example (Example 2.5), it would be appropriate to condition on the sample size and quote the conditional variance given $n = 1000$ as a proper measure of accuracy. Note that n is ancillary, its distribution is free of θ, so conditioning on it doesn't change the likelihood. In Welch's example (Example 2.6), we could give the conditional probability of covering θ, conditional on $X_1 - X_2 = 1$. Note that $X_1 - X_2$ is also an ancillary statistic like n in Cox's example, it contains no information about θ – so fixing it would not change the likelihood – but its value, like the value of n, gives us some idea about how much information there is in the data. You are asked to carry out Fisher's suggestion in Problem 4.

Suppose you are a classical statistician and faced with this example you are ready to make conditional inference as recommended by Fisher. Unfortunately, there is a catch. Classical statistics also recommends that inference be based on minimal sufficient statistics. These two principles, namely the conditionality principle (CP) and sufficiency principle (SP) together have a far reaching implication. Birnbaum (1962) proved that they imply one must then follow the likelihood principle (LP), which requires that inference be based on the likelihood alone, ignoring the sample space. A precise statement and proof are given in Appendix B.

Bayesian analysis satisfies the likelihood principle since the posterior depends on the data only through the likelihood. Most classical inference procedures violate the likelihood principle.

Closely related to the violation of LP is the stopping rule paradox in classical inference. There is a hilarious example due to Pratt (Berger, 1985a, pp. 30-31).

2.5 Elements of Bayesian Decision Theory

We can approach problems of inference in a mathematically more formal way through statistical decision theory. This would make the problems somewhat abstract and divorced from the real-life connotations but, on the other hand, provides a unified conceptual framework for handling very diverse problems.

A classical statistical decision problem, vide Section 1.5, has the following ingredients. It has as data the observed value of $\boldsymbol{X}$, the density $f(\boldsymbol{x}|\boldsymbol{\theta})$ where the parameter $\boldsymbol{\theta}$ lies in some subset Θ (known as the parameter space) of the p-dimensional Euclidean space $\mathcal{R}^p$. It also has a space $\mathcal{A}$ of actions or decisions $\boldsymbol{a}$ and a loss function $L(\boldsymbol{\theta}, \boldsymbol{a})$ which is the loss incurred when the parameter is $\boldsymbol{\theta}$ and the action taken is $\boldsymbol{a}$. The loss function is assumed to be bounded below so that integrals that appear later are well-defined. Typically, $L(\boldsymbol{\theta}, \boldsymbol{a})$ will be ≥ 0 for all $\boldsymbol{\theta}$ and $\boldsymbol{a}$. We treat actions and decisions as essentially the same in this framework though in non-statistical decision problems there will be some conceptual difference between a decision and the action it leads to. Finally it has a collection of decision functions or rules $\delta(\boldsymbol{x})$ that take values in $\mathcal{A}$. Suppose $\delta(\boldsymbol{x}) = \boldsymbol{a}$ for given $\boldsymbol{x}$. Then the statistician who follows this particular rule $\delta(\boldsymbol{x})$ will choose action $\boldsymbol{a}$ given this particular data and incur the loss $L(\boldsymbol{\theta}, \boldsymbol{a})$.

Both estimation and testing are special cases. Suppose the object is to estimate $\tau(\boldsymbol{\theta})$, a real-valued function of $\boldsymbol{\theta}$. Then $\mathcal{A} = \mathcal{R}$, $L(\boldsymbol{\theta}, a) = (a - \tau(\boldsymbol{\theta}))^2$ and a decision function $\delta(\boldsymbol{x})$ is an estimate of $\tau(\boldsymbol{\theta})$. If it is a problem of testing $H_0 : \boldsymbol{\theta} = \boldsymbol{\theta}_0$ versus $H_1 : \boldsymbol{\theta} \neq \boldsymbol{\theta}_0$, say, then $\mathcal{A} = \{a_0, a_1\}$ where a_j means the decision to accept H_j, $L(\boldsymbol{\theta}, a_j) = 0$ if $\boldsymbol{\theta}$ satisfies H_j and $L(\boldsymbol{\theta}, a_j) = 1$ otherwise. If $I(\boldsymbol{x})$ is the indicator of a rejection region for H_0, then the corresponding $\delta(\boldsymbol{x})$ is equal to a_j if $I(\boldsymbol{x}) = j$, $j = 0, 1$.

We recall also how one evaluates the performance of $\delta(\boldsymbol{x})$ in classical statistics through the average loss or risk function

$$R(\boldsymbol{\theta}, \delta) = E_{\boldsymbol{\theta}}(L(\boldsymbol{\theta}, \delta(\boldsymbol{X}))).$$

If δ is an estimate of $\tau(\boldsymbol{\theta})$ in an estimation problem, then $R(\boldsymbol{\theta}, \delta) = E_{\boldsymbol{\theta}}(\tau(\boldsymbol{\theta}) - \delta(\boldsymbol{X}))^2$ is the MSE (mean squared error). If δ is the indicator function of an H_0-rejection region, then $R(\boldsymbol{\theta}, \delta)$ is the probability of error of first kind if $\boldsymbol{\theta} \in \Theta_0$ and probability of error of second kind if $\boldsymbol{\theta} \in \Theta_1$.

For a Bayesian, $\boldsymbol{\theta}$ is a random variable with prior distribution $\pi(\boldsymbol{\theta})$ before seeing the data, for example, at the planning stage of an experiment. The relevant risk at this stage is the so-called preposterior risk

$$\int_{\Theta} R(\boldsymbol{\theta}, \delta)\pi(\boldsymbol{\theta})d\boldsymbol{\theta} = R(\pi, \delta).$$

It depends on δ and the prior. On the other hand, after the data are in hand, the relevant distribution of $\boldsymbol{\theta}$ is given by the posterior density $\pi(\boldsymbol{\theta}|\boldsymbol{x})$ and the relevant risk is the posterior risk

$$E(L(\boldsymbol{\theta}, \boldsymbol{a})|\boldsymbol{x}) = \psi(\boldsymbol{x}, \boldsymbol{a}).$$

The posterior risk associated with δ is $\psi(\boldsymbol{x}, \delta(\boldsymbol{x}))$. So, in principle, there are two Bayesian decision problems.

A. Given $\boldsymbol{X} = \boldsymbol{x}$, choose an optimal $\boldsymbol{a}$, i.e., choose an $\boldsymbol{a}$ to minimize $\psi(\boldsymbol{x}, \boldsymbol{a})$.
B. At the planning stage, choose an optimal $\delta(\boldsymbol{X})$, denoted as δ_{π} and called the Bayes decision rule or simply the Bayes rule, to minimize $R(\pi, \delta)$.

We have the following pleasant fact, which shows in a sense both problems give the same answer for a given $\boldsymbol{X}$.

Theorem 2.7. *(a) For any δ,*

$$R(\pi, \delta) = E(\psi(\boldsymbol{X}, \delta(\boldsymbol{X}))).$$

b) Suppose $\boldsymbol{a}(\boldsymbol{x})$ minimizes $\psi(\boldsymbol{x}, \boldsymbol{a})$, i.e.

$$\psi(\boldsymbol{x}, \boldsymbol{a}(\boldsymbol{x})) = \inf_{\boldsymbol{a}} \psi(\boldsymbol{x}, \boldsymbol{a}).$$

Then the decision function $\boldsymbol{a}(\boldsymbol{x})$ minimizes $R(\pi, \delta)$.

Proof. (a) Because $E(L(\theta, \delta(\boldsymbol{X}))|\boldsymbol{X}) = \psi(\boldsymbol{X}, \delta(\boldsymbol{X}))$ by definition of ψ, the result follows by taking expectations on both sides.

(b) Let $\boldsymbol{a}(\boldsymbol{x})$, as defined in the theorem, be denoted by δ_0. Then, by part (a), and definition of $\boldsymbol{a}(\boldsymbol{x})$,

$$\begin{aligned}
R(\pi, \delta_0) &= E(\psi(\boldsymbol{X}, \boldsymbol{a}(\boldsymbol{X}))) \\
&\leq E(\psi(\boldsymbol{X}, \delta(\boldsymbol{X}))) \text{ for any } \delta, \\
&= R(\pi, \delta),
\end{aligned}$$

so that $R(\pi, \delta_0) = \inf_\delta R(\pi, \delta)$, as claimed. $\square$

This fact will be used below in the sections on estimation, interval estimation and testing.

2.6 Improper Priors

For point and interval estimates and to some extent in testing, objective priors are often improper. We have considered an improper prior for μ in $N(\mu, \sigma^2)$ earlier in Examples 2.1 and 2.4 but somewhat indirectly. Also, one of the Beta priors in Example 2.2 was improper. We discuss a few basic facts about improper priors. We follow Berger (1985a).

An improper prior density $\pi(\boldsymbol{\theta})$ is non-negative for all $\boldsymbol{\theta}$ but

$$\int_\Theta \pi(\boldsymbol{\theta}) d(\boldsymbol{\theta}) = \infty.$$

Such an improper prior can be used in the Bayes formula for calculating the posterior, provided the denominator is finite for all $\boldsymbol{x}$ (or all but a set of $\boldsymbol{x}$ with zero probability for all $\boldsymbol{\theta}$), i.e.,

$$\int_\Theta \pi(\boldsymbol{\theta}) f(\boldsymbol{x}|\boldsymbol{\theta}) d\boldsymbol{\theta} < \infty.$$

Then the posterior density $\pi(\boldsymbol{\theta}|\boldsymbol{X} = \boldsymbol{x})$ is a proper probability density function and can be used at least in inference problems or the posterior decision problem where we define and minimize $\psi(\boldsymbol{x}, \boldsymbol{a})$. However, for improper priors usually $R(\pi, \delta)$ is not used.

The most common improper priors are

$$\pi_1(\mu) = C, \quad -\infty < \mu < \infty,$$
$$\pi_2(\sigma) = \frac{1}{\sigma}, \quad 0 < \sigma < \infty,$$

for location and scale parameters. Both the improper priors may be interpreted as a sort of limit of the proper priors:

$$\pi_{1,L}(\mu) = \begin{cases} 1/(2L) & \text{if } -L < \mu < L; \\ 0 & \text{otherwise,} \end{cases}$$

$$\pi_{2,L}(\sigma) = \begin{cases} A/\sigma & \text{if } 0 < 1/L < \sigma < L; \\ 0 & \text{otherwise,} \end{cases}$$

where $A = 1/(2\log L)$, in the sense that the posteriors for π_1 and π_2 may be obtained by making $L \to \infty$ in $\pi_{i,L}(\boldsymbol{\theta}|\boldsymbol{X})$. Also, as pointed out by Heath and Sudderth (1978), the posteriors for π_i are same as the posteriors for suitably chosen proper but finitely additive priors.

2.7 Common Problems of Bayesian Inference

There are three common problems, as in classical statistics, namely, point estimation, interval estimation, and testing. We have already seen examples of point estimates and tests of one-sided hypotheses, so we begin with these two problems and then turn to interval estimates (credible intervals) and testing of a sharp null hypothesis. Testing a sharp null hypothesis will be illustrated with a popular Bayes test for the normal mean due to Jeffreys. We also discuss prediction and a few other topics related to testing and interval estimation.

Because the differences between Bayesian inference and Bayesian decision theory is mainly one of nuances, we do not make any sharp distinctions between the two approaches. So our treatment of these three problems as well as other problems later includes elements of both – loss functions from decision theory as well as evidential descriptive measures from inference. A full Bayesian study of a problem consists of two stages, the planning or preposterior stage followed by posterior Bayesian analysis of data collected. At the planning stage one would have problems of choosing optimum design and optimum sample size. Then the integrated Bayes risk $R(\pi) = \inf_\delta R(\pi, \delta)$ plays a central role.

In this book we concentrate on the posterior Bayes analysis of data.

2.7.1 Point Estimates

For a real valued θ, standard Bayes estimates are the posterior mean or the posterior median. The posterior mean is the Bayes estimate corresponding with squared error loss and the posterior median is the Bayes estimate for absolute deviation loss. Along with the posterior mean one reports the posterior variance or its square root, the posterior standard deviation of θ. If one chooses to work with the posterior median, it would be convenient to report a couple of other posterior quantiles to give an idea of the posterior variability of θ. One could report at least the first and third posterior quartiles.

If the posterior is unimodal then the posterior mode is another choice. It is similar to the MLE of classical statistics. Indeed if the prior is uniform, both are identical. Along with the posterior mode one can report a suitable highest posterior density (HPD) credible interval as a measure of posterior variability. If the parameter is a vector, common choices for reporting are the posterior mean vector and the posterior dispersion matrix. Again if the posterior is unimodal, one can report the posterior mode with a suitable HPD credible set. Problem 14 illustrates this with a multivariate normal model with known dispersion matrix and a multivariate normal or uniform prior for the normal mean vector.

2.7.2 Testing

We want to test

$$H_0 : \boldsymbol{\theta} \in \Theta_0 \quad \text{versus} \quad H_1 : \boldsymbol{\theta} \in \Theta_1. \tag{2.7}$$

If Θ_0 and Θ_1 are of the same dimension as for one-sided null and alternative hypotheses, it is convenient and easy to choose a prior density that assigns positive prior probability to Θ_0 and Θ_1. One then calculates the posterior probabilities $P\{\Theta_i | \boldsymbol{x}\}$ as well as the posterior odds ratio (or simply *posterior odds*), namely,

$$P\{\Theta_0 | \boldsymbol{x}\}/P\{\Theta_1 | \boldsymbol{x}\}$$

that most people prefer. One would then find a threshold like $1/9$ or $1/19$, etc. to decide what constitutes evidence against H_0. The Bayes rule for 0-1 loss is to choose the hypothesis with higher posterior probability.

There is a conceptual problem with this approach. If the prior is improper, then the prior probabilities may be undefined — they are, strictly speaking, undefined in the example with one-sided null and alternatives. Even if the prior is proper, the prior probabilities assigned to Θ_i, i.e., $P(\Theta_i)$ may not be carefully chosen and so may not be satisfactory. Surely, if our attitude to H_0 is still as in classical Statistics, namely, that it should not be rejected unless there is compelling evidence to the contrary, then it would be unreasonable to assign less prior probability to Θ_0 than Θ_1. In fact an objective or impartial choice would be to assign equal probabilities. These things can be done better if we use the following alternative way of specifying the prior.

Let π_0 and $1 - \pi_0$ be the prior probabilities of Θ_0 and Θ_1. Let $g_i(\boldsymbol{\theta})$ be the prior p.d.f. of $\boldsymbol{\theta}$ under Θ_i, so that

$$\int_{\Theta_i} g_i(\boldsymbol{\theta}) d\boldsymbol{\theta} = 1.$$

The prior in the previous approach is nothing but

$$\pi(\boldsymbol{\theta}) = \pi_0 g_0(\boldsymbol{\theta}) I\{\boldsymbol{\theta} \in \Theta_0\} + (1 - \pi_0) g_1(\boldsymbol{\theta}) I\{\boldsymbol{\theta} \in \Theta_1\}. \tag{2.8}$$

We do not require any longer that Θ_0 and Θ_1 are of the same dimension. So in principle, sharp null hypotheses are also covered. We can now proceed as before and report posterior probabilities or posterior odds. To compute these posterior quantities, note that the marginal density of X under the prior π can be expressed as

$$m_\pi(x) = \int_\Theta f(x|\boldsymbol{\theta})\pi(\boldsymbol{\theta})\, d\boldsymbol{\theta}$$

$$= \pi_0 \int_{\Theta_0} f(x|\boldsymbol{\theta})g_0(\boldsymbol{\theta})\, d\boldsymbol{\theta} + (1 - \pi_0) \int_{\Theta_1} f(x|\boldsymbol{\theta})g_1(\boldsymbol{\theta})\, d\boldsymbol{\theta} \qquad (2.9)$$

and hence the posterior density of $\boldsymbol{\theta}$ given the data $X = x$ as

$$\pi(\boldsymbol{\theta}|x) = \frac{f(x|\boldsymbol{\theta})\pi(\boldsymbol{\theta})}{m_\pi(x)} = \begin{cases} \pi_0 f(x|\boldsymbol{\theta})g_0(\boldsymbol{\theta})/m_\pi(x) & \text{if } \boldsymbol{\theta} \in \Theta_0; \\ (1 - \pi_0)f(x|\boldsymbol{\theta})g_1(\boldsymbol{\theta})/m_\pi(x) & \text{if } \boldsymbol{\theta} \in \Theta_1. \end{cases} \qquad (2.10)$$

It follows then that

$$P^\pi(H_0|x) = P^\pi(\Theta_0|x) = \frac{\pi_0}{m_\pi(x)} \int_{\Theta_0} f(x|\boldsymbol{\theta})g_0(\boldsymbol{\theta})\, d\boldsymbol{\theta}$$

$$= \frac{\pi_0 \int_{\Theta_0} f(x|\boldsymbol{\theta})g_0(\boldsymbol{\theta})\, d\boldsymbol{\theta}}{\pi_0 \int_{\Theta_0} f(x|\boldsymbol{\theta})g_0(\boldsymbol{\theta})\, d\boldsymbol{\theta} + (1 - \pi_0) \int_{\Theta_1} f(x|\boldsymbol{\theta})g_1(\boldsymbol{\theta})\, d\boldsymbol{\theta}} \quad \text{and}$$

$$P^\pi(H_1|x) = P^\pi(\Theta_1|x) = \frac{(1 - \pi_0)}{m_\pi(x)} \int_{\Theta_1} f(x|\boldsymbol{\theta})g_1(\boldsymbol{\theta})\, d\boldsymbol{\theta}$$

$$= \frac{(1 - \pi_0) \int_{\Theta_1} f(x|\boldsymbol{\theta})g_1(\boldsymbol{\theta})\, d\boldsymbol{\theta}}{\pi_0 \int_{\Theta_0} f(x|\boldsymbol{\theta})g_0(\boldsymbol{\theta})\, d\boldsymbol{\theta} + (1 - \pi_0) \int_{\Theta_1} f(x|\boldsymbol{\theta})g_1(\boldsymbol{\theta})\, d\boldsymbol{\theta}}.$$

One may also report the *Bayes factor*, which does not depend on π_0. The Bayes factor of H_0 relative to H_1 is defined as

$$BF_{01} = \frac{\int_{\Theta_0} f(\boldsymbol{x}|\boldsymbol{\theta})g_0(\boldsymbol{\theta})\, d\boldsymbol{\theta}}{\int_{\Theta_1} f(\boldsymbol{x}|\boldsymbol{\theta})g_1(\boldsymbol{\theta})\, d\boldsymbol{\theta}}. \qquad (2.11)$$

Clearly, $BF_{10} = 1/BF_{01}$. The posterior odds ratio of H_0 relative to H_1 is

$$\left(\frac{\pi_0}{1 - \pi_0}\right) BF_{01},$$

which reduces to BF_{01} if $\pi_0 = \frac{1}{2}$. Thus, BF_{01} is an important evidential measure that is free of π_0. The smaller the value of BF_{01}, the stronger the evidence against H_0.

Let us consider an example to illustrate some of these measures. It will be extended to include the well-known Jeffreys' analysis later.

Example 2.8. Consider a blood test conducted for determining the sugar level of a person with diabetes two hours after he had his breakfast. It is of interest to see if his medication has controlled his blood sugar levels. Assume that the test result X is $N(\theta, 100)$, where θ is the true level. In the appropriate

population (diabetic but under this treatment), θ is distributed according to a $N(100, 900)$. Then, marginally X is $N(100, 1000)$, and the posterior distribution of θ given $X = x$ is normal with
mean $= \frac{900}{1000} x + \frac{100}{1000} 100 = 0.9x + 10$ and variance $= \frac{100 \times 900}{1000} = 90$.
Suppose we want to test $H_0 : \theta \leq 130$ versus $H_1 : \theta > 130$. If the blood test shows a sugar level of 130, what can be concluded? Note that, given this test result, the true mean blood sugar level (θ) may be assumed to be $N(127, 90)$. Consequently, we obtain,

$$P(\theta \leq 130 | X = 130) = \Phi\left(\frac{130 - 127}{\sqrt{90}}\right) = \Phi(.316) = 0.624, \text{ and hence}$$

$$P(\theta > 130 | X = 130) = 0.376. \text{ Therefore,}$$

$$\text{Posterior odds ratio} = 0.624/0.376 = 1.66.$$

Because $\pi_0 = P^{\pi}(\theta \leq 130) = \Phi(\frac{130-100}{30}) = \Phi(1)$, the prior odds ratio is $\Phi(1)/(1 - \Phi(1)) = .8413/.1587 = 5.3$, and thus the Bayes factor turns out to be $1.66/5.3 = .313$.

It can also be noted here that in one-sided testing situations when a continuous prior π can be specified readily for the entire parameter space, there is no need to express it in the form of $\pi(\theta) = \pi_0 g_0(\theta) I\{\theta \in \Theta_0\} + (1 - \pi_0) g_1(\theta) I\{\theta \in \Theta_1\}$. However, the problem of testing a point null hypothesis turns out to be quite different as shown below.

Testing a Point Null Hypothesis

The problem is to test

$$H_0 : \theta = \theta_0 \text{ versus } H_1 : \theta \neq \theta_0. \tag{2.12}$$

Consider the following examples, which indicate when we need to consider point nulls and when we need not.

Example 2.9. In a statistical quality control situation, θ is the size of a unit and acceptable units are with $\theta \in (\theta_0 - \delta, \theta_0 + \delta)$. Then one would like to test

$$H_{0\delta} : |\theta - \theta_0| \leq \delta.$$

In this problem the length of the interval, 2δ, can be explicitly specified. On the other hand, this is not the case in the following.

Example 2.10. (i) Suppose we want to test the hypothesis,

$$H_0 : \text{ Vitamin C has no effect on the common cold.}$$

Clearly this is not meant to be thought of as an exact point null; surely vitamin C has *some* effect, though perhaps a very minuscule effect. Thus, in reality, this is still the case of an interval null hypothesis, with a very small unspecified interval. However, it would be better represented as a point null hypothesis.

(ii) On the other hand, a hypothesis such as

$$H_0 : \text{ Astrology cannot predict the future}$$

can perhaps be represented as an exact point null.

Since these issues are important, we summarize the main points below. If the interval in an interval null hypothesis, along with π_0, g_0, and g_1 can be specified, it is best to treat the problem as an interval null hypothesis problem and proceed accordingly. However, when the interval around θ_0 is small but unspecified, and g_0 is difficult to specify, it is best to approximate the interval null by a point null. Conceptually testing a point null is not a different problem, but there are complications. First of all, it is not possible to use a continuous prior density because any such prior will necessarily assign prior probability zero to the null hypothesis. Consequently, the posterior probability of the null hypothesis will also be zero. Intuitively, this is clear: if the null hypothesis is *a priori* impossible, it will remain so *a posteriori* also. Therefore, a prior probability of $\pi_0 > 0$ needs to be assigned to the point θ_0 and the remaining probability of $\pi_1 = 1 - \pi_0$ will be spread over $\{\theta \neq \theta_0\}$ using a density g_1. Simply take g_0 to be a point mass at θ_0 in (2.8). If the point null hypothesis approximates an interval null hypothesis, $H_0 : \theta \in (\theta_0 - \epsilon, \theta_0 + \epsilon)$, then π_0 is the probability assigned to the interval $(\theta_0 - \epsilon, \theta_0 + \epsilon)$ by a continuous prior. The complication now is that the prior π is of the form

$$\pi(\theta) = \pi_0 I\{\theta = \theta_0\} + (1 - \pi_0)g_1(\theta)I\{\theta \neq \theta_0\} \qquad (2.13)$$

and hence has both discrete and continuous parts. However, (2.9) and (2.10) yield,

$$m(x) = \pi_0 f(x|\theta_0) + (1 - \pi_0)m_1(x), \qquad (2.14)$$

where

$$m_1(x) = \int_{\theta \neq \theta_0} f(x|\theta)g_1(\theta)\, d\theta.$$

Therefore, from (2.10),

$$\begin{aligned}
\pi(\theta_0|x) &= \frac{f(x|\theta_0)\pi_0}{m(x)} \\
&= \frac{\pi_0 f(x|\theta_0)}{\pi_0 f(x|\theta_0) + (1 - \pi_0)m_1(x)} \\
&= \left\{ 1 + \frac{1 - \pi_0}{\pi_0}\frac{m_1(x)}{f(x|\theta_0)} \right\}^{-1}. \qquad (2.15)
\end{aligned}$$

It follows then that the posterior odds ratio is given by

$$\frac{\pi(\theta_0|x)}{1 - \pi(\theta_0|x)} = \frac{\pi_0}{(1 - \pi_0)} \frac{f(x|\theta_0)}{m_1(x)},$$

and hence the Bayes factor of H_0 relative to H_1 (which is the ratio of the above posterior odds ratio to the prior odds ratio of $\pi_0/(1 - \pi_0)$) is

$$B = B(x) = BF_{01}(x) = \frac{f(x|\theta_0)}{m_1(x)}. \tag{2.16}$$

Thus, (2.15) can be expressed as

$$\pi(\theta_0|x) = \left\{ 1 + \frac{1 - \pi_0}{\pi_0} BF_{01}^{-1}(x) \right\}^{-1}. \tag{2.17}$$

Example 2.11. Suppose $X \sim \mathrm{B}(n, \theta)$ and we want to test $H_0 : \theta = \theta_0$ versus $H_1 : \theta \neq \theta_0$, a problem similar to checking whether a given coin is biased based on n independent tosses (where θ_0 will be taken to be 0.5). Under the alternative hypothesis, suppose θ is distributed as $\mathrm{Beta}(\alpha, \beta)$. Then $m_1(x)$ is given by

$$m_1(x) = \binom{n}{x} \frac{\Gamma(\alpha + \beta)}{\Gamma(\alpha)\Gamma(\beta)} \frac{\Gamma(\alpha + x)\Gamma(\beta + n - x)}{\Gamma(\alpha + \beta + n)},$$

so that

$$BF_{01}(x) = \binom{n}{x} \theta_0^x (1 - \theta_0)^{n-x} \Big/ \left(\binom{n}{x} \frac{\Gamma(\alpha + \beta)}{\Gamma(\alpha)\Gamma(\beta)} \frac{\Gamma(\alpha + x)\Gamma(\beta + n - x)}{\Gamma(\alpha + \beta + n)} \right)$$

$$= \theta_0^x (1 - \theta_0)^{n-x} \Big/ \left(\frac{\Gamma(\alpha + \beta)}{\Gamma(\alpha)\Gamma(\beta)} \frac{\Gamma(\alpha + x)\Gamma(\beta + n - x)}{\Gamma(\alpha + \beta + n)} \right)$$

$$= \frac{\Gamma(\alpha)\Gamma(\beta)}{\Gamma(\alpha + \beta)} \frac{\Gamma(\alpha + \beta + n)}{\Gamma(\alpha + x)\Gamma(\beta + n - x)} \theta_0^x (1 - \theta_0)^{n-x}.$$

Hence, we obtain,

$$\pi(\theta_0|x) = \left\{ 1 + \frac{1 - \pi_0}{\pi_0} BF_{01}^{-1}(x) \right\}^{-1}$$

$$= \left\{ 1 + \frac{1 - \pi_0}{\pi_0} \frac{\frac{\Gamma(\alpha+\beta)}{\Gamma(\alpha)\Gamma(\beta)} \frac{\Gamma(\alpha+x)\Gamma(\beta+n-x)}{\Gamma(\alpha+\beta+n)}}{\theta_0^x (1 - \theta_0)^{n-x}} \right\}^{-1}.$$

Further discussion on hypothesis testing will be deferred to Chapter 6 where basic aspects of model selection will also be considered.

Jeffreys Test for Normal Mean with Unknown σ^2

Suppose the data consist of i.i.d. observations $X_1, X_2, \ldots, X_n$ from a normal $N(\mu, \sigma^2)$ distribution. We want to test $H_0 : \mu = \mu_0$ versus $H_1 : \mu \neq \mu_0$, where

μ_0 is some specified number. Without loss of generality, we assume $\mu_0 = 0$. Note that the parameter σ^2 is common in the two models corresponding to H_0 and H_1 and μ occurs only in H_1. Also, in this example μ and σ^2 are orthogonal parameters in the sense that the Fisher information matrix is orthogonal. In such situations, Jeffreys (1961) suggests using (improper) objective prior only for the common parameter and using default proper prior for the other parameter that occurs only in one model. Let us consider the following priors in our example. We take the prior $g_0(\sigma) = 1/\sigma$ for σ under H_0. Under H_1, we take the same prior for σ and add a conditional prior for μ given σ, namely

$$g_1(\mu|\sigma) = \frac{1}{\sigma}g_2\left(\frac{\mu}{\sigma}\right).$$

where $g_2(\cdot)$ is a p.d.f. An initial natural choice for g_2 is $N(0, c^2)$. Thus the prior conditional variance of μ is calibrated with respect to σ^2 as recommended by Jeffreys. Usually, one takes $c = 1$.

Jeffreys points out that one would expect the Bayes factor BF_{01} should tend to zero if $\bar{x} \to \infty$ and $s^2 = \frac{1}{n-1}\sum(x_i - \bar{x})^2$ is bounded. He gives an argument that implies that unless g_1 has no finite moments, this will not happen. In particular, with $g_2 = $ normal, it can be verified (Problem 12) directly that BF_{01} doesn't tend to zero as above. Jeffreys suggested we should take g_2 to be Cauchy. So the priors recommended by Jeffreys are

$$g_0(\sigma) = \frac{1}{\sigma} \qquad \text{under } H_0$$

and

$$g_1(\mu, \sigma) = \frac{1}{\sigma}g_1(\mu|\sigma) = \frac{1}{\sigma}\frac{1}{\sigma\pi(1 + \mu^2/\sigma^2)} \qquad \text{under } H_1.$$

One may now find the Bayes factor BF_{01} using (2.11). Let the joint density of $X_1, \ldots, X_n$ under $N(\mu, \sigma^2)$ model be denoted by $f(x_1, \ldots, x_n|\mu, \sigma^2)$. Then BF_{01} is given by

$$BF_{01} = \frac{\int f(x_1, \ldots, x_n|0, \sigma^2)g_0(\sigma)d\sigma}{\int f(x_1, \ldots, x_n|\mu, \sigma^2)g_1(\mu, \sigma)\,d\mu\,d\sigma},$$

where $g_0(\sigma)$ and $g_1(\mu, \sigma)$ are as given above. The integral in the numerator of BF_{01} can be obtained in closed form. However, no closed form is available for the denominator. To calculate this one can proceed as follows. The Cauchy density $g_1(\mu|\sigma)$ can be written as a Gamma scale mixture of normals

$$g_1(\mu|\sigma) = \int_0^{\infty} \frac{\sigma}{\sqrt{2\pi}}\tau^{-1/2}e^{-\sigma^2\tau/2}\left(\frac{\sqrt{\tau}}{\sqrt{2\pi}}e^{-\tau\mu^2/2}\right)d\tau$$

where τ is the mixing Gamma variable. Then to calculate the denominator of BF_{01}, one can integrate over μ and σ in closed form. Finally, one has a one-dimensional integral over τ left.

Example 2.12. Einstein's theory of gravitation predicts that light is deflected by gravitation and specifies the amount of deflection. Einstein predicted that light of stars would deflect under gravitational pull of the sun on the nearby stars, but the effect would be visible only during a total solar eclipse when the deflection can be measured through apparent change in a star's position. A famous experiment by a team led by British astrophysicist Eddington, immediately after the First World War (see Gardner, 1997), led to acceptance of Einstein's theory. Though many other better designed experiments have confirmed Einstein's theory since then, Eddington's expedition remains historically important. There are four observations, two collected in 1919 in Eddington's expedition, and two more collected by other groups in 1922 and 1929. The observations are $x_1 = 1.98, x_2 = 1.61, x_3 = 1.18, x_4 = 2.24$ (all in seconds as measures of angular deflection). Suppose they are normally distributed around their predicted value μ. Then $X_1, \cdots, X_4$ are independent and identically distributed as $N(\mu, \sigma^2)$. Einstein's prediction is $\mu = 1.75$. We will test $H_0 : \mu = 1.75$ versus $H_1 : \mu \neq 1.75$, where σ^2 is unknown.

If we use the conventional priors of Jeffreys to calculate the Bayes factor BF_{01} in this example, it turns out to be 2.98 (Problem 7). Thus the calculations with the given data lend some support to Einstein's prediction. However, the evidence in the data isn't very strong. This particular experiment has not been repeated because of unavoidable experimental errors. There are now better confirmations of Einstein's theory, vide Gardner (1997).

2.7.3 Credible Intervals

Bayesian interval estimates for θ are similar to confidence intervals of classical inference. They are called credible intervals or sets.

Definition 2.13. *For $0 < \alpha < 1$, a $100(1 - \alpha)\%$ credible set for θ is a subset $C \subset \Theta$ such that*

$$P\{C|X = x\} = 1 - \alpha.$$

Usually C is taken to be an interval. Let θ be a continuous random variable, $\theta^{(1)}, \theta^{(2)}$ be $100\alpha_1\%$ and $100(1 - \alpha_2)\%$ quantiles with $\alpha_1 + \alpha_2 = \alpha$. Let $C = [\theta^{(1)}, \theta^{(2)}]$. Then $P(C|X = x) = 1 - \alpha$. Usually equal tailed intervals are chosen so $\alpha_1 = \alpha_2 = \alpha/2$.

If θ is discrete, usually it would be difficult to find an interval with exact posterior probability $1 - \alpha$. There the condition is relaxed to

$$P(C|X = x) \geq 1 - \alpha$$

with the inequality being as close to an equality as possible. In general, one may use a conservative inequality like this in the continuous case also if exact posterior probability $1 - \alpha$ is difficult to attain.

Whereas the (frequentist) confidence statements do not apply to whether a given interval for a given x covers the "true" θ, this is not the case with credible intervals. The credibility $1-\alpha$ of a credible set does answer a layman's question on whether the given set covers the "true" θ with probability $1-\alpha$. This is because in the Bayesian approach, "true" θ is a random variable with a data dependent probability distribution, namely, the posterior distribution.

For arbitrary priors, these probabilities will usually not have any frequency interpretation over repetitions like confidence statements. But for common objective priors, such statements are usually approximately true because of the normal approximation to the posterior distribution (see Chapter 4). Moreover, the approximations are surprisingly accurate for the Jeffreys prior. You are invited to verify this in Problem 8. Some explanation of this comes from the discussion of probability matching priors (Chapter 5).

The equal tailed credible interval need not have the smallest size, namely, length or area or volume whichever is appropriate. For that one needs an HPD (Highest Posterior Density) interval.

Definition 2.14. *Suppose the posterior density for θ is unimodal. Then the HPD interval for θ is the interval*

$$C = \{\theta : \pi(\theta|X = x) \geq k\},$$

where k is chosen such that

$$P(C|X = x) = 1 - \alpha.$$

Example 2.15. Consider a normal prior for mean of a normal population with known variance σ^2. The posterior is normal with mean and variance given by equations (2.2) and (2.3). The HPD interval is the same as the equal tailed interval centered at the posterior mean,

$$C = \text{ posterior mean } \pm z_{\alpha/2} \text{ posterior s.d.}$$

Credible intervals are very easy to calculate unlike confidence intervals, the construction of which requires pivotal quantities or inversion of a family of tests (Chapter 1, Section 1.4.3).

For a vector $\boldsymbol{\theta}$, one may consider a HPD credible set, specially if the posterior is unimodal. Alternatively, one may have credible intervals for each component. One may also report the probability of simultaneous coverage of all components.

2.7.4 Testing of a Sharp Null Hypothesis Through Credible Intervals

Some Bayesians are in favor of testing, say, $H_0 : \theta = \theta_0$ versus $H_1 : \theta \neq \theta_0$ by accepting H_0 if θ_0 belongs to a chosen credible set. This is similar to the relation between confidence intervals and classical testing, except that there the tests are inverted to get confidence intervals. This must be thought of as a very informal way of testing. If one really believes that the sharp null is a well-formulated theory and deserves to be tested, one would surely want to attach a posterior probability to it. That is not possible in this approach.

Because the inference based on credible intervals often has good frequency properties, a test based on them also is similar to a classical test. This is in sharp contrast with inference based on Bayes factors or posterior odds (Section 2.7.2 and Chapter 6).

2.8 Prediction of a Future Observation

We have already done this informally earlier. Suppose the data are $x_1, \cdots, x_n$, where $X_1, \ldots, X_n$ are i.i.d. with density $f(x|\theta)$, e.g., $N(\mu, \sigma^2)$ with σ^2 known. We want to predict the unobserved X_{n+1} or set up a predictive credible interval for X_{n+1}.

Prediction by a single number $t(x_1, \cdots, x_n)$ based on $x_1, \cdots, x_n$ with squared error loss amounts to considering prediction loss

$$
E\left\{(X_{n+1} - t)^2 | \boldsymbol{x}\right\} = E\left[\left\{(X_{n+1} - E(X_{n+1}|\boldsymbol{x})) - (t - E(X_{n+1}|\boldsymbol{x}))\right\}^2 | \boldsymbol{x}\right]
$$

$$
= E\left\{(X_{n+1} - E(X_{n+1}|\boldsymbol{x}))^2 | \boldsymbol{x}\right\} + (t - E(X_{n+1}|\boldsymbol{x}))^2
$$

which is minimum at

$$
t = E(X_{n+1}|\boldsymbol{x}).
$$

To calculate the predictor we need to calculate the predictive distribution

$$
\pi(x_{n+1}|\boldsymbol{x}) = \int_\Theta \pi(x_{n+1}|\boldsymbol{x}, \boldsymbol{\theta})\pi(\boldsymbol{\theta}|\boldsymbol{x})\, d\boldsymbol{\theta}
$$

$$
= \int_\Theta f(x_{n+1}|\boldsymbol{\theta})\pi(\boldsymbol{\theta}|\boldsymbol{x})\, d\boldsymbol{\theta}.
$$

Let $\mu(\boldsymbol{\theta}) = \int_{-\infty}^{\infty} x f(x|\boldsymbol{\theta})\, dx$. It can be shown that

$$
E(X_{n+1}|\boldsymbol{x}) = E(\mu(\boldsymbol{\theta})|\boldsymbol{x}) = \int_\Theta \mu(\boldsymbol{\theta})\pi(\boldsymbol{\theta}|\boldsymbol{x})\, d\boldsymbol{\theta}
$$

and hence for the normal problem the predictor is $\int_{-\infty}^{\infty} \mu\pi(\mu|\boldsymbol{x})\, d\mu = $ posterior mean of μ.

Similarly in Example 2.2, the predictive probability that the next ball is red is

$$
E(X_{n+1}|\boldsymbol{x}) = E(p|\boldsymbol{x}) = \frac{\alpha + r}{\alpha + \beta + n}
$$

where $r = \sum_1^n x_i$.

A predictive credible interval for X_{n+1} is (c, d) where c and d are $100\alpha_1\%$ and $100(1 - \alpha_2)\%$ quantiles of the predictive distribution of X_{n+1} given $\boldsymbol{x}$. Usually, one takes $\alpha_1 = \alpha_2 = \alpha/2$ as for credible intervals.

2.9 Examples of Cox and Welch Revisited

In both these problems (see Examples 2.5 and 2.6), the parameter is a location parameter. A common objective prior is

$$\pi(\theta) = \text{ constant } , -\infty < \theta < \infty.$$

You can verify (Problem 11) that the objective Bayesian answers, namely, posterior variance in Cox's example and posterior probability in Welch's example, agree with the corresponding conditional frequentist answers recommended by Fisher. This would typically be the case for location and scale parameters.

2.10 Elimination of Nuisance Parameters

In problems of testing and estimation, the main object of interest may be not the full vector $\boldsymbol{\theta}$ but one of its components. Which component is important will depend on the context. To fix ideas let $\boldsymbol{\theta} = (\theta_1, \boldsymbol{\theta}_2)$ and θ_1 be the parameter of importance. The unimportant parameters $\boldsymbol{\theta}_2$ are called nuisance parameters.

Classical statistics has three ways of eliminating nuisance parameters $\boldsymbol{\theta}_2$ and thus simplifying the problem of inference about θ_1. We explain through three examples.

Example 2.16. Suppose X_1 and X_2 are independent Poisson with mean λ_1, λ_2. You want to test $H_0 : \lambda_1 = \lambda_2$. We can reparameterize (λ_1, λ_2) as $\theta_1 = \frac{\lambda_1}{\lambda_1 + \lambda_2}$, $\theta_2 = \lambda_1 + \lambda_2$. Then θ_1 is the parameter of interest. Under H_0, $\theta_1 = \frac{1}{2}$, only θ_2 is the unknown parameter. $T = X_1 + X_2$ is sufficient for $\lambda_1 + \lambda_2$ and the conditional distribution of X_1 given T is binomial$(n = T, p = 1/2)$, which can be used to construct a conditional test.

Example 2.17. In the second example we use an invariance argument. Consider a sample from $N(\mu, \sigma^2)$. We want to test $H_0 : \mu = 0$ against say $H_1 : \mu > 0$, which can be reformulated as $H_0 : \mu/\sigma = 0$ and $H_1 : \mu/\sigma > 0$. Again reparameterize as $(\theta_1 = \mu/\sigma, \theta_2 = \sigma)$. Note that $(\bar{X}, S^2 = \frac{1}{n-1}\sum(X_i - \bar{X})^2)$ is a sufficient statistic and $\bar{X}/S$ is invariant under the transformation

$$X_i \to cX_i, \quad i = 1, 2, \cdots, n.$$

So $\bar{X}/S = \bar{Z}/S_z$, where $Z_i = X_i/\sigma$ depends only θ_1. The usual t-test is based on $\bar{X}/S$.

Example 2.18. In the third method one constructs what is called a profile likelihood for θ_1 by maximizing the joint likelihood with respect to $\boldsymbol{\theta}_2$ and then using it as a sort of likelihood for θ_1. Thus the profile likelihood is

$$L_p(\theta_1) = \sup_{\boldsymbol{\theta}_2} f(\boldsymbol{x}|\theta_1, \boldsymbol{\theta}_2) = f(\boldsymbol{x}|\theta_1, \hat{\boldsymbol{\theta}}_2(\theta_1))$$

where $\hat{\boldsymbol{\theta}}_2(\theta_1)$ is the MLE of $\boldsymbol{\theta}_2$ if θ_1 is given.

In a full Bayesian approach, a nuisance parameter causes no problem. One simply integrates it out in the joint posterior. Suppose, however, that one does not want to do a full Bayesian analysis but rather construct a Bayesian analogue of profile likelihood, on the basis of which some exploratory Bayesian analysis for θ_1 will be done. Once again, this is easy. One uses

$$L(\theta_1) = \int f(\boldsymbol{x}|\theta_1, \boldsymbol{\theta}_2)\pi(\boldsymbol{\theta}_2|\theta_1)\, d\boldsymbol{\theta}_2.$$

We give an example to indicate that integration makes better sense than treating the unknown $\boldsymbol{\theta}_2$ as known and equal to the conditional MLE $\hat{\boldsymbol{\theta}}_2(\theta_1)$.

Example 2.19. (due to Neyman and Scott). Let $X_{i1}, X_{i2}, i = 1, 2, \ldots, n$, be $2n$ independent normal random variables, with X_{i1}, X_{i2} being i.i.d. $N(\mu_i, \sigma^2)$. Here $\sigma^2 = \theta_1$ is the parameter of interest and $(\mu_1, \ldots, \mu_n) = \boldsymbol{\theta}_2$ is the nuisance parameter. One may think of a weighing machine with no bias but some variability; μ_i is the weight of ith object, X_{i1}, X_{i2} are two measurements of the weight of the ith object. The profile likelihood is

$$L_p(\sigma^2) \propto \sup_{\mu_1, \cdots, \mu_n} \sigma^{-2n} \exp\left(-\frac{1}{2\sigma^2}\sum_{i=1}^{n}\left\{(X_{i1} - \mu_i)^2 + (X_{i2} - \mu_i)^2\right\}\right)$$

$$\propto \sigma^{-2n} \exp\left(-\frac{1}{2\sigma^2}\sum_{i=1}^{n}\left\{(X_{i1} - \bar{X}_i)^2 + (X_{i2} - \bar{X}_i)^2\right\}\right),$$

where $\bar{X}_i = (X_{i1} + X_{i2})/2$. If one maximizes it to get an estimate of θ_1, it will be the usual MLE of σ^2, namely,

$$\frac{1}{2n}\sum_{i=1}^{n}\left\{(X_{i1} - \bar{X}_i)^2 + (X_{i2} - \bar{X}_i)^2\right\}.$$

It is easy to show (Problem 13) that the estimate is inconsistent; it converges in probability to $\sigma^2/2$. If one corrects it for its bias by dividing by n, instead of $2n$, it becomes consistent. To rectify problems with profile likelihood, Cox and Reid (1987) have considered an asymptotic conditional likelihood, which behaves better than profile likelihood.

The simple-minded Bayesian likelihood is

$$L(\sigma^2)$$

$$= \int \sigma^{-2n} \exp\left(-\frac{1}{2\sigma^2}\sum_{i=1}^{n}\left\{(X_{i1} - \mu_i)^2 + (X_{i2} - \mu_i)^2\right\}\right)\pi(\boldsymbol{\mu}|\sigma^2)\, d\boldsymbol{\mu}$$

$$\propto \sigma^{-n} \exp\left(-\frac{1}{2\sigma^2}\sum_{i=1}^{n}\left\{(X_{i1} - \bar{X}_i)^2 + (X_{i2} - \bar{X}_i)^2\right\}\right),$$

where $\boldsymbol{\mu} = (\mu_1, \ldots, \mu_n)$ has an improper uniform prior. Maximizing it one gets a consistent estimate of σ^2.

Berger et al. (1999) discuss many such examples with subtle problems of lack of identifiability of the parameters in the model. For example, if one has two binomials, $B(n_i, p_i)$, $i = 1, 2$ where n_i is large, p_i is small, and $n_1 p_1 = n_2 p_2 = \lambda$, then both will be well approximated by a Poisson with mean λ. So data would provide a lot of information on $\lambda = n_i p_i$ but not so on (n_i, p_i). If both parameters of binomial, namely, n and p are unknown, then they may have identifiability problems in this sense.

2.11 A High-dimensional Example

Examples discussed so far have one thing in common – the dimension of the parameter space is small. We refer to them as low-dimensional. Many of the new problems we have to solve today have a high-dimensional parameter space. We refer to them as high-dimensional. One such example appears below.

Example 2.20. New biological screening experiments, namely, microarrays, test simultaneously thousands of genes to identify their functions in a particular context (say, in producing a particular protein or a particular kind of tumor). On the basis of the data some genes, usually in hundreds, are considered "expressed" if they are thought to have this function. They are taken up for further study by more traditional and time-consuming techniques. Without going into the fascinating biochemistry behind these experiments, we provide a statistical formulation.

The data consist of $(\bar{X}_i, S_i)$, $i = 1, 2, \ldots, p$ where $\bar{X}_i$, S_i are the sample mean and s.d. based on raw data $X_{i1}, X_{i2}, \ldots, X_{ir}$ of size r on the ith gene. For fixed i, $X_{i1}, \ldots, X_{ir}$ are i.i.d. $N(\mu_i, \sigma_i^2)$. Further, $\mu_i = 0$ if the ith gene is not expressed and $\mu_i \neq 0$ if the gene is indeed expressed. Of course, we could carry out a separate t-test for each i but this ignores some additional information that we can get by considering all the genes together in a Bayesian way. Moreover, a simple-minded testing for each gene separately would increase enormously the number of false discoveries. For example, if one tests for each i with $\alpha = 0.05$, then even if no genes are really expressed there would be $N\alpha$ false rejections of the null hypothesis of "no expression". We put a prior on μ_i's and σ_i^2's as follows. We assume that (μ_i, σ_i^2), $i = 1, 2, \ldots, p$ are i.i.d. given certain hyper-parameters. The prior distribution for μ_i, given σ_i^2 is mixture of two normals $pN(0, c\sigma_i^2) + (1 - p)N(\theta, c\sigma_i^2)$ and σ_i^2 are i.i.d. inverse Gamma. The prior distribution has five (hyper) parameters, namely, p, c, θ and the shape and scale parameters. If the proportion of genes expected to be functional can be guessed, we would set p to be equal to this proportion. We would have to put a (second stage) prior on the remaining four parameters making this an example of hierarchical priors. A somewhat simpler approach (empirical Bayes) is to estimate the (hyper) parameters from data. We will see in Chapter 9 that there is a lot of information about them in the data. In either case, data about all the genes affect inference about each gene through these (hyper) parameters that are common to all the genes. Our prior is based on a judgment of exchangeability of (μ_i, σ_i^2), $i = 1, 2, \ldots, p$ and de Finetti's Theorem in Section 2.12.

The inference for each gene is quite simple in the PEB (parametric empirical Bayes) approach. It is more complicated in the hierarchical Bayes setup but doable. Both are discussed in Chapter 9.

2.12 Exchangeability

One may often be able to judge that a set of parameters $(\theta_1, \ldots, \theta_p)$ or a set of observables like $(X_1, X_2, \ldots, X_n)$ are exchangeable, i.e., their joint distribution function is left unaltered if the arguments are permuted. Thus if

$$P\{X_1 \leq x_1, \cdots, X_n \leq x_n\} = P\{X_1 \leq x_{i_1}, \cdots, X_n \leq x_{i_n}\}$$

for all $n!$ permutations $x_{i_1}, \ldots, x_{i_n}$ of $x_1, \ldots, x_n$, one says $X_1, \ldots, X_n$ are exchangeable. A simple way of generating exchangeable random variables is to choose an indexing random parameter η and have the random variables conditionally i.i.d. given η. In many cases the converse is also true, as shown by de Finetti (1974, 1975), and Hewitt and Savage (1955). We only discuss de Finetti's theorem.

We say X_i, $i = 1, 2, \ldots, n, n+1, \ldots$, is a sequence of exchangeable random variables if $\forall n > 1$, $X_1, X_2, \ldots, X_n$ are exchangeable.

Theorem 2.21. *(de Finetti). Suppose X_i's constitute an exchangeable sequence and each X_i takes only values 0 or 1. Then, for some π,*

$$P\{X_1 = x_1, \cdots, X_n = x_n\} = \int_0^1 \eta^{\sum_{i=1}^n x_i} (1 - \eta)^{n - \sum_{i=1}^n x_i} \, d\pi(\eta),$$

$\forall n$, $\forall x_1, \ldots, x_n$ *equal to 0 or 1, i.e., given η, $X_1, \ldots, X_n$ are conditionally i.i.d. Bernoulli with parameter η and η has distribution π.*

A Bayesian may interpret this as follows. The subjective judgment of exchangeability leads to both a Bernoulli likelihood and the existence of a prior π. If one has also a prediction rule as in Problem 18, π can be specified. Thus at least in this interpretation the prior and the likelihood have the same logical status, both arise from a subjective judgment about observables.

Hewitt and Savage (1955) show that even if the random variables take values in $\mathcal{R}^p$, or more generally in a nice measurable space, then a similar representation as conditionally i.i.d. random variables holds. See Schervish (1995) for a statement and proof.

In many practical cases, vide Example 2.20, one may perceive certain parameters $\theta_1, \ldots, \theta_p$ as exchangeable. Even if the parameters do not form an infinite sequence, it is convenient to represent them as conditionally i.i.d. given a hyperparameter. Often as in Example 2.20, the form of $\pi(\theta|\eta)$ is also dictated by operational convenience. We show in Chapter 5 we can check if this form is validated by the data.

2.13 Normative and Descriptive Aspects of Bayesian Analysis, Elicitation of Probability

Do most people faced with uncertainty make a decision as if they were Bayesian, each with her subjective prior and utility? The answer is generally *No*. The Bayesian approach is not claimed to be a description of how people tend to make a decision. On the other hand Bayesians believe, on the basis of various sets of rationality axioms and their consequences (as discussed in Chapter 3), people should act as if they have a prior and utility. The Bayesian approach is normative rather than descriptive. There have been empirical as well as philosophical studies of these issues. We refer the interested reader to Raiffa and Schlaiffer (1961) and French and Ríos Insua (2000). We explore tentatively a couple of issues related to this.

It is an odd fact in our intellectual history that the concept of probability, which is so fundamental both in daily life and science, was developed only during the European Renaissance. It is tempting to speculate that our current inability to behave rationally under uncertainty is related to the late arrival of probability on the intellectual scene. Most Bayesians hope the situation will improve with the passage of time and attempts to educate ourselves to act rationally.

Related to these facts is the inability of most people to express their uncertainty in terms of a well calibrated probability. Probability is still most easily calculated in gambling or similar problems where outcomes are equally likely, in problems like life or medical insurance, where empirical calculations based on repetitions is possible or under exchangeability. Most examples of successful elicitation of subjective probability involve exchangeability in some form. However, there have been some progress in elicitation. Some of these examples are discussed in Chapter 5 .

These examples and attempts notwithstanding, full elicitation of subjective probability is still quite rare. Most priors used in practice are at least partly nonsubjective. They are obtained through some objective, i.e., nonsubjective algorithms. In some sense they are uniform distributions that take into account what is known, namely some prior moments or quartiles and the geometry in the parameter space. We discuss objective priors and Bayesian analysis based on them in the next section.

2.14 Objective Priors and Objective Bayesian Analysis

We refer to the Bayesian analysis based on objective priors as *objective Bayesian analysis*. One would expect that as elicitation improves, *subjective Bayesian analysis* would be used increasingly in place of objective Bayesian analysis. All Bayesians agree that wherever prior information is available, one should try to use a prior reflecting that as far as possible. In fact, one of the attractions of the Bayesian paradigm is that use of prior expert information is a possibility. Incorporation of prior expert opinion would strengthen considerably purely data based analysis in real-life decision problems as well as problems of statistical inference with small sample size or high or infinite dimensional parameter space. In this approach use of objective Bayesian analysis has no conflict with the subjectivist approach. It may also have a legitimate place in subjective Bayesian analysis as a reference point or origin with which to compare the role and importance of prior information in a particular Bayesian decision. In a similar spirit, it may also be used to report to general readers or to a group of Bayesians with different priors.

We discuss in Chapter 3 algorithms for generating common objective priors such as the Jeffreys or reference or probability matching priors. We also discuss there common criticisms, such as the fact that these priors are improper and depend on the experiment, as well as our answers to such criticisms.

In examples with low-dimensional Θ, objective Bayesian analysis has some similarities with frequentist answers, as in Examples 2.2 and 2.4, in that the estimate obtained or hypothesis accepted tends to be very close to what a frequentist would have done. However, the objective Bayesian has a posterior distribution and a data based evaluation of the error or risk associated with inference or decision, namely, the posterior variance or posterior error or posterior risk.

In high-dimensional problems, e.g., Example 2.20, it is common to use hierarchical priors with objective prior of the above type used at the highest level of the hierarchy. One then typically uses MCMC without always checking whether the posteriors are proper – in fact checking mathematically may be very difficult. Truncation of the prior, with careful variation of the stability of the posterior provides good numerical insight. However, this is not the only place where an objective prior is used in the hierarchy. In fact, in Example 2.20, the prior for (μ_i, σ_i^2) arises from a subjective assumption of exchangeability but the particular form taken is for convenience. This is a non-subjective choice but, as indicated in Chapter 9, some data based validation is possible.

The objective Bayesian analysis in high-dimensional problems is also close in spirit to frequentist answers to such problems. Indeed it is a pleasant fact that, as in low-dimensional problems but for different reasons, the frequentist answers are almost identical to the Bayesian answers. The frequentist answers are based on the parametric empirical Bayes (PEB) approach, in which the parameters in the last stage of hierarchical priors are estimated from data rather than given an objective prior. As in the low-dimensional case, the objective Bayesian analysis has some advantages over frequentist analysis. The PEB approach used by frequentists tends to underestimate the posterior risk.

Though it is implicit in the above discussion, it is worth pointing out that Bayesian analysis can be based on an improper prior only if the posterior is proper. Somewhat surprisingly, the posterior is usually proper when one uses the Jeffreys prior or a reference prior, but counter-examples exist; see Ghosh (1997) and Section 7.4.7.

Because the objective priors are improper, the usual type of preposterior analysis cannot be made at the planning stage. In particular one cannot compare different designs for an experiment and make an optimal choice. For the same reason choosing optimal sample size is a problem. It is suggested in Chapter 6 that a partial solution is to take a few observations, say the minimum number of observations needed to make the posterior proper, and use the proper posterior as a proper prior. The additional data can be used to update it. For an application of these ideas, see Ghosh et al. (2002). Unfortunately, when all the data have been collected at the stage of formulating the prior, one would need to modify the above simple procedure.

2.15 Other Paradigms

In earlier sections, we have discussed several aspects of the Bayesian paradigm and its logical advantages. In this context we have also discussed in some detail various problems with the classical frequentist approach.

Some of these problems of classical statistics can be resolved, or at least mitigated by appropriate conditioning. Even though Birnbaum's theorem shows extensive conditioning and restriction to minimal sufficiency would lead to fundamental changes in the classical paradigm and it may be quite awkward to find a suitable conditioning, the idea of conditioning makes it possible to reconcile a lot of objective Bayesian analysis and classical statistics if suitable conditioning is made. At least this makes communication relatively easy between the paradigms.

There have also been attempts to create a new paradigm of inference based on sufficiency, conditioning and likelihood. An excellent treatment is available is Sprott (2000). Some of our reservations are listed in Ghosh (2002).

One should also mention belief functions and upper and lower probabilities of Dempster and Schafer (see Dempster (1967), Shafer (1976) and Shafer (1987)). Wasserman and Kadane (1990) have shown that under certain axioms, their approach may be identified with a robust Bayesian point of view. Problems of foundations of probability and inference remain an active area.

An entirely different popular approach is data analysis. Data analysis makes few assumptions, it is very innovative and yet easy to communicate, However, it is rather ad hoc and cannot quite be called a paradigm. If machine or statistical learning emerges as a new alternative paradigm for learning from data, then data analysis would find in it the paradigm it currently lacks.

2.16 Remarks

Even though there are several different paradigms, we believe the Bayesian approach is not only the most logical but also very flexible and easy to communicate. Many innovations in computation have led to wide applicability as well as wide acceptance from not only statisticians but other scientists. Within the Bayesian paradigm it is relatively easy to use information as well as solve real-life decision problems. Also, we can now construct our priors, with a fair amount of confidence as to what they represent, to what extent they use subjective prior information and to what extent they are part of an algorithm to produce a posterior.

The fact that there are no paradoxes or counter-examples suggests the logical foundations are secure in spite of a rapid, vigorous growth, specially in the past two decades. The advantage of a strong logical foundation is that it makes the subject a discipline rather than a collection of methods, like data analysis. It also allows new problems to be approached systematically and therefore with relative ease.

Though based on subjective ideas, the paradigm accepts likelihood, and frequentist validation in the real world as well as consequent calibration of probabilities, utilities, likelihood based methods.

In other words, it seems to combine many of the conceptual and methodological strengths of both classical statistics and data analysis, but is free from the logical weaknesses of both.

Ultimately, each reader has to make up her own mind but hopefully, even a reader, not completely convinced of the superiority of Bayesian analysis, will learn much that would be useful to her in the paradigm of her choice. This book is offered in a spirit of reconciliation and exploration of paradigms, even though from a Bayesian point of view. In many ways current mutual interaction between the three paradigms is reminiscent of the periods of similar rapid growth in the eighteenth, nineteenth, and early twentieth centuries. We have in mind specially the history of *least squares*, which began as a data analytic tool, then got itself a probabilistic model in the hands of Gauss, Laplace, and others. The associated inferential probabilities were simultaneously subjective and frequentist. The interested reader may also want to browse through von Mises (1957).

2.17 Exercises

1. (a) (French (1986)) Three prisoners, **A**, **B**, and **C**, are each held in solitary confinement. **A** knows that two of them will be hanged and one will be set free but he does not know who will go free. Therefore, he reasons that he has $\frac{1}{3}$ chance of survival. He asks the guard who will go free, but has no success there. Being an intelligent person, he comes up with the following question for the guard:

   *If two of us must die, then I know that either **B** or **C** must die and possibly both. Therefore, if you tell me the name of one who is to die, I learn nothing about my own fate; further, because we are kept apart, I cannot reveal it to them. So tell me the name of one of them who is to* *die.*

The guard likes this logic and tells **A** that **C** will be hanged. **A** now argues that either he or **B** will go free, and so now he has $\frac{1}{2}$ chance of survival. Is this reasoning correct?

b) There are three chambers, one of which has a prize. The master of ceremonies will give the prize to you if you guess the right chamber correctly. You first make a random guess. Then he shows you one chamber which is empty. You have an option to stick to your original guess or switch to the remaining other chamber. (The chamber you guessed first has not been opened). What should you do?

2. Suppose $X|\mu \sim N(\mu, \sigma^2)$, σ^2 known and $\mu \sim N(\eta, \tau^2)$, η and τ^2 known.

(a) Show that the joint density $g(x, \mu)$ of X and μ can be written as

$$g(x, \mu) = \pi(\mu)f(x|\mu) = \frac{1}{2\pi\sigma\tau} \exp\left\{-\frac{1}{2}\left[\frac{(\mu - \eta)^2}{\tau^2} + \frac{(x - \mu)^2}{\sigma^2}\right]\right\}$$

$$= \frac{1}{\sqrt{2\pi(\tau^2 + \sigma^2)}} \exp\left(-\frac{(x - \eta)^2}{2(\tau^2 + \sigma^2)}\right)$$

$$\times \sqrt{\frac{\tau^2 + \sigma^2}{2\pi\tau^2\sigma^2}} \exp\left\{-\frac{\tau^2 + \sigma^2}{2\tau^2\sigma^2}\left(\mu - \frac{\tau^2\sigma^2}{\tau^2 + \sigma^2}\left(\frac{\eta}{\tau^2} + \frac{x}{\sigma^2}\right)\right)^2\right\}.$$

(b) From (a) show that the marginal density $m(x)$ of X is

$$m(x) = \frac{1}{\sqrt{2\pi(\tau^2 + \sigma^2)}} \exp\left(-\frac{(x - \eta)^2}{2(\tau^2 + \sigma^2)}\right),$$

and the posterior density $\pi(\mu|x)$ of $\mu|X = x$ is

$$\pi(\mu|x) = \sqrt{\frac{\tau^2 + \sigma^2}{2\pi\tau^2\sigma^2}} \exp\left\{-\frac{\tau^2 + \sigma^2}{2\tau^2\sigma^2}\left(\mu - \frac{\tau^2\sigma^2}{\tau^2 + \sigma^2}\left(\frac{\eta}{\tau^2} + \frac{x}{\sigma^2}\right)\right)^2\right\}.$$

(c) What are the posterior mean and posterior s.d. of μ given $X = x$?

(d) Instead of a single observation X as above, consider a random sample $X_1, \ldots, X_n$. What is the minimal sufficient statistic and what is the likelihood function for μ now? Work out (b) and (c) in this case.

3. Let $X_1, \ldots, X_n$ be i.i.d. $N(\mu, \sigma^2)$, σ^2 known. Consider testing

$$H_0 : \mu \leq \mu_0 \text{ versus } H_1 : \mu > \mu_0.$$

(a) Compute the P-value. Compare it with the posterior probability of H_0 when μ is assumed to have the uniform prior.

(b) Do the same for a sharp H_0.

4. Refer to Welch's problem, Example 2.6. Follow Fisher's suggestion and calculate $P\{\text{CI covers } \theta | X_1 - X_2\}$ and verify it agrees with the objective Bayes solution with improper uniform prior for θ.

5. (Berger's version of Welch's problem, see Berger (1985b)). Suppose X_1 and X_2 are i.i.d. having the discrete distribution:

$$X = \begin{cases} \theta - 1/2 & \text{with probability } 1/2; \\ \theta + 1/2 & \text{with probability } 1/2, \end{cases}$$

where θ is an unknown real number.
(a) Show that the set C given by

$$C = \begin{cases} \{(X_1 + X_2)/2\} & \text{if } X_1 \neq X_2; \\ \{X_1 - 1\} & \text{if } X_1 = X_2, \end{cases}$$

is a 75% confidence set for θ.
(b) Calculate $P\{C \text{ covers } \theta | X_1 - X_2\}$.
6. Can the Welch paradox occur if X_1, X_2 are i.i.d. $N(\theta, 1)$?
7. (Newton versus Einstein). In Example 2.12 calculate the Bayes factor, BF_{01} for the given data using Jeffreys prior.
8. Let $X_1, \ldots, X_n$ be i.i.d. Bernoulli(p) (i.e., $B(1, p)$).
(a) Assume p has Jeffreys prior. Construct the $100(1 - \alpha)\%$ HPD credible interval for p.
(b) Suppose $n = 10$ and $\alpha = 0.05$. Calculate the frequentist coverage probability of the interval in (a) using simulation.
9. Consider the same model as in Problem 8. Derive the minimax estimate of p under the square error loss. Plot and compare the mean square error of this estimate with that of $\bar{X}$ for $n = 10, 50, 100$, and 400. (The minimax estimate seems to do better at least upto $n = 100$.)
10. Let $X_1, \ldots, X_n$ be i.i.d. $N(\mu, \sigma^2)$, σ^2 known. Suppose μ has the $N(\eta, \tau^2)$ prior distribution with known η and τ^2.
(a) Construct the $100(1 - \alpha)\%$ HPD credible interval for μ.
(b) Construct a $100(1 - \alpha)\%$ predictive interval for X_{n+1}.
(c) Consider the uniform prior for this problem by letting $\tau^2 \longrightarrow \infty$. Work out (a) and (b) in this case.
11. (a) Refer to Example 2.6. Let $C(X_1, X_2)$ denote the $100(1 - \alpha)\%$ confidence interval for θ. Assume that θ has Jeffreys prior. Then show that

$$P\{C(X_1, X_2) \text{ covers } \theta | X_1 - X_2\} = P\{\theta \in C(X_1, X_2) | X_1, X_2\}.$$

(b) Recall Example 2.5. Assume that μ has Jeffreys prior. Then show that

$$\text{Var}(\mu | \boldsymbol{x}) = \text{Var}(\bar{X} | n).$$

12. Let $X_1, \ldots, X_n$ be i.i.d. $N(\mu, \sigma^2)$, σ^2 unknown. Consider Jeffreys test (Section 2.7.2) for testing $H_0 : \mu = \mu_0$ versus $H_1 : \mu \neq \mu_0$. Consider both the normal and Cauchy priors for $\mu | \sigma^2$ under H_1. Suppose $\bar{X} \to \infty$ and s^2 is bounded. Compute BF_{01} under both the priors and show that BF_{01} converges to zero for Cauchy prior but does not converge to zero for normal prior.

13. (a) Refer to Example 2.19. Show that the usual MLE of σ^2, namely,

$$\frac{1}{2n} \sum_{i=1}^{n} \left\{ (X_{i1} - \bar{X}_i)^2 + (X_{i2} - \bar{X}_i)^2 \right\}$$

is inconsistent. Correct it to get a consistent estimate.

(b) Suppose $X_1, \ldots, X_k$ are i.i.d. B(n, p). Both n and p are unknown, but only n is of interest, so p is a nuisance parameter (see Berger et al. (1999)). Derive the following likelihoods for n: (i) profile likelihood, (ii) conditional likelihood, i.e., that obtained from the conditional distribution of X_i's given their sum (and n), (iii) integrated likelihood with respect to the uniform prior, and (iv) integrated likelihood with respect to Jeffreys prior.

(c) Suppose the observations are (17, 19, 21, 28, 30). Plot and compare the different likelihoods in b) above, and comment.

14. Suppose $\boldsymbol{X}|\boldsymbol{\mu} \sim N_p(\boldsymbol{\mu}, \Sigma)$, Σ known and $\boldsymbol{\mu} \sim N_p(\boldsymbol{\eta}, \Gamma)$, $\boldsymbol{\eta}$ and Γ known.

(a) Show that the above probability structure is equivalent to $\boldsymbol{X} = \boldsymbol{\mu} + \boldsymbol{\epsilon}$, $\boldsymbol{\epsilon} \sim N_p(\boldsymbol{0}, \Sigma)$, $\boldsymbol{\mu} \sim N_p(\boldsymbol{\eta}, \Gamma)$, $\boldsymbol{\epsilon}$ and $\boldsymbol{\mu}$ are independent and Σ, $\boldsymbol{\eta}$, Γ are known.

(b) From (a) show that the joint distribution of $\boldsymbol{X}$ and $\boldsymbol{\mu}$ is

$$\begin{pmatrix} \boldsymbol{X} \\ \boldsymbol{\mu} \end{pmatrix} \sim N_{2p} \left(\begin{pmatrix} \boldsymbol{\eta} \\ \boldsymbol{\eta} \end{pmatrix}, \begin{pmatrix} \Sigma + \Gamma & \Gamma \\ \Gamma & \Gamma \end{pmatrix} \right).$$

(c) From (b) and using multivariate normal theory, show that

$$\boldsymbol{\mu}|\boldsymbol{X} = \boldsymbol{x} \sim N_p \left(\Gamma(\Sigma + \Gamma)^{-1}\boldsymbol{x} + \Sigma(\Sigma + \Gamma)^{-1}\boldsymbol{\eta}, \Gamma - \Gamma(\Sigma + \Gamma)^{-1}\Gamma \right).$$

(d) What are the posterior mean and posterior dispersion matrix of $\boldsymbol{\mu}$? Construct a $100(1 - \alpha)\%$ HPD credible set for $\boldsymbol{\mu}$.

(e) Work out (d) with a uniform prior.

15. Let $X_1, \ldots, X_m$ and $Y_1, \ldots, Y_n$ be independent random samples, respectively, from $N(\mu_1, \sigma^2)$ and $N(\mu_2, \sigma^2)$, where σ^2 is known. Construct a $100(1 - \alpha)\%$ credible interval for $(\mu_1 - \mu_2)$ assuming a uniform prior on (μ_1, μ_2).

16. Let $X_1, \ldots, X_m$ and $Y_1, \ldots, Y_n$ be independent random samples, respectively, from $N(\mu, \sigma_1^2)$ and $N(\mu, \sigma_2^2)$, where both σ_1^2 and σ_2^2 are known. Construct a $100(1 - \alpha)\%$ credible interval for the common mean μ assuming a uniform prior. Show that the frequentist $100(1 - \alpha)\%$ confidence interval leads to the same answer.

17. (Behrens-Fisher problem) Let $X_1, \ldots, X_m$ and $Y_1, \ldots, Y_n$ be independent random samples, respectively, from $N(\mu_1, \sigma_1^2)$ and $N(\mu_2, \sigma_2^2)$, where all the four parameters are unknown, but inference on $\mu_1 - \mu_2$ is of interest. To derive a confidence interval for $\mu_1 - \mu_2$ and also test $H_0 : \mu_1 = \mu_2$, the Behrens-Fisher solution is to use the statistic

$$T = \frac{\bar{X} - \bar{Y}}{\sqrt{s_1^2/m + s_2^2/n}},$$

where $s_1^2 = \sum_{i=1}^{m}(X_i - \bar{X})^2/(m-1)$ and $s_2^2 = \sum_{i=1}^{n}(Y_i - \bar{Y})^2/(n-1)$.
(a) Show that

$$\frac{s_1^2/m + s_2^2/n}{\sigma_1^2/m + \sigma_2^2/n} \sim \frac{\chi_\nu^2}{\nu},$$

approximately, where ν can be estimated by

$$\hat{\nu} = \frac{(s_1^2/m + s_2^2/n)^2}{s_1^4/(m^2(m-1)) + s_2^4/(n^2(n-1))}.$$

(Hint: If we want to approximate the weighted sum, $\sum_{i=1}^{k} a_i V_i$ of independent $\chi_{r_i}^2$, by a χ_ν^2/ν, then a method of moment estimate for ν is available, see Satterwaite (1946) and Welch (1949).)
(b) Using (a), justify that T is approximately distributed like a Student's t with $\hat{\nu}$ degrees of freedom under H_0.
(c) Show numerically that the $100(1-\alpha)\%$ confidence interval for $\mu_1 - \mu_2$ derived using T is conservative, i.e., its confidence coefficient will always be $\geq 1-\alpha$. (See Robinson (1976). A Bayesian solution to the Behrens-Fisher problem is discussed in Chapter 8.)

18. Suppose $X_1, X_2, \ldots, X_n$ are i.i.d. Bernoulli(p) and the prediction loss is squared error. Further, suppose that for all $n \geq 1$, the Bayes prediction rule is given by

$$E(X_{n+1}|X_1, \ldots, X_n) = \frac{\alpha + \sum_{i=1}^{n} X_i}{\alpha + \beta + n},$$

for some $\alpha > 0$ and $\beta > 0$. Show that this is possible iff the prior on p is Beta(α, β).

19. Suppose $(N_1, \ldots, N_k)$ have the multinomial distribution with density

$$f(n_1, \ldots, n_k|\boldsymbol{p}) = \frac{n!}{n_1! n_2! \cdots n_k!} \prod_{j=1}^{k} p_j^{n_j}.$$

Let $\boldsymbol{p}$ have the Dirichlet prior with density

$$f(\boldsymbol{p}|\boldsymbol{\alpha}) = \frac{\Gamma(\sum_{i=1}^{k} \alpha_i)}{\prod_{i=1}^{k} \Gamma(\alpha_i)} \prod_{i=1}^{k} p_i^{\alpha_i - 1}.$$

(a) Find the posterior distribution of $\boldsymbol{p}$.
(b) Find the posterior mean vector and the dispersion matrix of $\boldsymbol{p}$.
(c) Construct a $100(1-\alpha)\%$ HPD credible interval for p_1 and also for $p_1 + p_2$.

20. Let p denote the probability of success with a particular drug for some disease. Consider two different experiments to estimate p. In the first experiment, n randomly chosen patients are treated with this drug and let X denote the number of successes. In the other experiment patients are treated with this drug, one after the other until r successes are observed. In this experiment, let Y denote the total number of patients treated with this drug.

    (a) Construct $100(1-\alpha)\%$ HPD credible intervals for p under $U(0,1)$ and Jeffreys prior when $X = x$ is observed.

    (b) Construct $100(1-\alpha)\%$ HPD credible intervals for p under $U(0,1)$ and Jeffreys prior when $Y = y$ is observed.

    (c) Suppose $n = 16$, $x = 6$, $r = 6$, and $y = 16$. Now compare (a) and (b) and comment with reference to LP.

21. Let $X_1, \ldots, X_n$ be i.i.d. $N(\mu, \sigma^2)$, σ^2 known. Suppose we want to test $H_0 : \mu = \mu_0$ versus $H_1 : \mu \neq \mu_0$. Let $\pi_0 = P(H_0) = 1/2$ and under H_1, let $\mu \sim N(\mu_0, \tau^2)$. Show that, unlike in the case of a one-sided alternative, P-value and the posterior probability of H_0 can be drastically different here.

22. Let $X_1, \ldots, X_n$ be i.i.d. $N(\mu, \sigma^2)$, where both μ and σ^2 are unknown. Take the prior $\pi(\mu, \sigma^2) \propto \sigma^{-2}$. Consider testing

$$H_0 : \mu \leq \mu_0 \text{ versus } H_1 : \mu > \mu_0.$$

    Compute the P-value. Compare it with the posterior probability of H_0. Compute the Bayes factor BF_{01}.

3

Utility, Prior, and Bayesian Robustness

We begin this chapter with a discussion of rationality axioms for preference and how one may deduce the existence of a utility and prior. Later we explore how robust or sensitive is Bayesian analysis to the choice of prior, utility, and model. In the process, we introduce and examine various quantitative evaluations of robustness.

3.1 Utility, Prior, and Rational Preference

We have introduced in Chapter 2 problems of estimation and testing as Bayesian decision problems. We recall the components of a general Bayesian decision problem.

Let $\mathcal{X}$ be the sample space, Θ the parameter space, $f(x|\theta)$ the density of X and $\pi(\theta)$ prior probability density. Moreover, there is a space $\mathcal{A}$ of actions "a" and a loss function $L(\theta, a)$. The decision maker (DM) chooses "a" to minimize the posterior risk

$$\psi(a|x) = \int L(\theta, a)\pi(\theta|x)\, d\theta, \tag{3.1}$$

where $\pi(\theta|x)$ is the posterior density of θ given x. Note that given the loss function and the prior, there is a natural preference ordering $a_1 \preceq a_2$ (i.e., a_2 is at least as good as a_1) iff $\psi(a_2|x) \leq \psi(a_1|x)$.

There is a long tradition of foundational study dating back to Ramsey (1926), in which one starts with such a preference relation on $\mathcal{A} \times \mathcal{A}$ satisfying certain rational axioms (i.e., axioms modeling rational behavior) like transitivity. It can then be shown that such a relation can only be induced as above via a loss function and a prior. i.e., $\exists L$ and π such that

$$a_1 \preceq a_2 \quad \text{iff} \quad \int L(\theta, a_2)\pi(\theta)\, d\theta \leq \int L(\theta, a_1)\pi(\theta)\, d\theta. \tag{3.2}$$

In other words, from an objectively verifiable rational preference relation, one can recover the subjective loss function and prior. If there is no sample data, then π would qualify as a subjective prior for the DM. If we have data x, a likelihood function $f(x|\theta)$ is given and we are examining a preference relation given x, then also one can deduce the existence of L and π such that

$$a_1 \preceq a_2 \;\; \text{iff} \;\; \int L(\theta, a_2)\pi(\theta|x)\,d\theta \leq \int L(\theta, a_1)\pi(\theta|x)\,d\theta \qquad (3.3)$$

under appropriate axioms.

In Section 3.2, we explore the elicitation or construction of a loss function given certain rational preference relations. In the next couple of sections, we discuss a result that shows we must have a (subjective) prior if our preference among actions satisfies certain axioms about rational behavior. Together, they justify (3.2) and throw some light on (3.3). In the remaining sections we examine different aspects of sensitivity of Bayesian analysis with respect to the prior. Suppose one thinks of the prior as only an approximate quantification of prior belief. In principle, one would have a whole family of such priors, all approximately quantifying one's prior belief. How much would the Bayesian analysis change as the prior varies over this class? This is a basic question in the study of Baycsian robustness.

It turns out that there are some preference relations weaker than those of Section 3.3 that lead to a situation like what was mentioned above. i.e., one can show the existence of a class of priors such that

$$a_1 \preceq a_2 \;\; \text{iff} \;\; \int L(\theta, a_2)\pi(\theta)\,d\theta \leq \int L(\theta, a_1)\pi(\theta)\,d\theta \qquad (3.4)$$

for all π in the given class. This preference relation is only a partial ordering, i.e., not all pairs a_1, a_2 can be ordered.

The Bayes rule $a(x)$ minimizing $\psi(a|x)$ also minimizes the integrated risk of decision rules $\delta(x)$,

$$r(\pi, \delta) = \int_{\Theta} R(\theta, \delta)\pi(\theta)\,d\theta,$$

where $R(\theta, \delta)$ is the risk of δ under θ, namely, $\int_{\mathcal{X}} L(\theta, \delta(x))f(x|\theta)\,dx$. Given a pair of decision rules, we can define a preference relation

$$a_1 \preceq a_2 \;\; \text{iff} \;\; r(\pi, a_2(.)) \leq r(\pi, a_1(.)). \qquad (3.5)$$

One can examine a converse of (3.5) in the same way as we did with (3.2) through (3.4). One can start with a preference relation that orders decision rules (rather than actions) and look for rationality axioms which would guarantee existence of L and π. For (3.2), (3.3) and (3.5) a good reference is Schervish (1995) or Ferguson (1967). Classic references are Savage (1954) and DeGroot (1970); other references are given later. For (3.4) a good reference is Kadane et al. (1999).

A similar but different approach to subjective probability is via coherence, due to de Finetti (1972). We take this up in Section 3.4.

3.2 Utility and Loss

It is tempting to think of the loss function $L(\theta, a)$ and a utility function $u(\theta, a) = -L(\theta, a)$ as conceptually a mirror image of each other. French and Ríos Insua (2000) point out that there can be important differences that depend on the context.

In most statistical problems the DM (decision maker) is really trying to learn from data rather than implement a decision in the real world that has monetary consequences. For convenience we refer to these as decision problems of Type 1 and Type 2. In Type 1 problems, i.e., problems without monetary consequences (see Examples 2.1–2.3) for each θ there is usually a correct decision $a(\theta)$ that depends on θ, and $L(\theta, a)$ is a measure of how far "a" is away from $a(\theta)$ or a penalty for deviation from $a(\theta)$. In a problem of estimating θ, the correct decision $a(\theta)$ is θ itself. Common losses are $(a - \theta)^2$, $|a - \theta|$, etc. In the problem of estimating $\tau(\theta)$, $a(\theta)$ equals $\tau(\theta)$ and common losses are $(\tau(\theta) - a)^2$, $|\tau(\theta) - a|$, etc. In testing a null hypothesis H_0 against H_1, the 0-1 loss assigns no penalty for a correct decision and a unit penalty for an incorrect decision. In Type 2 problems, there is a similarity with gambles where one must evaluate the consequence of a risky decision. Historically, in such contexts one talks of utility rather than loss, even though either could be used. We consider below an axiomatic approach to existence of a utility for Type 2 problems but we use the notations for a statistical decision problem by way of illustration. We follow Ferguson (1967) here as well as in the next section.

Let $\mathcal{P}$ denote the space of all consequences like (θ, a). It is customary to regard them as non-numerical pay-offs. Let $\mathcal{P}^*$ be the space of all probability distributions on $\mathcal{P}$ that put mass on a finite number of points. The set $\mathcal{P}^*$ represents risky decisions with uncertainty quantified by a known element of $\mathcal{P}^*$. Suppose the DM has a preference relation on $\mathcal{P}^*$, namely a total order, i.e., given any pair $p_1, p_2 \in \mathcal{P}^*$, either $p_1 \preceq p_2$ (p_2 is preferred) or $p_2 \preceq p_1$ (p_1 is preferred) or both. Suppose also the preference relation is transitive, i.e., if $p_1 \preceq p_2$ and $p_2 \preceq p_3$, then $p_1 \preceq p_3$. We refer the reader to French and Ríos Insua (2000) for a discussion of how compelling are these conditions. It is clear that one can embed $\mathcal{P}$ as subset of $\mathcal{P}^*$ by treating each element of $\mathcal{P}$ as a degenerate element of $\mathcal{P}^*$. Thus the preference relation is also well-defined on $\mathcal{P}$. Suppose the relation satisfies axioms $\mathbf{H}_1$ and $\mathbf{H}_2$.

$\mathbf{H}_1$ If p_1, p_2 and $q \in \mathcal{P}^*$ and $0 < \lambda \leq 1$, then $p_1 \preceq p_2$ if and only if $\lambda p_1 + (1 - \lambda)q \preceq \lambda p_2 + (1 - \lambda)q$.

$\mathbf{H}_2$ If p_1, p_2, p_3 are in $\mathcal{P}^*$ and $p_1 \prec p_2 \prec p_3$, then there exist numbers $0 < \lambda < 1$, $0 < \mu < 1$, such that

$$\lambda p_3 + (1 - \lambda)p_1 \prec p_2 \prec \mu p_3 + (1 - \mu)p_1.$$

Ferguson (1967) shows that if $\mathbf{H}_1$ and $\mathbf{H}_2$ hold then there exists a utility $u(.)$ on $\mathcal{P}^*$ such that $p_1 \preceq p_2$ if and only if $u(p_1) \leq u(p_2)$, where for $p^* =$

$\sum_{i=1}^{m} \lambda_i p_i$, with $\lambda_i \geq 0$ and $\sum_{i=1}^{m} \lambda_i = 1$, $u(p^*)$ is defined to be the average

$$u(p^*) = \sum_{i=1}^{m} \lambda_i u(p_i). \qquad (3.6)$$

The main idea of the proof, which may also be used for eliciting the utility, is to start with a pair $p_1^* \prec p_2^*$, i.e., $p_1^* \preceq p_2^*$ but $p_1^* \not\sim p_2^*$. (Here $\sim$ denotes the equivalence relation that the DM is indifferent between the two elements.) Consider all $p_1^* \preceq p^* \preceq p_2^*$. Then by the assumptions $\mathbf{H}_1$ and $\mathbf{H}_2$, one can find $0 \leq \lambda^* \leq 1$ such that the DM would be indifferent between p^* and $(1-\lambda^*)p_1^* + \lambda^* p_2^*$. One can write $\lambda^* = u(p^*)$ and verify that $p_1^* \preceq p_3^* \preceq p_4^* \preceq p_2^*$ iff $\lambda_3^* \equiv u(p_3^*) \leq u(p_4^*) \equiv \lambda_4^*$ as well as the relation (3.6) above. For $p_2^* \preceq p^*$, by a similar argument one can find a $0 \leq \mu^* \leq 1$ such that

$$p_2^* \sim (1 - \mu^*)p_1^* + \mu^* p^*$$

from which one gets

$$p^* \sim (1 - \lambda^*)p_1^* + \lambda^* p_2^*,$$

where $\lambda^* = 1/\mu^*$. Set $\lambda^* = u(p^*)$ as before. In a similar way, one can find a λ^* for $p^* \preceq p_1^*$ and set $u(p^*) = \lambda^*$.

In principle, λ^* can be elicited for each p^*. Incidentally, utility is not unique. It is unique up to a change in origin and scale. Our version is chosen so that $u(p_1^*) = 0$, $u(p_2^*) = 1$.

French and Ríos Insua (2000) point out that most axiomatic approaches to the existence of a utility first exhibit a utility on $\mathcal{P}^*$ and then restrict it to $\mathcal{P}$, whereas intuitively, one would want to define $u(.)$ first on $\mathcal{P}$ and then extend it to $\mathcal{P}^*$. They discuss how this can be done.

3.3 Rationality Axioms Leading to the Bayesian Approach[1]

Consider a decision problem with all the ingredients discussed in Section 3.1 except the prior. If the sample space and the action space are finite, then the number of decision functions (i.e., functions from $\mathcal{X}$ to $\mathcal{A}$) is finite. In this case, the decision maker (DM) may be able to order any pair of given decision functions according to her rational preference of one to the other taking into account consequences of actions and all inherent uncertainties. Consider a randomized decision rule defined by

$$\delta = p_1 \delta_1 + p_2 \delta_2 + \cdots + p_k \delta_k,$$

where $\delta_1, \delta_2, \ldots, \delta_k$ constitute a listing of all the non-randomized decision functions and $(p_1, p_2, \ldots, p_k)$ is a probability vector, i.e., $p_i \geq 0$ and $\sum_{i=1}^{k} p_i =$

---

[1] Section 3.3 may be omitted at first reading.

1. The representation means that for each x, the probability distribution $\delta(x)$ in the action space is the same as choosing the action $\delta_i(x)$ with probability p_i. Suppose the DM can order any pair of randomized decision functions also in a rational way and it reduces to her earlier ordering if the randomized decision functions being compared are in fact non-randomized with one p_i equal to 1 and other p_i's equal to zero. Under certain axioms that we explore below, there exists a prior $\pi(\theta)$ such that $\delta_1^* \preceq \delta_2^*$, i.e., the DM prefers δ_1^* to δ_2^* if and only if

$$r(\pi, \delta_1^*) = \sum_{\theta,a} \pi(\theta) P_\theta(a|\delta_1^*) L(\theta, a) \leq \sum_{\theta,a} \pi(\theta) P_\theta(a|\delta_2^*) L(\theta, a) = r(\pi, \delta_2^*),$$

where $P_\theta(a|\delta^*)$ is the probability of choosing the action "a" when θ is the value of the parameter and δ^* is used, i.e., using the representation $\delta^* = \sum_i p_i^* \delta_i$,

$$P_\theta(a|\delta^*) = \sum_x P_\theta(x) \sum_i p_i^* I_i(x),$$

and I_i is the indicator function

$$I_i(x) = \begin{cases} 1 & \text{if } \delta_i(x) = a; \\ 0 & \text{if } \delta_i(x) \neq a. \end{cases}$$

We need to work a little to move from here to the starting point of Ferguson (1967).

As far as the preference is concerned, it is only the risk function of δ that matters. Also δ appears in the risk function only through $P_\theta(a|\delta)$ which, for each δ, is a probability distribution on the action space. Somewhat trivially, for each $\theta_0 \in \Theta$, one can also think of it as a probability distribution q on the space $\mathcal{P}$ of all (θ, a), $\theta \in \Theta$, $a \in \mathcal{A}$ such that

$$q(\theta, a) = \begin{cases} P_{\theta_0}(a|\delta) & \text{if } \theta = \theta_0; \\ 0 & \text{if } \theta \neq \theta_0. \end{cases}$$

As in Section 3.2, let the set of probability distributions putting probability on a finite number of points in $\mathcal{P}$ be $\mathcal{P}^*$. The DM can think of the choice of a δ as a somewhat abstract gamble with pay-off $(P_\theta(a_1|\delta), P_\theta(a_2|\delta), \cdots)$ if θ is true. This pay-off sits on $(\theta, a_1), (\theta, a_2)\ldots$. Let $\mathcal{G}$ be the set of all gambles of this form $[p_1, \ldots, p_m]$ where p_i is a probability distribution on $\mathcal{P}$ that is the pay-off corresponding to the ith point θ_i in $\Theta = \{\theta_1, \theta_2, \ldots, \theta_m\}$. Further, let $\mathcal{G}^*$ be the set of all probability distributions putting mass on a finite number of points in $\mathcal{G}$. The DM can embed her δ in $\mathcal{G}$ and suppose she can extend her preference relation to $\mathcal{G}$ and $\mathcal{G}^*$. If axioms $\mathbf{H}_1$ and $\mathbf{H}_2$ of Section 3.2 hold, then there exists a utility function u_g on $\mathcal{G}^*$ that induces the preference relation $\preceq_g$ on $\mathcal{G}^*$. We assume the preference relations $\preceq$ on $\mathcal{P}^*$ and $\preceq_g$ on $\mathcal{G}^*$ are connected as follows vide Ferguson (1967).

$\mathbf{A}_1$ If $p_i \preceq p_i'$, $i = 1, \ldots, m$, then $[p_1, \ldots, p_m] \preceq_g [p_1', \ldots, p_m']$.

$\mathbf{A}_2$ If $p \prec p'$, then $[p, \ldots, p] \prec_g [p', \ldots, p']$.

To proceed further, we need one more assumption, $\mathbf{A}_3$ of Ferguson (1967). If $p_1, \ldots, p_k$ are elements of $\mathcal{P}$ and $\lambda_1, \ldots, \lambda_k$ are non-negative numbers adding up to 1, let $(\lambda_1 p_1, \ldots, \lambda_k p_k)$ denote the element of $\mathcal{P}^*$ that chooses pay-off p_i with probability λ_i, $1 \leq i \leq k$. Then $\mathbf{A}_3$ is given by

$\mathbf{A}_3$ $(\lambda_1 [p_{11}, \ldots, p_{1m}], \cdots, \lambda_k [p_{k1}, \ldots, p_{km}])$

$$\sim_g [(\lambda_1 p_{11}, \ldots, \lambda_k p_{k1}), \cdots, (\lambda_1 p_{1m}, \ldots, \lambda_k p_{km})],$$

where $\sim_g$ denotes equivalence under the preference relation on $\mathcal{G}^*$.

Then, under these three assumptions, it is shown by Ferguson that $\preceq_g$ is induced by a prior $\pi(\theta)$ and the loss function $L(\theta, a)$ as indicated in Section 3.1.

The need to extend the preference relation on the space of decision functions to all pairs of elements of $\mathcal{G}^*$ is somewhat artificial. It is of course true that in many practical decision problems the space $\mathcal{G}^*$ would occur naturally. For example, even in a statistical problem, if the loss or utility arising from the combination (θ, a) doesn't depend on θ, then the extension to $\mathcal{G}^*$ would be relatively natural. An illuminating and penetrating discussion of various sets of axioms leading to existence of utility and prior appears in Chapter 2 of French and Ríos Insua (2000). They also provide references to a huge literature and a brief survey.

3.4 Coherence

There is an alternative way of justifying a Bayesian approach to decision making on the basis of the notion of *coherence* as modified by Freedman and Purves (1969) and Heath and Sudderth (1978). Coherence was originally introduced by de Finetti to show any quantification of uncertainty that does not satisfy the axioms of a (finitely additive) probability distribution would lead to sure loss in suitably chosen gambles. This is treated in Appendix C.

To return to coherence in the context of decision making, suppose A stands for a set in the space of θ and x values, and $A_x = \{\theta : (\theta, x) \in A\}$. Given x, the DM's uncertainty about A is given by $q(x, A_x)$. An MC (master of ceremonies) chooses a betting system (A, b), where A is as above and b is a bounded real valued function of x. The DM accepts the gamble with pay-off

$$\psi(\theta, x) = b(x) \left[I_A(\theta, x) - q(x, A_x) \right].$$

She gets $b(x)q(x, A_x)$ or pays $b(x)[1 - q(x, A_x)]$ depending on whether θ lies in A_x or not. The expected pay-off is

$$E(\theta) = \int \psi(\theta, x) \, p(dx|\theta).$$

If she accepts k such gambles defined as above by $(A_{(1)}, b_{(1)}), \ldots, (A_{(k)}, b_{(k)})$, then her expected pay-off is the sum of the k expected pay-offs. She will face sure loss if

$$\inf_\theta \left(\sum_{i=1}^{k} E_i(\theta) \right) > 0.$$

The idea is that if q reflects her uncertainty about θ, then this combination of bets is fair and so acceptable to her. However any rational choice of q should avoid sure loss as defined above. Such a choice is said to be coherent if no finite combination of acceptable bets can lead to sure loss. The basic result of Freedman and Purves (1969) and Heath and Sudderth (1978) is that in order to be coherent, the DM must act like a Bayesian with a (finitely additive) prior and q must be the resulting posterior. A similar result is proved by Berti et al. (1991).

3.5 Bayesian Analysis with Subjective Prior

We have already discussed basics of subjective prior Bayesian inference in Chapter 2. In the following, we shall concentrate on some issues related to robustness of Bayesian inference. The notations used will be mostly as given in Chapter 2, but some of those will be recalled and a few additional notations will be introduced here as needed.

Let $\mathcal{X}$ be the sample space and Θ be the parameter space. As before, suppose X has (model) density $f(x|\theta)$ and θ has (prior) probability density $\pi(\theta)$. Then the joint density of (X, θ), for $x \in \mathcal{X}$ and $\theta \in \Theta$, is

$$h(x, \theta) = f(x|\theta)\pi(\theta).$$

The marginal density of X corresponding with this joint density is

$$m_\pi(x) = m(x|\pi) = \int_\Theta f(x|\theta\, d\pi(\theta).$$

Note that this can be expressed as

$$m_\pi(x) = \begin{cases} \int_\Theta f(x|\theta)\pi(\theta)\, d\theta & \text{if } X \text{ is continuous,} \\ \sum_\Theta f(x|\theta)\pi(\theta) & \text{if } X \text{ is discrete.} \end{cases}$$

Often we shall use $m(x)$ for $m_\pi(x)$, especially if the prior π which is being used is clear from the context. Recall that the posterior density of θ given x is given by

$$\pi(\theta|x) = \frac{h(x, \theta)}{m_\pi(x)} = \frac{f(x|\theta)\pi(\theta)}{m_\pi(x)}.$$

The posterior mean and posterior variance with respect to prior π will be denoted by $E^\pi(\theta|x)$ and $V^\pi(\theta|x)$, respectively. Similarly, the posterior probability of a set $A \subset \Theta$ given x will be denoted by $P^\pi(A|x)$.

3.6 Robustness and Sensitivity

Intuitively, robustness means lack of sensitivity of the decision or inference to assumptions in the analysis that may involve a certain degree of uncertainty. In an inference problem, the assumptions usually involve choice of the model and prior, whereas in a decision problem there is the additional assumption involving the choice of the loss or utility function. An analysis to measure the sensitivity is called sensitivity analysis. Clearly, robustness with respect to all three of these components is desirable. That is to say that reasonable variations from the choice used in the analysis for the model, prior, and loss function do not lead to unreasonable variations in the conclusions arrived at. We shall not, however, discuss robustness with respect to model and loss function here in any great detail. Instead, we would like to mention that there is substantial literature on this and references can be found in sources such as Berger (1984, 1985a, 1990, 1994), Berger et al. (1996), Kadane (1984), Leamer (1978), Ríos Insua and Ruggeri (2000), and Wasserman (1992).

Because justification from the viewpoint of rational behavior is usually desired for inferential procedures, we would like to cite the work of Nobel laureate Kahneman on Bayesian robustness here. In his joint paper with Tversky (see Tversky et al. (1981) and Kahneman et al. (1982)), it was shown in psychological studies that seemingly inconsequential changes in the formulation of choice problems caused significance shifts of preference. These 'inconsistencies' were traced to all the components of decision making. This probably means that robustness of inference cannot be taken for granted but needs to be earned.

The following example illustrates why sensitivity to the choice of prior can be an important consideration.

Example 3.1. Suppose we observe X, which follows Poisson(θ) distribution. Further, it is felt *a priori* that θ has a continuous distribution with median 2 and upper quartile 4. i.e. $P^{\pi}(\theta \leq 2) = 0.5 = P^{\pi}(\theta \geq 2)$ and $P^{\pi}(\theta \geq 4) = 0.25$. If these are the only prior inputs available, the following three are candidates for such a prior:

(i) $\pi_1 : \theta \sim \text{exponential}(a)$ with $a = \log(2)/2$;

(ii) $\pi_2 : \log(\theta) \sim N(\log(2), (\log(2)/z_{.25})^2)$; and

(iii) $\pi_3 : \log(\theta) \sim \text{Cauchy}(\log(2), \log(2))$.

Then (i) under π_1, $\theta|x \sim \text{Gamma}(a + 1, x + 1)$, so that the posterior mean is $(a + 1)/(x + 1)$;

(ii) under π_2, if we let $\gamma = \log(\theta)$, and $\tau = \log(2)/z_{.25} = \log(2)/0.675$, we obtain

$$E^{\pi_2}(\theta|x) = E^{\pi_2}(\exp(\gamma)|x)$$

$$= \frac{\int_{-\infty}^{\infty} \exp(-e^{\gamma}) \exp(\gamma(x + 1)) \exp(-(\gamma - \log(2))^2/(2\tau^2)) \, d\gamma}{\int_{-\infty}^{\infty} \exp(-e^{\gamma}) \exp(\gamma x) \exp(-(\gamma - \log(2))^2/(2\tau^2)) \, d\gamma};$$

and (iii) under π_3, again if let $\gamma = \log(\theta)$, we get

Table 3.1. Posterior Means under π_1, π_2, and π_3

					x					
π	0	1	2	3	4	5	10	15	20	50
π_1	.749	1.485	2.228	2.971	3.713	4.456	8.169	11.882	15.595	37.874
π_2	.950	1.480	2.106	2.806	3.559	4.353	8.660	13.241	17.945	47.017
π_3	.761	1.562	2.094	2.633	3.250	3.980	8.867	14.067	19.178	49.402

$$
E^{\pi_3}(\theta|x) = E^{\pi_3}(\exp(\gamma)|x)
$$

$$
= \frac{\int_{-\infty}^{\infty} \exp(-e^{\gamma}) \exp(\gamma(x+1)) \left[1 + \left(\frac{\gamma-\log(2)}{\log(2)}\right)^2\right]^{-1} d\gamma}{\int_{-\infty}^{\infty} \exp(-e^{\gamma}) \exp(\gamma x) \left[1 + \left(\frac{\gamma-\log(2)}{\log(2)}\right)^2\right]^{-1} d\gamma}.
$$

To see if the choice of prior matters, simply examine the posterior means under the three different priors in Table 3.1.

For small or moderate x ($x \leq 10$), there is robustness: the choice of prior does not seem to matter too much. For large values of x, the choice does matter. The inference that a conjugate prior obtains then is quite different from what a heavier tailed prior would obtain. It is now clear that there are situations where it does matter what prior one chooses from a class of priors, each of which is considered reasonable given the available prior information.

The above example indicates that there is no escape from investigating prior robustness formally. How does one then reconcile this with the single prior Bayesian argument? It is certainly true that if one has a utility/loss function and a prior distribution there are compelling reasons for a Bayesian analysis using these. However, this assumes the existence of these two entities, and so it is of interest to know if one can justify the Bayesian viewpoint for statistics without this assumption. Various axiomatic systems for statistics can be developed (see Fishburn (1981)) involving a preference ordering for statistical procedures together with a set of axioms that any 'coherent' preference ordering must satisfy. Justification for the Bayesian approach then follows from the fact that any rational or coherent preference ordering corresponds to a Bayesian preference ordering (see Berger (1985a)). This means that there must be a loss function and a prior distribution such that this axiom system is compatible with the Bayesian approach corresponding to these. However, even then there are no compelling reasons to be a die-hard single prior Bayesian. The reason is that it is impractical to arrive at a total preference ordering. If we stop short of this and we are only able to come up with a partial preference ordering (see Seidenfeld et al. (1995) and Kadane et al. (1999)), the result will be a Bayesian analysis (again) using a class of prior distributions (and a class of utilities). This is the philosophical justification for a "robust Bayesian" as noted in Berger's book (Berger (1985a)). One could, of course, argue that a second stage of prior on the class Γ of possible priors is the natural solution to arrive at a single prior, but it is not clear how to arrive at this second stage prior.

3.7 Classes of Priors

There is a vast literature on how to choose a class, Γ of priors to model prior uncertainty appropriately. The goals (see Berger (1994)) are clearly
(i) to ensure that as many 'reasonable' priors as possible are included,
(ii) to try to eliminate 'unreasonable' priors,
(iii) to ensure that Γ does not require prior information which is difficult to elicit, and
(iv) to be able to compute measures of robustness without much difficulty.

As can be seen, (i) is needed to ensure robustness and (ii) to ensure that one does not erroneously conclude lack of robustness. The above mentioned are competing goals and hence can only be given weights which are appropriate in the given context. The following example from Berger (1994) is illuminating.

Example 3.2. Suppose θ is a real-valued parameter, prior beliefs about which indicate that it should have a continuous prior distribution, symmetric about 0 and having the third quartile, Q_3, between 1 and 2. Consider, then
$\Gamma_1 = \left\{ N(0, \tau^2), 2.19 < \tau^2 < 8.76 \right\}$ and
$\Gamma_2 = \{ \text{ symmetric priors with } 1 < Q_3 < 2 \}$.
Even though Γ_1 can be appropriate in some cases, it will mostly be considered "rather small" because it contains only sharp-tailed distributions. On the other hand, Γ_2 will typically be "too large," containing priors, shapes of some of which will be considered unreasonable. Starting with Γ_2 and imposing reasonable constraints such as unimodality on the priors can lead to sensible classes such as

$$\Gamma_3 = \{ \text{ unimodal symmetric priors with } 1 < Q_3 < 2 \} \supset \Gamma_1.$$

It will be seen that computing measures of robustness is not very difficult for any of these three classes.

3.7.1 Conjugate Class

The class consisting of conjugate priors (discussed in some detail in Chapter 5) is one of the easiest classes of priors to work with. If $X \sim N(\theta, \sigma^2)$ with known σ^2, the conjugate priors for θ are the normal priors $N(\mu, \tau^2)$. So one could consider

$$\Gamma_C = \left\{ N(\mu, \tau^2), \mu_1 \le \mu \le \mu_2, \tau_1^2 \le \tau^2 \le \tau_2^2 \right\}$$

for some specified values of μ_1, μ_2, τ_1^2, and τ_2^2. The advantage with the conjugate class is that posterior quantities can be calculated in closed form (for natural conjugate priors). In the above case, if $\theta \sim N(\mu, \tau^2)$, then $\theta|X = x \sim N(\mu^*(x), \delta^2)$, where $\mu^*(x) = (\tau^2/(\tau^2 + \sigma^2))x + (\sigma^2/(\tau^2 + \sigma^2))\mu$ and $\delta^2 = \tau^2\sigma^2/(\tau^2 + \sigma^2)$. Minimizing and maximizing posterior quantities then becomes an easy task (see Leamer (1978), Leamer (1982), and Polasek (1985)). The crucial drawback of the conjugate class is that it is usually "too small" to provide robustness. Further, tails of these prior densities are similar to those of the likelihood function, and hence prior moments greatly influence posterior inferences. Thus, even when the data is in conflict with the specified prior information the conjugate priors used can have very pronounced effect (which may be undesirable if data is to be trusted more). Details on this can be found in Berger (1984, 1985a, 1994). It must be added here that mixtures of conjugate priors, on the other hand, can provide robust inferences. In particular, the Student's t prior, which is a scale mixture of normals, having flat tails can be a good choice in some cases. We discuss some of these details later (see Section 3.9).

3.7.2 Neighborhood Class

If π_0 is a single elicited prior, then uncertainty in this elicitation can be modeled using the class

$$\Gamma_N = \{\pi \text{ which are in the neighborhood of } \pi_0\}.$$

A natural and well studied class is the ϵ-contamination class,

$$\Gamma_\epsilon = \{\pi : \pi = (1 - \epsilon)\pi_0 + \epsilon q, q \in Q\},$$

ϵ reflecting the uncertainty in π_0 and Q specifying the contaminations. Some choices for Q are, all distributions q, all unimodal distributions with mode θ_0, and all unimodal symmetric distributions with mode θ_0. The ϵ-contamination class with appropriate choice of Q can provide good robustness as we will see later.

3.7.3 Density Ratio Class

Assuming the existence of densities for all the priors in the class, the density ratio class is defined as

$$\Gamma_{DR} = \{\pi : L(\theta) \leq \alpha\pi(\theta) \leq U(\theta) \text{ for some } \alpha > 0\}$$
$$= \left\{\pi : \frac{L(\theta)}{U(\theta')} \leq \frac{\pi(\theta)}{\pi(\theta')} \leq \frac{U(\theta)}{L(\theta')} \text{ for all } \theta, \theta'\right\}, \tag{3.7}$$

for specified non-negative functions L and U (see DeRobertis and Hartigan (1981)). If we take $L \equiv 1$ and $U \equiv c$, then we get

$$\Gamma_{DR} = \left\{\pi : c^{-1} \leq \frac{\pi(\theta)}{\pi(\theta')} \leq c \text{ for all } \theta, \theta'\right\}.$$

Some other classes have also been studied. For example, the sub-sigma field class is obtained by defining the prior on a sub-sigma field of sets. See Berger (1990) for details and references. Because many distributions are determined by their moments, once the distributional form is specified, sometimes bounds are specified on their moments to arrive at a class of priors (see Berger (1990)).

3.8 Posterior Robustness: Measures and Techniques

Measures of sensitivity are needed to examine the robustness of inference procedures (or decisions) when a class Γ of priors are under consideration. In recent years two types of these measures have been studied. Global measures of sensitivity such as the range of posterior quantities and local measures such as the derivatives (in a sense to be made clear later) of these quantities. Attempts have also been made to derive robust priors and robust procedures using these measures.

3.8.1 Global Measures of Sensitivity

Example 3.3. Suppose $X_1, X_2, \ldots, X_n$ are i.i.d. $N(\theta, \sigma^2)$, with σ^2 known and let Γ be all $N(0, \tau^2)$, $\tau^2 > 0$, priors for θ. Then the variation in the posterior mean is simply $\left(\inf_{\tau^2>0} E(\theta|\overline{x}), \sup_{\tau^2>0} E(\theta|\overline{x})\right)$. Because, for fixed τ^2, $E(\theta|\overline{x}) = (\tau^2/(\tau^2 + \sigma^2))\overline{x}$, this range can easily be seen to be $(0, \overline{x})$ or $(\overline{x}, 0)$ according as $\overline{x} \geq 0$ or $\overline{x} < 0$. If $\overline{x}$ is small in magnitude, this range will be small. Thus the robustness of the procedure of using posterior mean as the Bayes estimate of θ will depend crucially on the magnitude of the observed value of $\overline{x}$.

As can be seen from the above example, a natural global measure of sensitivity of the Bayesian quantity to the choice of prior is the range of this quantity as the prior varies in the class of priors of interest. Further, as explained in Berger (1990), typically there are three categories of Bayesian quantities of interest.

(i) Linear functionals of the prior: $\rho(\pi) = \int_\Theta h(\theta)\,\pi(d\theta)$, where h is a given function.

If h is taken to be the likelihood function l, we get an important linear functional, the marginal density of data, i.e., $m(\pi) = \int_\Theta l(\theta)\,\pi(d\theta)$.

(ii) Ratio of linear functionals of the prior: $\rho(\pi) = \frac{1}{m(\pi)}\int_\Theta h(\theta)l(\theta)\,\pi(d\theta)$ for some given function h.

If we take $h(\theta) = \theta$, $\rho(\pi)$ is the posterior mean. For $h(\theta) = I_C(\theta)$, the indicator function of the set C, we get the posterior probability of C.

(iii) Ratio of nonlinear functionals: $\rho(\pi) = \frac{1}{m(\pi)}\int_\Theta h(\theta, \phi(\pi))l(\theta)\,\pi(d\theta)$ for some given h. For $h(\theta, \phi(\pi)) = (\theta - \mu(\pi))^2$, where $\mu(\pi)$ is the posterior mean, we get $\rho(\pi) =$ the posterior variance.

Note that extreme values of linear functionals of the prior as it varies in a class Γ are easy to compute if the extreme points of Γ can be identified.

Example 3.4. Suppose $X \sim N(\theta, \sigma^2)$, with σ^2 known and the class Γ of interest is

$$\Gamma_{SU} = \{ \text{ all symmetric unimodal distributions with mode } \theta_0 \}.$$

Then ϕ denoting the standard normal density, $m(\pi) = \int_{-\infty}^{\infty} \frac{1}{\sigma}\phi(\frac{x-\theta}{\sigma})\pi(\theta)\, d\theta$. Note that any unimodal symmetric (about θ_0) density π is a mixture of uniform densities symmetric about θ_0. Thus the extreme points of Γ_{SU} are $U(\theta_0 - r, \theta_0 + r)$ distributions. Therefore,

$$\inf_{\pi \in \Gamma_{SU}} m(\pi) = \inf_{r>0} \frac{1}{2r} \int_{\theta_0-r}^{\theta_0+r} \frac{1}{\sigma}\phi(\frac{x-\theta}{\sigma})\, d\theta$$

$$= \inf_{r>0} \frac{1}{2r} \left\{ \Phi(\frac{\theta_0 + r - x}{\sigma}) - \Phi(\frac{\theta_0 - r - x}{\sigma}) \right\}, \qquad (3.8)$$

$$\sup_{\pi \in \Gamma_{SU}} m(\pi) = \sup_{r>0} \frac{1}{2r} \int_{\theta_0-r}^{\theta_0+r} \frac{1}{\sigma}\phi(\frac{x-\theta}{\sigma})\, d\theta$$

$$= \sup_{r>0} \frac{1}{2r} \left\{ \Phi(\frac{\theta_0 + r - x}{\sigma}) - \Phi(\frac{\theta_0 - r - x}{\sigma}) \right\}. \qquad (3.9)$$

In empirical Bayes problems (to be discussed later), for example, maximization of the above kind is needed to select a prior. This is called *Type II maximum likelihood* (see Good (1965)).

To study ratio-linear functionals the following results from Sivaganesan and Berger (1989) are useful.

Lemma 3.5. *Suppose C_T is a set of probability measures on the real line given by $C_T = \{\nu_t : t \in T\}$, $T \subset \mathcal{R}^d$, and let $\mathcal{C}$ be the convex hull of C_T. Further suppose h_1 and h_2 are real-valued functions defined on $\mathcal{R}$ such that $\int |h_1(x)|\, dF(x) < \infty$ for all $F \in \mathcal{C}$, and $K + h_2(x) > 0$ for all x for some constant K. Then, for any k,*

$$\sup_{F \in \mathcal{C}} \frac{k + \int h_1(x)\, dF(x)}{K + \int h_2(x)\, dF(x)} = \sup_{t \in T} \frac{k + \int h_1(x)\nu_t(dx)}{K + \int h_2(x)\nu_t(dx)}, \qquad (3.10)$$

$$\inf_{F \in \mathcal{C}} \frac{k + \int h_1(x)\, dF(x)}{K + \int h_2(x)\, dF(x)} = \inf_{t \in T} \frac{k + \int h_1(x)\nu_t(dx)}{K + \int h_2(x)\nu_t(dx)}. \qquad (3.11)$$

Proof. Because $\int h_1(x)\, dF(x) = \int h_1(x) \int_T \nu_t(dx)\mu(dt)$, for some probability measure μ on T, using Fubini's theorem,

$$k + \int h_1(x)\, dF(x) = \int (k + h_1(x)) \int_T \nu_t(dx)\mu(dt)$$

$$= \int_T \left(\int (k + h_1(x))\nu_t(dx) \right) \mu(dt)$$

$$= \int_T \left[\left(\frac{\int (k + h_1(x))\nu_t(dx)}{\int (K + h_2(x))\nu_t(dx)} \right) \int (K + h_2(x))\nu_t(dx) \right] \mu(dt)$$

$$\leq \left(\sup_{t \in T} \frac{\int (k + h_1(x))\nu_t(dx)}{\int (K + h_2(x))\nu_t(dx)} \right) \left(K + \int h_2(x)\, dF(x) \right).$$

Therefore,

$$\sup_{F \in \mathcal{C}} \frac{k + \int h_1(x)\, dF(x)}{K + \int h_2(x)\, dF(x)} \leq \sup_{t \in T} \frac{\int (k + h_1(x))\nu_t(dx)}{\int (K + h_2(x))\nu_t(dx)}.$$

However, because $\mathcal{C} \supset \mathcal{C}_T$,

$$\sup_{F \in \mathcal{C}} \frac{k + \int h_1(x)\, dF(x)}{K + \int h_2(x)\, dF(x)} \geq \sup_{t \in T} \frac{\int (k + h_1(x))\nu_t(dx)}{\int (K + h_2(x))\nu_t(dx)}.$$

Hence the proof for the supremum, and the proof for the infimum is along the same lines. $\square$

Theorem 3.6. *Consider the class Γ_{SU} of all symmetric unimodal prior distributions with mode θ_0. Then it follows that*

$$\sup_{\pi \in \Gamma_{SU}} E^{\pi}(g(\theta)|x) = \sup_{r>0} \frac{\frac{1}{2r} \int_{\theta_0-r}^{\theta_0+r} g(\theta)f(x|\theta)\, d\theta}{\frac{1}{2r} \int_{\theta_0-r}^{\theta_0+r} f(x|\theta)\, d\theta}, \tag{3.12}$$

$$\inf_{\pi \in \Gamma_{SU}} E^{\pi}(g(\theta)|x) = \inf_{r>0} \frac{\frac{1}{2r} \int_{\theta_0-r}^{\theta_0+r} g(\theta)f(x|\theta)\, d\theta}{\frac{1}{2r} \int_{\theta_0-r}^{\theta_0+r} f(x|\theta)\, d\theta}. \tag{3.13}$$

Proof. Note that $E^{\pi}(g(\theta)|x) = \dfrac{\int g(\theta)f(x|\theta)\, d\pi(\theta)}{\int f(x|\theta)\, d\pi(\theta)}$, where $f(x|\theta)$ is the density of the data x. Now Lemma 3.5 can be applied by recalling that any unimodal symmetric distribution is a mixture of symmetric uniform distributions. $\square$

Example 3.7. Suppose $X|\theta \sim N(\theta, \sigma^2)$ and robustness of the posterior mean with respect to Γ_{SU} is of interest. Then, range of posterior mean over this class can be easily computed using Theorem 3.6. We thus obtain,

$$\sup_{\pi \in \Gamma_{SU}} E^{\pi}(\theta|x) = \sup_{r>0} \frac{\frac{1}{2r} \int_{\theta_0-r}^{\theta_0+r} \frac{\theta}{\sigma}\phi\left(\frac{x-\theta}{\sigma}\right) d\theta}{\frac{1}{2r} \int_{\theta_0-r}^{\theta_0+r} \frac{1}{\sigma}\phi\left(\frac{x-\theta}{\sigma}\right) d\theta}$$

$$= x + \sup_{r>0} \frac{\phi\left(\frac{\theta_0-r-x}{\sigma}\right) - \phi\left(\frac{\theta_0+r-x}{\sigma}\right)}{\Phi\left(\frac{\theta_0+r-x}{\sigma}\right) - \Phi\left(\frac{\theta_0-r-x}{\sigma}\right)},$$

$$\inf_{\pi \in \Gamma_{SU}} E^{\pi}(\theta|x) = \inf_{r>0} \frac{\frac{1}{2r} \int_{\theta_0-r}^{\theta_0+r} \frac{\theta}{\sigma}\phi\left(\frac{x-\theta}{\sigma}\right) d\theta}{\frac{1}{2r} \int_{\theta_0-r}^{\theta_0+r} \frac{1}{\sigma}\phi\left(\frac{x-\theta}{\sigma}\right) d\theta}$$

$$= x + \inf_{r>0} \frac{\phi\left(\frac{\theta_0-r-x}{\sigma}\right) - \phi\left(\frac{\theta_0+r-x}{\sigma}\right)}{\Phi\left(\frac{\theta_0+r-x}{\sigma}\right) - \Phi\left(\frac{\theta_0-r-x}{\sigma}\right)}.$$

Example 3.8. Suppose $X|\theta \sim N(\theta, \sigma^2)$ and it is of interest to test $H_0 : \theta \leq \theta_0$ versus $H_1 : \theta > \theta_0$. Again, suppose that Γ_{SU} is the class of priors to be considered and robustness of this class is to be examined. Because

$$P^\pi(H_0|x) = P^\pi(\theta \le \theta_0|x)$$
$$= \frac{\int_{-\infty}^{\infty} I_{(-\infty,\theta_0]}(\theta) f(x|\theta)\, d\pi(\theta)}{\int_{-\infty}^{\infty} f(x|\theta)\, d\pi(\theta)},$$

we can apply Theorem 3.6 here as well. We get,

$$\sup_{\pi \in \Gamma_{SU}} P^\pi(H_0|x) = \sup_{r>0} \frac{\frac{1}{2r}\int_{\theta_0-r}^{\theta_0} \frac{1}{\sigma}\phi(\frac{x-\theta}{\sigma})\, d\theta}{\frac{1}{2r}\int_{\theta_0-r}^{\theta_0+r} \frac{1}{\sigma}\phi(\frac{x-\theta}{\sigma})\, d\theta}$$
$$= \sup_{r>0} \frac{\Phi(\frac{\theta_0-x}{\sigma}) - \Phi(\frac{\theta_0-r-x}{\sigma})}{\Phi(\frac{\theta_0+r-x}{\sigma}) - \Phi(\frac{\theta_0-r-x}{\sigma})},$$

and similarly,

$$\inf_{\pi \in \Gamma_{SU}} P^\pi(H_0|x) = \inf_{r>0} \frac{\Phi(\frac{\theta_0-x}{\sigma}) - \Phi(\frac{\theta_0-r-x}{\sigma})}{\Phi(\frac{\theta_0+r-x}{\sigma}) - \Phi(\frac{\theta_0-r-x}{\sigma})}.$$

It can be seen that the above bounds are, respectively, 0.5 and α, where $\alpha = \Phi(\frac{x-\theta_0}{\sigma})$, the P-value.

We shall now consider the density-ratio class that was mentioned earlier in (3.7) and is given by

$$\Gamma_{DR} = \{\pi : L(\theta) \le \alpha\pi(\theta) \le U(\theta) \text{ for some } \alpha > 0\},$$

for specified non-negative functions L and U. For $\pi \in \Gamma_{DR}$ and any real-valued π-integrable function h on the parameter space Θ, let $\pi(h) = \int_\Theta h(\theta)\pi(d\theta)$. Further, let $h \equiv h^+ - h^-$ be the usual decomposition of h into its positive and negative parts, i.e., $h^+(u) = \max\{h(x),0\}$ and $h^-(u) = \max\{-h(x),0\}$. Then we have the following theorem (see DeRobertis and Hartigan (1981)).

Theorem 3.9. *For U-integrable functions h_1 and h_2, with h_2 positive a.s. with respect to all $\pi \in \Gamma_{DR}$,*

$$\inf_{\pi \in \Gamma_{DR}} \frac{\pi(h_1)}{\pi(h_2)} \quad \text{is the unique solution } \lambda \text{ of}$$

$$U(h_1 - \lambda h_2)^- + L(h_1 - \lambda h_2)^+ = 0, \tag{3.14}$$

$$\sup_{\pi \in \Gamma_{DR}} \frac{\pi(h_1)}{\pi(h_2)} \quad \text{is the unique solution } \lambda \text{ of}$$

$$U(h_1 - \lambda h_2)^+ + L(h_1 - \lambda h_2)^- = 0. \tag{3.15}$$

Proof. Let $\lambda_0 = \inf_{\pi \in \Gamma_{DR}} \frac{\pi(h_1)}{\pi(h_2)}$, $c_1 = \inf_{\pi \in \Gamma_{DR}} \pi(h_2)$ and $c_2 = \sup_{\pi \in \Gamma_{DR}} \pi(h_2)$. Then $0 < c_1 < c_2 < \infty$, and $|\lambda_0| < \infty$. Because $U(h_1 - \lambda h_2)^- + L(h_1 - \lambda h_2)^+ = \inf_{\pi \in \Gamma_{DR}} \pi(h_1 - \lambda h_2)$ for any λ, note that $\lambda_0 \ge \lambda$ if and only if $U(h_1 - \lambda h_2)^- + L(h_1 - \lambda h_2)^+ \ge 0$. However, $\lambda_0 > \lambda$ if and only if $U(h_1 - \lambda h_2)^- + L(h_1 - \lambda h_2)^+ > 0$. A similar argument for the supremum. $\quad\square$

Example 3.10. Suppose $X \sim N(\theta, \sigma^2)$, with σ^2 known. Consider the class Γ_{DR} with L being the Lebesgue measure and $U = kL$, $k > 1$. Because the posterior mean is

$$\frac{\int \theta f(x|\theta)\, d\pi(\theta)}{\int f(x|\theta)\, d\pi(\theta)} = \frac{\pi(\theta f(x|\theta))}{\pi(f(x|\theta))},$$

in the notation of Theorem 3.9, we have that $\inf_{\pi \in \Gamma_{DR}} E^\pi(\theta|x)$ is the unique solution λ of

$$k \int_{-\infty}^{\lambda} (\theta - \lambda) f(x|\theta)\, d\theta + \int_{\lambda}^{\infty} (\theta - \lambda) f(x|\theta)\, d\theta = 0, \tag{3.16}$$

and similarly, $\sup_{\pi \in \Gamma_{DR}} E^\pi(\theta|x)$ is the unique solution λ of

$$\int_{-\infty}^{\lambda} (\theta - \lambda) f(x|\theta)\, d\theta + k \int_{\lambda}^{\infty} (\theta - \lambda) f(x|\theta)\, d\theta = 0. \tag{3.17}$$

Noting that $f(x|\theta) = \frac{1}{\sigma}\phi(\frac{x-\theta}{\sigma}) = \frac{1}{\sigma}\phi(\frac{\theta-x}{\sigma})$, and letting λ_1 be the minimum and λ_2 the maximum, the above equations may be rewritten as

$$(k-1)\left[(\frac{\lambda_1 - x}{\sigma})\Phi(\frac{\lambda_1 - x}{\sigma}) + \phi(\frac{\lambda_1 - x}{\sigma})\right] = \frac{\lambda_1 - x}{\sigma}, \tag{3.18}$$

$$(k-1)\left[(\frac{\lambda_2 - x}{\sigma})\Phi(\frac{\lambda_2 - x}{\sigma}) + \phi(\frac{\lambda_2 - x}{\sigma})\right] = k(\frac{\lambda_2 - x}{\sigma}). \tag{3.19}$$

Now let $k(\frac{\lambda_1 - x}{\sigma}) = \gamma$. Then $\lambda_2 = x + \sigma\frac{\gamma}{k}$. Put $\lambda_0 = x - \sigma\frac{\gamma}{k}$, or $\frac{\lambda_0 - x}{\sigma} = -\frac{\gamma}{k}$. Then we see from the second equation above that

$$(k-1)\left[(\frac{\lambda_0 - x}{\sigma})\Phi(\frac{\lambda_0 - x}{\sigma}) + \phi(\frac{\lambda_0 - x}{\sigma})\right]$$
$$= (k-1)\left[-\frac{\gamma}{k}\Phi(-\frac{\gamma}{k}) + \phi(-\frac{\gamma}{k})\right]$$
$$= (k-1)\left[-\frac{\gamma}{k}(1 - \Phi(\frac{\gamma}{k})) + \phi(\frac{\gamma}{k})\right]$$
$$= (k-1)\left[\frac{\gamma}{k}\Phi(\frac{\gamma}{k}) + \phi(\frac{\gamma}{k})\right] - (k-1)\frac{\gamma}{k}$$
$$= 0,$$

implying that once λ_2 is obtained, say $\lambda_2 = x + \sigma\frac{\gamma}{k}$, the solution for λ_1 is simply $x - \sigma\frac{\gamma}{k}$. Table 3.2 tabulates $\gamma = \gamma(k)$ for various values of k. What one

Table 3.2. Values of $\gamma(k)$ for Some Values of k

k	1	1.25	1.5	2	3	4	5	10
$\gamma(k)$	0	0.089	0.162	0.276	0.436	0.549	0.636	0.901

can easily see from this table is that, if, for example, the prior density ratio between two parameter points is sure to be between 0.5 and 2, the posterior mean is sure to be within 0.276 standard deviation of x, and if instead the ratio is certain to be between 0.1 and 10, the range is certain to be no more than 1 s.d. either side.

3.8.2 Belief Functions

An entirely different approach to global Bayesian robustness is available, and this is through *belief functions* and *plausibility functions*. This originated with the introduction of *upper and lower probabilities* by Dempster (1967, 1968) but further evolved in various directions as can be seen from Shafer (1976, 1979), Wasserman (1990), Wasserman and Kadane (1990), and Walley (1991). The terminology of *infinitely alternating Choquet capacity* is also used in the literature. *Imprecise probability* is a generic term used in this context, which includes *fuzzy logic* as well as *upper and lower previsions*.

Recall that robust Bayesian inference uses a class of plausible prior probability measures. It turns out that associated with a *belief function* is a convex set of probability measures, of which the *belief function* is a lower bound, and the *plausibility function* an upper bound. Thus a belief function and a plausibility function can naturally be used to construct a class of prior probability distributions. Some specific details are given below skipping technical details and some generality.

Suppose the parameter space Θ is a Euclidean space and D is a convex, compact subset of a Euclidean space. Let μ be a probability measure on D and T be a map taking points in D to nonempty closed subsets of Θ. Then for each $A \subset \Theta$, define

$$A_* = \{d \in D : T(d) \subset A\}, \text{and}$$
$$A^* = \{d \in D : T(d) \cap A \neq \phi\}.$$

Define Bel and Pl on Θ by

$$Bel(A) = \mu(A_*) \text{ and } Pl(A) = \mu(A^*). \tag{3.20}$$

Then Bel is called a *belief function* and Pl, a *plausibility function* with source (D, μ, T). Note that $0 \leq Bel(A) \leq Pl(A) \leq 1$, $Bel(A) = 1 - Pl(A^c)$ for any A, and $Bel(\Theta) = Pl(\Theta) = 1$, $Bel(\phi) = Pl(\phi) = 0$. The above definition may be given the following meaning. If evidence comes from a random draw from D, then $Bel(A)$ may be interpreted to be the probability that this evidence implies A is true, whereas $Pl(A)$ can be thought of as the probability that this evidence is consistent with A being true. It can be checked that Bel is a probability measure iff $Bel(A) = Pl(A)$ for all A, or equivalently, $T(d)$ is almost surely a singleton set.

Example 3.11. Suppose it is known that the true value of θ lies in a fixed set $\Theta_0 \subset \Theta$. Set $T(d) = \Theta_0$ for all $d \in D$. Then $Bel(A) = 1$ if $\Theta_0 \subset A$; $Bel(A) = 0$ otherwise.

Example 3.12. Suppose P is a probability measure on Θ. Then P is also a belief function with source (Θ, P, T), where $T(\theta) = \{\theta\}$.

A probability measure P is said to be *compatible* with Bel and Pl if for each A, $Bel(A) \leq P(A) \leq Pl(A)$. Let $\mathcal{C}$ be the set of all probability measures *compatible* with Bel and Pl. Then $\mathcal{C} \neq \phi$ and for each A,

$$Bel(A) = \inf_{P \in \mathcal{C}} P(A) \text{ and } Pl(A) = \sup_{P \in \mathcal{C}} P(A).$$

This indicates that we can use Bel and Pl to construct prior envelopes. In particular, if Bel and Pl arise from any available partial prior information, then the set of *compatible* probability measures, $\mathcal{C}$, is exactly the class of prior distributions that a robust Bayesian analysis requires (compare with (3.4)).

Let $h : \Theta \to \mathcal{R}$ be any bounded, measurable function. Define its upper and lower expectations by

$$E^*(h) = \sup_{P \in \mathcal{C}} E_P(h) \text{ and } E_*(h) = \inf_{P \in \mathcal{C}} E_P(h), \tag{3.21}$$

where $E_P(h) = \int_{\Theta} h(\theta)\, P(d\theta)$. If we let

$$h^*(d) = \sup_{\theta \in T(d)} h(\theta) \text{ and } h_*(d) = \inf_{\theta \in T(d)} h(\theta),$$

then it can be shown that

$$E^*(h) = \int_D h^*(u)\, \mu(du) \text{ and } E_*(h) = \int_D h_*(u)\, \mu(du). \tag{3.22}$$

Details on these may be found in Wasserman (1990). Based on these ideas, some new techniques for Bayesian robustness measures can be derived when the prior envelopes arise from belief functions.

Suppose Bel is a belief function on Θ with source (D, μ, T) and $\mathcal{C}$ is the class of all prior probability measures *compatible* with Bel. Let $L(\theta) = f(x|\theta)$ be the likelihood function of θ given the data x, and let $L_A(\theta) = L(\theta)I_A(\theta)$, where I_A is the indicator function of $A \subset \Theta$. Then we have the following result and its application from Wasserman (1990).

Theorem 3.13. *If $L(\theta)$ is bounded and $A \subset \Theta$, then*

$$\inf_{\pi \in \mathcal{C}} \pi(A|x) = \frac{E_*(L_A)}{E_*(L_A) + E^*(L_{A^c})} = \frac{E_\mu((L_A)_*)}{E_\mu((L_A)_*) + E_\mu(L_{A^c}^*)}, \tag{3.23}$$

$$\sup_{\pi \in \mathcal{C}} \pi(A|x) = \frac{E^*(L_A)}{E^*(L_A) + E_*(L_{A^c})} = \frac{E_\mu(L_A^*)}{E_\mu(L_A^*) + E_\mu((L_{A^c})_*)}. \tag{3.24}$$

Example 3.14. Consider the class of ϵ-contamination priors:

$$\mathcal{C} = \{\pi : \pi = (1 - \epsilon)\pi_0 + \epsilon q, q \in Q\},$$

where Q is the class of all probability measures on Θ. This neighborhood class $\mathcal{C}$ corresponds to the belief function with source (D, μ, T), where $D = \Theta \cup \{d_0\}$, $\mu = (1 - \epsilon)\pi'_0 + \epsilon\delta$, and

$$T(d) = \begin{cases} \{d\} & \text{if } d \in \Theta; \\ \Theta & \text{if } d = d_0. \end{cases}$$

Here δ is a point mass on d_0 and π'_0 is a probability measure on D giving zero probability to d_0 and is identical to π_0 on $D - \{d_0\}$. Then from Theorem 3.13 above,

$$\sup_{\pi \in \mathcal{C}} \pi(A|x) = \frac{(1 - \epsilon) \int_A L(\theta)\pi_0(d\theta) + \epsilon \sup_{\theta \in A} L(\theta)}{(1 - \epsilon) \int_\Theta L(\theta)\pi_0(d\theta) + \epsilon \sup_{\theta \in A} L(\theta)}, \tag{3.25}$$

$$\inf_{\pi \in \mathcal{C}} \pi(A|x) = \frac{(1 - \epsilon) \int_A L(\theta)\pi_0(d\theta)}{(1 - \epsilon) \int_\Theta L(\theta)\pi_0(d\theta) + \epsilon \sup_{\theta \in A^c} L(\theta)}. \tag{3.26}$$

It may be noted that this is a different proof for the same result of Berger and Berliner (1986).

3.8.3 Interactive Robust Bayesian Analysis

Following Berger (1994), an interactive scheme for robust Bayesian analysis can be suggested according to the diagram Figure 3.1. The point to note is that, if lack of robustness is evident, then the class Γ of priors obtained from initial prior inputs has to be shrunk using further prior elicitation. Details on such an approach for shrinking a large quantile class of priors is described in Liseo et al. (1996).

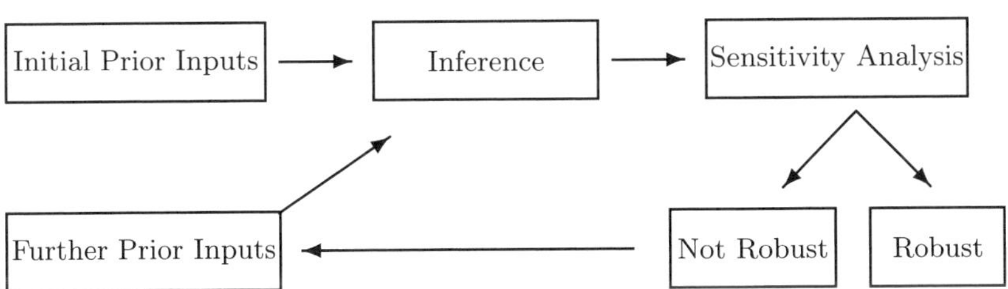

Fig. 3.1. Interactive robust Bayesian scheme.

3.8.4 Other Global Measures

As seen earlier, interpretation of the size of the range of posterior quantities needs to be done within the given context. However, some efforts have been made to derive certain generic measures also. Ruggeri and Sivaganesan (2000) suggest a scaled version of the range for this purpose. Suppose π_0 is a baseline prior, and let the sensitivity of the posterior mean of a target quantity $h(\theta)$ to deviations from π_0 be of interest. Let Γ be a class of plausible priors π on θ. Assume the following notation of $\rho^\pi(x) = E^\pi(h(\theta)|x)$, $\rho^0(x) = E^{\pi_0}(h(\theta)|x)$, and $V^\pi(x)$ denoting the posterior variance of $h(\theta)$ under prior π. Then the *relative sensitivity*, denoted by R_π, is defined as

$$R_\pi(x) = \frac{(\rho^\pi(x) - \rho^0(x))^2}{V^\pi(x)}. \tag{3.27}$$

The motivation for considering R_π is that the posterior variance V^π is a measure of accuracy in estimation of $h(\theta)$, and hence if the squared distance of $\rho^\pi(x)$ from $\rho^0(x)$ relative to this is not too large, robustness can be expected. The following example which is essentially from Ruggeri and Sivaganesan (2000) illustrates this idea.

Example 3.15. Let X have the $N(\theta, 1)$ distribution, and under π_0, let θ be $N(0, 2)$. Consider the class Γ of all $N(0, \tau^2)$ priors with $1 \le \tau^2 \le 10$. Consider sensitivity of posterior inferences about $h(\theta) = \theta$ when $x > 0$ is observed. Because the posterior distribution (under the prior $N(0, \tau^2)$) of θ given x is normal with mean $\tau^2 x/(\tau^2 + 1)$ and variance $\tau^2/(\tau^2 + 1)$, note that

$$\rho^\pi(x) - \rho^0(x) = \left(\frac{\tau^2}{\tau^2 + 1} - \frac{2}{3}\right) x \text{ and } R_\pi(x) = \frac{(\tau^2 - 2)^2 x^2}{9\tau^2(\tau^2 + 1)}.$$

It can then be easily checked that the range of $\rho^\pi(x) - \rho^0(x)$ is $8x/33$ and $\sup R_\pi(x) = 6.4x^2/99$. Thus, robustness can be expected when the observation x lies in the range $0 \le x \le 4$, but certainly not when $x = 10$.

3.8.5 Local Measures of Sensitivity

As can be noted from the previous section, unless the class Γ of possible priors is a 'nice' parametric class, or a class whose set of extreme points is easy to work with, computational complexity of global measures of robustness is high. Furthermore, this 'global' approach can become quite unfeasible for very complicated models. If, for example, $X \sim P_\theta$, and θ is p-dimensional, $p > 1$, then the range of posterior mean of θ_i may well depend on prior inputs on θ_j for $j \ne i$ also. If such is the case, global measures of robustness will involve computing ranges of posterior quantities of general functions $g(\theta)$ over classes of joint prior distributions of θ.

The alternative, which has attracted a lot of attention in recent years, is that of trying to study the effects of small perturbations to the prior. This is called local sensitivity. In this approach also, one may either study the sensitivity of the entire posterior distribution or that of some specified posterior quantity. Let us first consider the former as in Gustafson and Wasserman (1995). A different set of notations as given below are needed in this section. Let π be a prior probability measure and let π^x denote its corresponding posterior probability measure given the data x, i.e., $\pi^x(d\theta) = f(x|\theta)\pi(d\theta)/m_\pi(x)$ where $m_\pi(x) = \int_\Theta f(x|\theta)\pi(d\theta)$ is the marginal density of the data. Let $\mathcal{P}$ be the set of all probability measures on the probability space $(\Theta, \mathcal{B})$. A distance function $d : \mathcal{P} \to \mathcal{P}$ is needed to quantify changes in prior and posterior measures. Let ν_ϵ be a perturbation of π in the direction of a measure ν. Then the local sensitivity of $\mathcal{P}$ in the direction of ν can be defined (see Gustafson and Wasserman (1995)) by

$$s(\pi, \nu; x) = \lim_{\epsilon \downarrow 0} \frac{d(\pi^x, \nu_\epsilon^x)}{d(\pi, \nu_\epsilon)}. \tag{3.28}$$

Two different types of perturbations ν_ϵ have been considered. The linear perturbation is defined as $\nu_\epsilon = (1-\epsilon)\pi + \epsilon\nu$, and the geometric perturbation as $d\nu_\epsilon \propto (\frac{d\nu}{d\pi})^\epsilon d\pi$. (See Gelfand and Dey (1991) for details.) The local sensitivity $s(\pi, \nu; x)$ is simply the rate at which the perturbed posterior ν_ϵ^x tends to the 'initial' posterior π^x relative to the change in the prior. As a measure of overall sensitivity of a class Γ of priors one may take

$$s(\pi, \Gamma; x) = \sup_{\nu \in \Gamma} s(\pi, \nu; x).$$

There are many possible choices for d, the distance measure.

(i) $d_{TV}(\pi, \nu) = \sup_{A \in \mathcal{B}} |\pi(A) - \nu(A)|$, the total variation distance. In this case $s(\pi, \nu; x)$ for linear perturbations turns out to be the norm of the Fréchet derivative. To see this one needs to start with the Gateaux differential of the posterior. To define the Gateaux differential, let $\delta = \pi - \nu$, $||\delta|| = d_{TV}(\pi, \nu)$ and define $T : \mathcal{P} \to \mathcal{P}$ by $T(\pi) = \pi^x$. The Gateaux differential of T is then

$$\dot{T}_\pi(\delta) = \lim_{\epsilon \downarrow 0} \frac{d_{TV}(\pi^x, \nu_\epsilon^x)}{\epsilon} = \frac{m_\nu(x)}{m_\pi(x)} d_{TV}(\pi^x, \nu^x),$$

because

$$\nu_\epsilon^x = (1 - \lambda)\pi^x + \lambda\nu^x, \tag{3.29}$$

where $\lambda = \lambda(\epsilon) = \epsilon m_\nu(x)/\{(1 - \epsilon)m_\pi(x) + \epsilon m_\nu(x)\}$. Also, simply note that $d_{TV}(\pi^x, \nu_\epsilon^x) = \lambda(\epsilon)d_{TV}(\pi^x, \nu^x)$. Further, if the likelihood function $f(x|\theta)$ is bounded (in θ), then $\dot{T}_\pi(\delta)$ is a linear map on signed measures such that

$$T(\pi + \delta) = T(\pi) + \dot{T}_\pi(\delta) + o(||\delta||), \text{ as } ||\delta|| \to 0,$$

uniformly over all signed measures δ with mass 0 (see Diaconis and Freedman (1986)). Note then that

$$s(\pi, \Gamma; x) = \sup_{\substack{\delta = \nu - \pi \\ \nu \in \Gamma}} \frac{\|\dot{T}_\pi\|}{\|\delta\|}.$$

(ii) $d_\phi(\pi, \nu) = \int \phi\left(\frac{d\pi(\theta)}{d\nu(\theta)}\right) d\nu(\theta)$, where ϕ is a smooth convex function with bounded first and second derivatives near 1 and such that $\phi(1) = 0$. This is the ϕ-divergence measure of distance. (See Csiszär (1978), Goel (1983) and Goel (1986).) Several well-known divergence measures are special cases of ϕ-divergence measure for different convex functions. Listed in Table 3.3 are some such ϕ functions and the corresponding divergence measures obtained thereof. (See Rao (1982) for applications of many of these measures in statistics.)

Consider first the ϵ-contamination class of priors (or linear perturbations), and note that

$$s(\pi, \nu; x) = \lim_{\epsilon \downarrow 0} \frac{d(\pi^x, \nu_\epsilon^x)}{d(\pi, \nu_\epsilon)}.$$

Because $d_\phi(P, Q) = \int \phi\left(\frac{dP}{dQ}\right) dQ$, both $d_\phi(\pi, \nu_\epsilon)$ and $d_\phi(\pi^x, \nu_\epsilon^x)$ converge to 0 as $\epsilon \to 0$. In fact, we shall see that, $\frac{d}{d\epsilon} d_\phi(\pi, \nu_\epsilon)$ and $\frac{d}{d\epsilon} d_\phi(\pi^x, \nu_\epsilon^x)$ also converge to 0 as $\epsilon \to 0$, so that on applying the L̂ Hospital rule, we obtain

$$\begin{aligned}
s(\pi, \nu; x) &= \lim_{\epsilon \downarrow 0} \frac{d_\phi(\pi^x, \nu_\epsilon^x)}{d_\phi(\pi, \nu_\epsilon)} \\
&= \lim_{\epsilon \downarrow 0} \frac{\frac{d}{d\epsilon} d_\phi(\pi^x, \nu_\epsilon^x)}{\frac{d}{d\epsilon} d_\phi(\pi, \nu_\epsilon)} \\
&= \lim_{\epsilon \downarrow 0} \frac{\frac{d^2}{d\epsilon^2} d_\phi(\pi^x, \nu_\epsilon^x)}{\frac{d^2}{d\epsilon^2} d_\phi(\pi, \nu_\epsilon)}.
\end{aligned} \tag{3.30}$$

The following theorem then follows from Theorem 3.1 of Dey and Birmiwal (1994).

Table 3.3. ϕ Functions and the Corresponding Divergence Measures

$\phi(x)$	Divergence Measure
$x \log(x)$	Kullback-Leibler
$-\log(x)$	Directed divergence
$(x-1)\log(x)$	J-divergence
$\frac{1}{2}(\sqrt{x}-1)^2$	Hellinger distance or Kolmogorov's measure of distance
$1 - x^\alpha,\ 0 < \alpha < 1$	Generalized Bhattacharya measure
$(x-1)^2$	Chi-squared divergence or Kagan's measure of distance
$\frac{(x^\lambda - 1)}{\lambda(\lambda+1)},\ \lambda \neq 0, -1$	Power-weighted divergence

Theorem 3.16. *Suppose that $\int \frac{\nu^2(\theta)}{\pi(\theta)}\, d\theta < \infty$. Then*

$$s(\pi, \nu; x) = \frac{V_{\pi^x}\left(\frac{\nu(\theta)}{\pi(\theta)}\right)}{V_{\pi}\left(\frac{\nu(\theta)}{\pi(\theta)}\right)}, \tag{3.31}$$

where $V_q(h(\theta)) = \int h^2(\theta)\, dq(\theta) - \left(h(\theta)\, dq(\theta)\right)^2$.

Proof. In view of (3.30) above, it is enough to establish that

$$\frac{d^2}{d\epsilon^2} d_\phi(\pi^x, \nu_\epsilon^x)|_{\epsilon=0} = \phi''(1) V_{\pi^x}\left(\frac{\nu(\theta)}{\pi(\theta)}\right). \tag{3.32}$$

Recall from (3.29) that $\nu_\epsilon^x = h(\epsilon)\pi^x + (1 - h(\epsilon))\nu^x$, where $h(\epsilon) = (1 - \epsilon)m_\pi(x)/m_{\nu_\epsilon}(x) = (1 - \epsilon)m_\pi(x)/\{(1 - \epsilon)m_\pi(x) + \epsilon m_\nu(x)\}$. Now let $\gamma = \gamma_\epsilon(\theta, x) = \nu_\epsilon^x(\theta)/\pi^x(\theta)$, and note that

$$\gamma = \frac{h(\epsilon)\pi^x(\theta) + (1 - h(\epsilon))\nu^x(\theta)}{\pi^x(\theta)}$$

$$= h(\epsilon)\left(1 - \frac{\nu^x(\theta)}{\pi^x(\theta)}\right) + \frac{\nu^x(\theta)}{\pi^x(\theta)}.$$

Therefore,

$$\frac{d}{d\epsilon}\gamma = \left(1 - \frac{\nu^x(\theta)}{\pi^x(\theta)}\right)\frac{d}{d\epsilon}h(\epsilon)$$

$$= \left\{\frac{m_{\nu_\epsilon}(x)(-m_\pi(x)) - (1 - \epsilon)m_\pi(x)(-m_\pi(x) + m_{\nu_\epsilon}(x))}{m_{\nu_\epsilon}^2(x)}\right\}\left(1 - \frac{\nu^x(\theta)}{\pi^x(\theta)}\right)$$

$$= \frac{(1 - \epsilon)m_\pi^2(x) - m_\pi(x)m_{\nu_\epsilon}(x) - (1 - \epsilon)m_\pi(x)m_\nu(x)}{m_{\nu_\epsilon}^2(x)}\left(1 - \frac{\nu^x(\theta)}{\pi^x(\theta)}\right),$$

and hence

$$\frac{d}{d\epsilon}\gamma|_{\epsilon=0} = -\frac{m_\pi(x)m_\nu(x)}{m_\pi^2(x)}\left(1 - \frac{\nu^x}{\pi^x}\right)$$

$$= -\frac{m_\nu(x)}{m_\pi(x)}\left(1 - \frac{\nu^x(\theta)}{\pi^x(\theta)}\right),$$

and similarly,

$$\frac{d^2}{d\epsilon^2}\gamma|_{\epsilon=0} = -2\frac{m_\nu(x)(m_{\nu_\epsilon}(x) - m_\pi(x))}{m_\pi^2(x)}\left(1 - \frac{\nu^x(\theta)}{\pi^x(\theta)}\right).$$

Now because

$$d_\phi(\pi^x, \nu_\epsilon^x) = \int \phi\left(\frac{\nu_\epsilon^x(\theta)}{\pi^x(\theta)}\right)\pi^x(\theta)\, d\theta$$

$$= \int \phi\left(\gamma_\epsilon(\theta, x)\right)\pi^x(\theta)\, d\theta,$$

and because ϕ is a smooth function with bounded first and second derivatives near 1, applying the dominated convergence theorem (DCT), one obtains,

$$\frac{d}{d\epsilon}d_\phi(\pi^x,\nu^x_\epsilon) = \int \phi'(\gamma_\epsilon)\frac{d}{d\epsilon}\gamma\pi^x(\theta)\,d\theta, \text{ and}$$

$$\frac{d}{d\epsilon}d_\phi(\pi^x,\nu^x_\epsilon)|_{\epsilon=0} = -\phi'(1)\frac{m_\nu(x)}{m_\pi(x)}\int(1-\frac{\nu^x(\theta)}{\pi^x(\theta)})\pi^x(\theta)\,d\theta$$

$$= 0.$$

Further, noting that $\frac{d^2}{d\epsilon^2}d_\phi(\pi^x,\nu^x_\epsilon) = \frac{d}{d\epsilon}\int \phi'(\gamma_\epsilon)\frac{d}{d\epsilon}\gamma\pi^x(\theta)\,d\theta$, and applying DCT once again, one obtains,

$$\frac{d^2}{d\epsilon^2}d_\phi(\pi^x,\nu^x_\epsilon) = \int\{\phi'(\gamma)\frac{d^2}{d\epsilon^2}\gamma + \phi''(\gamma)(\frac{d}{d\epsilon}\gamma)^2\}\pi^x(\theta)\,d\theta, \text{ and}$$

$$\frac{d^2}{d\epsilon^2}d_\phi(\pi^x,\nu^x_\epsilon)|_{\epsilon=0} = \phi''(1)(\frac{m_\nu(x)}{m_\pi(x)})^2\int(1-\frac{\nu^x(\theta)}{\pi^x(\theta)})^2\pi^x(\theta)\,d\theta,$$

because

$$\int\left(\frac{d^2}{d\epsilon^2}\gamma|_{\epsilon=0}\right)\pi^x(\theta)\,d\theta = -2\frac{m_\nu(x)\big(m_{\nu_\epsilon}(x)-m_\pi(x)\big)}{m_\pi^2(x)}\int(1-\frac{\nu^x(\theta)}{\pi^x(\theta)})\pi^x(\theta)\,d\theta$$

$$= 0.$$

Further noting that

$$E_{\pi^x}\left(\frac{\nu(\theta)}{\pi(\theta)}\right) = \int\left(\frac{\nu(\theta)}{\pi(\theta)}\right)\frac{f(x|\theta\pi(\theta)}{m_\pi(x)}\,d\theta \tag{3.33}$$

$$= \frac{1}{m_\pi(x)}\int\nu(\theta)f(x|\theta\,d\theta \tag{3.34}$$

$$= \frac{m_\nu(x)}{m_\pi(x)}, \tag{3.35}$$

we get

$$(\frac{m_\nu(x)}{m_\pi(x)})^2\int(1-\frac{\nu^x(\theta)}{\pi^x(\theta)})^2\pi^x(\theta)\,d\theta$$

$$= \int\left(\frac{m_\nu(x)}{m_\pi(x)} - \frac{m_\nu(x)}{m_\pi(x)}\frac{\nu^x(\theta)}{\pi^x(\theta)}\right)^2\pi^x(\theta)\,d\theta$$

$$= \int\left(\frac{m_\nu(x)}{m_\pi(x)} - \frac{m_\nu(x)}{m_\pi(x)}\frac{f(x|\theta)\nu(\theta)}{m_\nu(x)}\frac{m_\pi(x)}{f(x|\theta)\pi(\theta)}\right)^2\pi^x(\theta)\,d\theta$$

$$= \int\left(\frac{\nu(\theta)}{\pi(\theta)} - \frac{m_\nu(x)}{m_\pi(x)}\right)^2\pi^x(\theta)\,d\theta$$

$$= V_{\pi^x}\left(\frac{\nu(\theta)}{\pi(\theta)}\right),$$

which concludes the proof. $\quad\square$

We consider next geometric perturbations. The following theorem then follows from Theorem 3.2 of Dey and Birmiwal (1994).

Theorem 3.17. *Suppose that*
(i) $\int \left(\log \frac{\nu(\theta)}{\pi(\theta)}\right)^2 \pi(\theta)\, d\theta < \infty$, *and*
(ii) $\int \left(\log \frac{\nu(\theta)}{\pi(\theta)}\right)^2 \left(\frac{\nu(\theta)}{\pi(\theta)}\right)^{\epsilon} \pi(\theta)\, d\theta < \infty$ *for some* $\epsilon > 0$. *Then*

$$s(\pi, \nu; x) = \frac{V_{\pi^x}\left(\log \frac{\nu(\theta)}{\pi(\theta)}\right)}{V_{\pi}\left(\log \frac{\nu(\theta)}{\pi(\theta)}\right)}. \tag{3.36}$$

Proof. As before in Theorem 3.16, it is enough to establish that

$$\frac{d^2}{d\epsilon^2} d_{\phi}(\pi^x, \nu_{\epsilon}^x)\big|_{\epsilon=0} = \phi''(1) V_{\pi^x}\left(\log \frac{\nu(\theta)}{\pi(\theta)}\right), \tag{3.37}$$

proving the desired result. $\square$

Applications of these results are similar to those of a related simpler approach as shown below. The other approach to local sensitivity analysis is simply to look at variation of the curvature of ϕ-divergence as discussed in Dey and Birmiwal (1994) and Delampady and Dey (1994). This turns out to be much easier also as shown below. Consider the class Γ of ϵ-contamination priors,

$$\Gamma_{\epsilon} = \{\pi : \pi = (1 - \epsilon)\pi_0 + \epsilon q, q \in Q\}.$$

Then the curvature $C(q)$ defined by $C(q) = \frac{d^2}{d\epsilon^2} \int \phi\left(\frac{\pi(\theta|x)}{\pi_0(\theta|x)}\right)\pi_0(\theta|x)\, d\theta$, under general regularity conditions has the form $C(q) = \phi''(1) V_{\pi_0(.|x)}\left(\frac{q(\theta)}{\pi_0(\theta)}\right)$ as seen previously. Similarly, if we consider the class

$$\Gamma_g = \left\{\pi : \pi = c(\epsilon)\pi_0^{1-\epsilon} q^{\epsilon}, q \in Q\right\},$$

then we have that $C(q) = \phi''(1) V_{\pi_0(.|x)}\left(\log \frac{q(\theta)}{\pi_0(\theta)}\right)$. Variation of these quantities over many parametric and nonparametric classes can be easily computed. The following example is from Dey and Birmiwal (1994).

Example 3.18. Consider $\mathbf{X}|\boldsymbol{\theta} \sim N_p(\boldsymbol{\theta}, I)$ and the class of Γ_g where under π_0, $\boldsymbol{\theta} \sim N(\boldsymbol{\mu}_0, \Sigma_0)$ and $Q = \{q : \boldsymbol{\theta}|q \sim N_p(\boldsymbol{\mu}_0, k\Sigma_0), k_1 < k < k_2\}$, with $k_1 < 1 < k_2$. Then the posterior distribution of $\boldsymbol{\theta}$ given $\mathbf{x}$ under π_0 is

$$N_p \left(\Sigma_0(I + \Sigma_0)^{-1}\mathbf{x} + (I + \Sigma_0)^{-1}\boldsymbol{\mu}_0, \Sigma_0(I + \Sigma_0)^{-1}\right),$$

and hence

$$V_{\pi_0(.|\mathbf{x})}\left(\log \frac{q(\boldsymbol{\theta})}{\pi_0(\boldsymbol{\theta})}\right)$$
$$= \left(\frac{k-1}{k}\right)^2 \left\{2\mathrm{trace}(I + \Sigma_0^{-1})^2 + 4(\mathbf{x} - \boldsymbol{\mu}_0)'\Sigma_0(I + \Sigma_0)^{-3}(\mathbf{x} - \boldsymbol{\mu}_0)\right\}.$$

It can then be shown that $C(q)$ attains its minimum at $k = 1$ and maximum at k_1 or k_2. The extent of robustness will of course depend on the data $\mathbf{x}$, smaller values of $C(q)$ indicating robustness.

Let Q_{US} be the class of unimodal spherically symmetric densities q such that $\max_\theta q(\theta) \leq h$ for some specified $h > 0$. Consider

$$\Gamma = \{\pi : \pi = (1 - \epsilon)\pi_0 + \epsilon q, q \in Q_{US}\}.$$

(See Sivaganesan (1989) for details on this class.) Then, under certain reasonable conditions (see Delampady and Dey (1994)),

$$\sup_{\pi \in \Gamma} C(q)$$

$$= \phi''(1) \sup_{q \in Q_{US}} V_{\pi_0(.|x)}\left(\frac{q(\theta)}{\pi_0(\theta)}\right)$$

$$= \frac{\phi''(1)}{m_{\pi_0}(x)} \sup_{V(r) \geq 1/h} \left\{ \int_{S(r)} \frac{f(x|\theta)}{\pi_0(\theta)}\, d\theta - \frac{1}{m_{\pi_0}(x)}\left[\int_{S(r)} f(x|\theta)\, d\theta\right]^2 \right\}, \quad (3.38)$$

where $S(r)$ is a sphere of radius r centered at 0 and $V(r)$ denotes its volume. The following example illustrates the use of this result.

Example 3.19. Let $X|\theta \sim N(\theta, 1)$, and under π_0, $\theta \sim N(0, \tau^2)$, $\tau^2 > 1$. Then m_{π_0} is the density of $N(0, \tau^2 + 1)$. Upper bounds for $C(q)$ (denoted by C^*) calculated using (3.38) are listed in Table 3.4 for selected values of τ and x. The extremely large values of C^* corresponding with $\tau = 1.1$ and $x = 3, 4$ indicate that these data are not compatible with π_0. However, the same data are compatible with π_0 if τ has a larger value, say 2.0. Some kind of calibration, however, is needed to precisely establish what magnitudes of curvature can be considered extreme.

Table 3.4. Bounds on Curvature for Different Values of τ and x

τ	x	C^*
1.1	2	909.3
	3	2.08225×10^8
	4	1.06395×10^{16}
1.5	2	1.0918
	3	13.7237
	4	454.3244
2.0	3	1.1186
	4	7.0946

Before we conclude this discussion, it should be mentioned that there is a large amount of literature on gamma minimax estimation, which is a frequentist approach to Bayesian robustness. The idea here is to look for the minimax estimator, but the class of priors considered (for minimaxity) being the one identified for Bayesian robustness consideration. Let us take a very brief look.

Recall that for any decision rule δ, its frequentist risk function is given by $R(\theta, \delta) = EL(\theta, \delta(X))$, where L is the loss function and the expectation is with respect to the distribution of $X|\theta$. If π is any prior distribution on θ, the Bayes risk of δ with respect to π is $r(\pi, \delta) = E^\pi R(\theta, \delta)$. The decision rule δ_π, which minimizes the Bayes risk $r(\pi, \delta)$, is the Bayes rule with respect to π. Under the minimax principle, the optimal decision rule δ^M (minimax rule) is that which minimizes the maximum of the frequentist risk $R(\theta, \delta)$. Equivalently, δ^M minimizes the maximum of the Bayes risk $r(\pi, \delta)$ over the class of all priors π. Under the gamma minimax principle, if π is constrained to lie in a class Γ, the optimal rule δ^g (gamma-minimax rule) minimizes $\sup_{\pi \in \Gamma} r(\pi, \delta)$.

Even though there are many attractive results in this topic, we will not be discussing them. Extensive discussion can be found in Berger (1984, 1985a), and further material in Ickstadt (1992) and Vidakovic (2000).

3.9 Inherently Robust Procedures

It is natural to look for priors and the resulting Bayesian procedures that are inherently robust. Adopting this approach will eliminate the need for checking robustness at the end by building robustness into the analysis at the beginning itself. Further, practitioners can demand "default" Bayesian procedures with built-in robustness that do not require specific sensitivity analyses requiring sophisticated tools.

Accumulated evidence indicates that priors with flatter tails than those of the likelihood tend to be more robust than easier choices such as conjugate priors. Literature here includes Dawid (1973), Box and Tiao (1973), Berger (1984, 1985a), O'Hagan (1988, 1990), Angers and Berger (1991), Fan and Berger (1992), and Geweke (1999). The following example from Berger (1994) illustrates some of these ideas.

Example 3.20. Let $X_1, \ldots, X_n$ be a random sample from a measurement error model, so that $X_i = \theta + \epsilon_i$, $i = 1, \ldots, n$ where ϵ_i are the measurement errors. ϵ_i's can then be reasonably assumed to be i.i.d. having a symmetric unimodal distribution with mean 0 and unknown variance σ^2. The location parameter θ is of inferential interest with the prior information that it is symmetric about 0 and has quartiles of ± 1, whereas σ^2 is a nuisance parameter with little prior information.

The simple "standard" analysis would assume that $X_i|\theta, \sigma^2$ are i.i.d. $N(\theta, \sigma^2)$ and $\pi(\theta, \sigma^2) \propto \frac{1}{\sigma^2} \pi_1(\theta)$ where under π_1, the prior distribution of

θ is $N(0, 2.19)$. (This may be contrasted with Jeffreys analysis discussed in Section 2.7.2.) This conjugate prior analysis suffers from nonrobustness as mentioned previously.

Instead, assume that $X_i | \theta, \sigma^2$ are i.i.d. $t_4(\theta, \sigma^2)$, and likewise assume that under π_1, the prior distribution of θ is Cauchy$(0, 1)$. This analysis would achieve certain robustness lacking in the previous approach. Any outliers in the data will be adequately handled by the Student's t model, and further, if the prior and the data are in conflict, the prior information will be mostly ignored. There are certain computational issues to be addressed here. The "standard" analysis is very easy whereas the robust approach is computationally intensive. However, the MCMC techniques that will be discussed later in the context of hierarchical Bayesian analysis can handle these problems.

O'Hagan (1990) and Angers (2000) discuss some of these issues formally using concepts that they call *credence* and *p-credence* that compare the tail behavior of the posterior distribution with that of heavy tailed distributions such as Student's t and exponential power density.

Further discussion of robust priors and robust procedures will be deferred to Chapters 4 and 5 where we shall consider default and reference priors that are improper priors.

3.10 Loss Robustness

Given the same decision problem, it is possible that different decision makers have different assessments for the consequences of their actions and hence may have different loss functions. In such a situation, it may be necessary to evaluate the sensitivity of Bayesian procedures to the choice of loss.

Example 3.21. Suppose X is Poisson(θ) and θ has the prior distribution of exponential with mean 1. Suppose $x = 0$ is observed. Then the posterior distribution of θ is exponential with mean $1/2$. Therefore, the Bayes estimator of θ under squared error loss is $1/2$ which is the posterior mean, whereas the Bayes estimator under absolute error loss is 0.3465, the posterior median. These are clearly different, and this difference may have some significant impact depending on the use to which the estimator is being put.

It is possible to provide a Bayesian approach to the study of loss robustness exactly as we have done for the prior distribution. In particular, if a class of loss functions is available, range of posterior expected losses can be computed and examined as was done in Dey et al. (1998) and Dey and Micheas (2000). There are also other approaches, such as that of computing non-dominated alternatives, which is outlined in Martín et al. (1998).

3.11 Model Robustness

The model for the observables is the most important component of statistical inference, and hence imprecisions in the specification of the model that can lead to inaccurate inferences must be viewed with great concern. There has been a lot of work in classical statistics in this regard, but most of that only addresses the problem of influence of outliers with respect to a specified target model. In principle, Bayesian approach to model robustness need not be any different from that for prior robustness or loss robustness. However, the problem gets complicated because the mapping of likelihood function to posterior density is not ratio-linear, and hence different techniques need to be employed to assess the sensitivity. If only a finite set of models need to be considered, the problem is a simple one and one simply needs to check the inferences obtained under the different models for the given data. It needs to be kept in mind that, even in this case, different models may be based on different parameters with different interpretations, and hence the specification of prior distributions may be a complicated problem. The following example which illustrates some of the possibilities is similar to Example 1 of Shyamalkumar (2000). (See Pericchi and Pérez (1994) and Berger et al. (2000) also.)

Example 3.22. Suppose the quantity of inferential interest is θ, the median of the model. Model uncertainty is represented by considering the set of two models,

$$\mathcal{M} = \{N(\theta, 1), \ \mathrm{Cauchy}(\theta, 0.675)\},$$

where 0.675 above is the scale parameter of the Cauchy distribution. In other words, X is either $N(\theta, 1)$ or $\mathrm{Cauchy}(\theta, 0.675)$. Since θ is the median of the model in either case, it is not difficult to specify its prior distribution. Suppose the prior π lies in the class Γ of $N(0, \tau^2)$, $1 \leq \tau^2 \leq 10$. The range of posterior means are as shown in Table 3.5.

As can be seen, model robustness is also dependent on the observed x, just like prior or loss robustness. In many situations, this robustness will be absent, and there is no solution other than providing further input on model refinements.

Model robustness does have a long history even though the material is not very extensive. Box and Tiao (1962) have considered this problem in a simple setup. Lavine (1991) and Fernández et al. (2001) have used a nonparametric class of models, and Bayarri and Berger (1998b) have studied robustness in selection models. These can be considered global robustness approaches as compared with the approach of local robustness adopted by Cuevas and Sanz (1988), Sivaganesan (1993), and Dey et al. (1996). Extrema of functional derivative of the posterior quantities are studied by these authors. This is similar to the local robustness approach for prior distributions. Some of the frequentist approaches such as Huber (1964, 1981) are also somewhat relevant.

Table 3.5. Range of Posterior Means for Different Models

Likelihood	$x = 2$		$x = 4$		$x = 6$	
	$\inf E(\theta|x)$	$\sup E(\theta|x)$	$\inf E(\theta|x)$	$\sup E(\theta|x)$	$\inf E(\theta|x)$	$\sup E(\theta|x)$
Normal	1.000	1.818	2.000	3.636	3.000	5.455
Cauchy	0.914	1.689	0.621	3.228	0.362	4.433

3.12 Exercises

1. **(St. Petersburg paradox).** Suppose you are invited to play the following game. A fair coin is tossed repeatedly until it comes up heads. The reward will be 2^n (in some unit of currency) if it takes n tosses until a head first appears. How much would you be willing to pay to play this game? Show that the expected monetary return is ∞, but few would be willing to pay very much to play the game.
2. Consider a lottery where it costs \$1 to buy a ticket. If you win the lottery you get \$1000. If the probability of winning the lottery is 0.0001, decide what you should do under each of the following utility functions, $u(x)$, x being the monetary gain:
   (a) $u(x) = x$; (b) $u(x) = \log_e(.3 + x)$; (c) $u(x) = \exp(1 + x/100)$.
3. A mango grower owns three orchards. Orchard I yields 50% of his total produce, II provides 30% and III provides the rest. Even though they are all of a single variety, 2% of the mangoes from I, 1% each from II and III are excessively sour tasting.
   (a) What is the probability that a mango randomly selected from the total produce is excessively sour?
   (b) What is the probability that a randomly selected mango that is found to be excessively sour came from orchard II?
   (c) Consider a box of 100 mangoes all of which came from a single orchard, but we don't know which one. A mango is selected randomly from this box and is found to be sour. What is the probability that a second mango randomly selected from the remaining 99 is also sour?
4. Show that the Student's t density can be expressed as a scale mixture of normal densities.
5. Refer to Example 3.1. Suppose that the prior for θ has median 1, and upper quartile 2. Consider the priors,
   (i) $\theta \sim$ exponential, (ii) $\log(\theta) \sim$ normal and (iii) $\log(\theta) \sim$ Cauchy.
   (a) Determine the hyperparameters of the three priors.
   (b) Plot the posterior mean $E^\pi(\theta|x)$ for the three priors when x lies in the range, $0 \le x \le 50$.
6. Let $X_1, X_2, \cdots, X_n$ be a random sample from Poisson(θ), where estimation of θ is of interest.
   (a) Derive the range of posterior means when the prior lies in the class of Gamma distributions with prior mean θ_0.

(b) Compute the range of posterior means if the prior density is known only to be a continuous non-increasing function.

7. Let $X_1, X_2, \ldots, X_n$ be i.i.d. $N(\theta, \sigma^2)$, σ^2 known. Consider the following class of conjugate priors for θ: $\Gamma = \{N(0, \tau^2), \tau^2 > 0\}$.
   (a) Find the range of posterior means.
   (b) Find the range of posterior variances.
   (c) Suppose $\bar{x} > 0$. Plot the range of 95% HPD credible intervals.
   (d) Suppose $\sigma^2 = 10$ and $n = 10$. Further, suppose that an $\bar{x}$ of large magnitude is observed. If, now, a $N(0, 1)$ prior is assumed (in which case prior mean is far from the sample mean but prior variance and sample variance are equal) show that the posterior mean and also the credible interval will show substantial shrinkage. Comment on this phenomenon of the prior not allowing the data to have more influence when the data and prior are in conflict. What would happen if instead a Cauchy$(0, 1)$ prior were to be used?

8. Let $X|\theta \sim N(\theta, 1)$ and let Γ_{SU} denote the class of unimodal priors which are symmetric about 0.
   (a) Plot $\{\inf_{\pi \in \Gamma_{SU}} m(x), \sup_{\pi \in \Gamma_{SU}} m(x)\}$ for $0 \leq x \leq 10$.
   (b) Plot $\{\inf_{\pi \in \Gamma_{SU}} E^{\pi}(\theta|x), \sup_{\pi \in \Gamma_{SU}} E^{\pi}(\theta|x)\}$ for $0 \leq x \leq 10$.

9. Let $X_1, X_2, \cdots, X_n$ be i.i.d. with density

$$f(x|\theta) = \exp(-(x - \theta)), \ x > \theta,$$

where $-\infty < \theta < \infty$. Consider the class of unimodal prior distributions on θ which are symmetric about 0. Compute the range of posterior means and that of the posterior probability that $\theta > 0$, for $n = 5$ and $\mathbf{x} = (0.1828, 0.0288, 0.2355, 1.6038, 0.4584)$.

10. Suppose $X_1, X_2, \ldots, X_n$ are i.i.d. $N(\theta, \sigma^2)$, where θ needs to be estimated, but σ^2 which is also unknown is a nuisance parameter. Let $\bar{x}$ denote the sample mean and $s_{n-1}^2 = \sum_{i=1}^{n}(x_i - \bar{x})^2/(n-1)$, the sample variance.
   (a) Show that under the prior $\pi(\theta, \sigma^2) \propto (\sigma^2)^{-1}$, the posterior distribution of θ is given by

$$\frac{\sqrt{n}(\theta - \bar{x})}{s_{n-1}} \sim t_{n-1}.$$

   (b) Using (a), justify the standard confidence interval

$$\bar{x} \pm t_{n-1}(\alpha/2)s_{n-1}/\sqrt{n}$$

as an HPD Bayesian credible interval of coefficient $100(1 - \alpha)\%$, where $t_{n-1}(\alpha/2)$ is the t_{n-1} quantile of order $(1 - \alpha/2)$.
   (c) If instead, $\theta|\sigma^2 \sim N(\mu, c\sigma^2)$ and $\pi(\sigma^2) \propto (\sigma^2)^{-1}$, for specified μ and c, what is the HPD Bayesian credible interval of coefficient $100(1 - \alpha)\%$?
   (d) In (c) above, suppose $c = 5$ and μ is not specified, but is known to lie in the interval $0 \leq \mu \leq 3$, $n = 9$, $\bar{x} = 0$ and $s_{n-1} = 1$. Investigate the

robustness of the credible interval given in (b) by computing the range of its posterior probability.

(e) Consider independent priors: $\theta \sim N(\mu, \tau^2)$, $\pi(\sigma^2) \propto (\sigma^2)^{-1}$, where $0 \leq \mu \leq 3$ and $5 \leq \tau^2 \leq 10$. Conduct a robustness study as in (d) now.

11. Let

$$\rho_\lambda(x) = \begin{cases} \frac{x^\lambda - 1}{\lambda} & \text{if } \lambda > 0; \\ \lim_{\lambda \to 0} \frac{x^\lambda - 1}{\lambda} = \log(x) & \text{if } \lambda = 0, \end{cases}$$

and consider the following family of probability densities introduced by Albert et al. (1991):

$$\pi(\theta | \mu, \phi, c, \lambda) = k(c, \lambda) \sqrt{\phi} \exp\left\{ -\frac{c}{2} \rho_\lambda \left(1 + \frac{\phi(\theta - \mu)^2}{c - 1} \right) \right\}, \quad (3.39)$$

where $k(c, \lambda)$ is the normalizing constant, $-\infty < \mu < \infty$, $\phi > 0$, $c > 1$, $\lambda \geq 0$.

(a) Show that π is unimodal symmetric about μ.

(b) Show that the family of densities defined by (3.39) contains many location-scale families.

(c) Show that normal densities are included in this family.

(d) Show that Student's t is a special case of this density when $\lambda = 0$.

(e) Show that (3.39) behaves like the double exponential when $\lambda = 1/2$.

(f) For $0 \leq \lambda \leq 1$, show that the density in (3.39) is a scale mixture of normal densities.

12. Suppose $X | \theta \sim N(\theta, \sigma^2)$, with θ being the parameter of interest. Explain how the family of prior densities given by (3.39) can be used to study the robustness of the posterior inferences in this case. In particular, explain what values of λ are expected to provide robustness over a large range of values of $X = x$.

13. Refer to the definition of *belief function*, Equation (3.20). Show that *Bel* is a probability measure iff $Bel(.) = Pl(.)$.

14. Show that any probability measure is also a belief function.

15. Refer to Example 3.14. Prove (3.25) and (3.26).

16. Refer to Example 3.14 again. Let $X | \theta \sim N(\theta, 1)$ and let π_0 denote $N(0, \tau^2)$ with $\tau^2 = 2$. Take $\epsilon = 0.2$ and suppose $x = 3.5$ is observed.

(a) Construct the 95% HPD credible interval for θ under π_0.

(b) Compute (3.25) and (3.26) for the interval in (a) now, and check whether robustness is present when the ϵ-contamination class of priors is considered.

17. (Dey and Birmiwal (1994)) Let $\mathbf{X} = (X_1, \ldots, X_k)$ have a multinomial distribution with probability mass function,

$P(X_1 = x_1, \cdots, X_k = x_k | \mathbf{p}) = \frac{n!}{\prod_{i=1}^{k} x_i!} \prod_{i=1}^{k} p_i^{x_i}$, with $n = \sum_{i=1}^{k} x_i$ and

$0 < p_i < 1$, $\sum_{i=1}^{k} p_i = 1$. Suppose under π_0, $\mathbf{p}$ has the Dirichlet distribution $\mathcal{D}(\boldsymbol{\alpha})$ with density

$\pi_0(\mathbf{p}) = \frac{\Gamma(\alpha_0)}{\prod_{i=1}^{k} \Gamma(\alpha_i)} \prod_{i=1}^{k} p_i^{\alpha_i - 1}$, with $\alpha_0 = \sum_{i=1}^{k} \alpha_i$ where $\alpha_i > 0$. Now consider the ϵ-contamination class of priors with $Q = \{\mathcal{D}(s\alpha), s \geq 1\}$. Derive the extreme values of the curvature $C(q)$.

4

Large Sample Methods

In order to make Bayesian inference about the parameter $\boldsymbol{\theta}$, given a model $f(\boldsymbol{x}|\boldsymbol{\theta})$, one needs to choose an appropriate prior distribution for θ. Given the data $\boldsymbol{x}$, the prior distribution is used to find the posterior distribution and various posterior summary measures, depending on the problem. Thus exact or approximate computation of the posterior is a major problem for a Bayesian. Under certain regularity conditions, the posterior can be approximated by a normal distribution with the maximum likelihood estimate (MLE) as the mean and inverse of the observed Fisher information matrix as the dispersion matrix, if the sample size is large. If more accuracy is needed, one may use the Kass-Kadane-Tierney or Edgeworth type refinements. Alternatively, one may sample from the approximate posterior and take resort to importance sampling. Posterior normality has an important philosophical implication, which we discuss below.

How the posterior inference is influenced by a particular prior depends on the relative magnitude of the amount of information in the data, which for i.i.d. observations may be measured by the sample size n or $nI(\boldsymbol{\theta})$ or observed Fisher information $\hat{\boldsymbol{I}}_n$ (defined in Section 4.1.2), and the amount of information in the prior, which is discussed in Chapter 5. As the sample size grows, the influence of the prior distribution diminishes. Thus for large samples, a precise mathematical specification of prior distribution is not necessary. In most cases of low-dimensional parameter space, the situation is like this. A Bayesian would refer to it as washing away of the prior by the data. There are several mathematical results embodying this phenomenon of which posterior normality is the most well-known.

This chapter deals with posterior normality and some of its refinements. We begin with a discussion on limiting behavior of posterior distribution in Section 4.1. A sketch of proof of asymptotic normality of posterior is given in this section. A more accurate posterior approximation based on Laplace's asymptotic method and its refinements by Tierney, Kass, and Kadane are the subjects of Section 4.3. A refinement of posterior normality is discussed in Section 4.2 where an asymptotic expansion of the posterior distribution with a leading normal term is outlined. Throughout this chapter, we consider only the case with a finite dimensional parameter. Also, $\boldsymbol{\theta}$ is assumed to be a "continuous" parameter with a prior density function.

We apply these results for determination of sample size in Section 4.2.1.

4.1 Limit of Posterior Distribution

In this section, we discuss the limiting behavior of posterior distributions as the sample size $n \to \infty$. The limiting results can be used as approximations if n is sufficiently large. They may be used also as a form of frequentist validation of Bayesian analysis. We begin with a discussion of posterior consistency in Section 4.1.1. Asymptotic normality of posterior distribution is the subject of Section 4.1.2.

4.1.1 Consistency of Posterior Distribution

Suppose a data sequence is generated as i.i.d. random variables with density $f(\boldsymbol{x}|\theta_0)$. Would a Bayesian analyzing this data with his prior $\pi(\theta)$ be able to learn about θ_0? Our prior knowledge about θ is updated into the posterior as we learn more from the data. Ideally, the updated knowledge about θ, represented by its posterior distribution, should become more and more concentrated near θ_0 as the sample size increases. This asymptotic property is known as consistency of the posterior distribution at θ_0. Let $X_1, \ldots, X_n$ be the observations at the nth stage, abbreviated as $\boldsymbol{X}_n$, having a density $f(\boldsymbol{x}_n \mid \boldsymbol{\theta})$, $\boldsymbol{\theta} \in \Theta \subset \mathcal{R}^p$. Let $\pi(\boldsymbol{\theta})$ be a prior density, $\pi(\boldsymbol{\theta} \mid \boldsymbol{X}_n)$ the posterior density as defined in (2.1), and $\Pi(. \mid \boldsymbol{X}_n)$ the corresponding posterior distribution.

Definition. *The sequence of posterior distributions $\Pi(. \mid \boldsymbol{X}_n)$ is said to be consistent at some $\boldsymbol{\theta}_0 \in \Theta$, if for every neighborhood U of $\boldsymbol{\theta}_0$, $\Pi(U \mid \boldsymbol{X}_n) \to 1$ as $n \to \infty$ with probability one with respect to the distribution under $\boldsymbol{\theta}_0$.*

The idea goes back to Laplace, who had shown the following. If $X_1, \ldots, X_n$ are i.i.d. Bernoulli with $P_\theta(X_1 = 1) = \theta$ and $\pi(\theta)$ is a prior density that is continuous and positive on $(0, 1)$, then the posterior is consistent at all θ_0 in $(0, 1)$. von Mises (1957) calls this the second fundamental law of large numbers; the first being Bernoulli's weak law of large numbers. Need for posterior consistency has been stressed by Freedman (1963, 1965) and Diaconis and Freedman (1986).

From the definition of convergence in distribution, it follows that consistency of $\Pi(. \mid \boldsymbol{X}_n)$ at $\boldsymbol{\theta}_o$ is equivalent to the fact that $\Pi(. \mid \boldsymbol{X}_n)$ converges to the distribution degenerate at $\boldsymbol{\theta}_0$ with probability one under $\boldsymbol{\theta}_0$.

Consistency of posterior distribution holds in the general case with a finite dimensional parameter under mild conditions. For general results see, for example, Ghosh and Ramamoorthi (2003). For a real parameter θ, consistency at θ_0 can be proved by showing $E(\theta \mid \boldsymbol{X}_n) \to \theta_0$ and $Var(\theta \mid \boldsymbol{X}_n) \to 0$ with probability one under θ_0. This follows from an application of Chebyshev's inequality.

Example 4.1. Let $X_1, X_2, \ldots, X_n$ be i.i.d. Bernoulli observations with $P_\theta(X_1 = 1) = \theta$. Consider a Beta (α, β) prior density for θ. The posterior density of θ given $X_1, X_2, \ldots, X_n$ is then a Beta $(\sum_{i=1}^{n} X_i + \alpha, \ n - \sum_{i=1}^{n} X_i + \beta)$ density with

$$E(\theta \mid X_1, \ldots, X_n) = \frac{\sum_{i=1}^{n} X_i + \alpha}{n + \alpha + \beta},$$

$$Var(\theta \mid X_1, \ldots, X_n) = \frac{(\sum_{i=1}^{n} X_i + \alpha)(n - \sum_{i=1}^{n} X_i + \beta)}{(\alpha + \beta + n)^2(\alpha + \beta + n + 1)}.$$

As $\frac{1}{n}\sum_{i=1}^{n} X_i \to \theta_0$ with P_{θ_0}-probability 1 by the law of large numbers, it follows that $E(\theta \mid X_1, \ldots, X_n) \to \theta_0$ and $Var(\theta \mid X_1, \ldots, X_n) \to 0$ with probability one under θ_0. Therefore, in view of the result mentioned in the previous paragraph, the posterior distribution of θ is consistent.

An important result related to consistency is the robustness of the posterior inference with respect to choice of prior. Let $X_1, \ldots, X_n$ be i.i.d. observations. Let π_1 and π_2 be two prior densities which are positive and continuous at $\boldsymbol{\theta}_0$, an interior point of Θ, such that the corresponding posterior distributions $\Pi_1(. \mid \boldsymbol{X}_n)$ and $\Pi_2(. \mid \boldsymbol{X}_n)$ are both consistent at $\boldsymbol{\theta}_0$. Then with probability one under $\boldsymbol{\theta}_0$

$$\int_\Theta \mid \pi_1(\boldsymbol{\theta} \mid \boldsymbol{X}_n) - \pi_2(\boldsymbol{\theta} \mid \boldsymbol{X}_n) \mid d\boldsymbol{\theta} \to 0$$

or equivalently,

$$\sup_A |\Pi_1(A \mid \boldsymbol{X}_n) - \Pi_2(A \mid \boldsymbol{X}_n)| \to 0.$$

Thus, two different choices of the prior distribution lead to approximately the same posterior distribution. A proof of this result is available in Ghosh et al. (1994) and Ghosh and Ramamoorthi (2003).

4.1.2 Asymptotic Normality of Posterior Distribution

Large sample Bayesian methods are primarily based on normal approximation to the posterior distribution of $\boldsymbol{\theta}$. As the sample size n increases, the posterior distribution approaches normality under certain regularity conditions and hence can be well approximated by an appropriate normal distribution if n is sufficiently large. When n is large, the posterior distribution becomes highly concentrated in a small neighborhood of the posterior mode. Suppose that the notations are as in Section 4.1.1, and $\tilde{\boldsymbol{\theta}}_n$ denotes the posterior mode. Under suitable regularity conditions, a Taylor expansion of $\log \pi(\boldsymbol{\theta} \mid \boldsymbol{X}_n)$ at $\tilde{\boldsymbol{\theta}}_n$ gives

$$\log \pi(\boldsymbol{\theta} \mid \boldsymbol{X}_n) = \log \pi(\tilde{\boldsymbol{\theta}}_n \mid \boldsymbol{X}_n) + (\boldsymbol{\theta} - \tilde{\boldsymbol{\theta}}_n)' \frac{\partial}{\partial \boldsymbol{\theta}} \log \pi(\boldsymbol{\theta} \mid \boldsymbol{X}_n)|_{\tilde{\boldsymbol{\theta}}_n}$$

$$- \frac{1}{2}(\boldsymbol{\theta} - \tilde{\boldsymbol{\theta}}_n)' \tilde{\boldsymbol{I}}_n (\boldsymbol{\theta} - \tilde{\boldsymbol{\theta}}_n) + \cdots$$

$$\approx \log \pi(\tilde{\boldsymbol{\theta}}_n \mid \boldsymbol{X}_n) - \frac{1}{2}(\boldsymbol{\theta} - \tilde{\boldsymbol{\theta}}_n)' \tilde{\boldsymbol{I}}_n (\boldsymbol{\theta} - \tilde{\boldsymbol{\theta}}_n) \qquad (4.1)$$

where $\tilde{\boldsymbol{I}}_n$ is a $p \times p$ matrix defined as

$$\tilde{\boldsymbol{I}}_n = (-\frac{\partial^2}{\partial \theta_i \partial \theta_j} \log \pi(\boldsymbol{\theta} \mid \boldsymbol{X}_n)) \mid_{\boldsymbol{\theta} = \tilde{\boldsymbol{\theta}}_n}$$

and may be called generalized observed Fisher information matrix. The term involving the first derivative is zero as the derivative is zero at the mode $\tilde{\boldsymbol{\theta}}_n$. Also, under suitable conditions the terms involving third and higher order derivatives can be shown to be asymptotically negligible as $\boldsymbol{\theta}$ is essentially close to $\tilde{\boldsymbol{\theta}}_n$. Because the first term in (4.1) is free of θ, $\pi(\boldsymbol{\theta}|\boldsymbol{X}_n)$, as a function of $\boldsymbol{\theta}$, is approximately represented as a density proportional to

$$\exp[-\frac{1}{2}(\boldsymbol{\theta} - \tilde{\boldsymbol{\theta}}_n)' \tilde{\boldsymbol{I}}_n (\boldsymbol{\theta} - \tilde{\boldsymbol{\theta}}_n)],$$

which is a $N_p(\tilde{\boldsymbol{\theta}}_n, \tilde{\boldsymbol{I}}_n^{-1})$ density (with p being the dimension of $\boldsymbol{\theta}$).

As the posterior distribution becomes highly concentrated in a small neighborhood of the posterior mode $\tilde{\boldsymbol{\theta}}_n$ where the prior density $\pi(\boldsymbol{\theta})$ is nearly constant, the posterior density $\pi(\boldsymbol{\theta} \mid \boldsymbol{X}_n)$ is essentially the same as the likelihood $f(\boldsymbol{X}_n \mid \boldsymbol{\theta})$. Therefore, in the above heuristics, we can replace $\tilde{\boldsymbol{\theta}}_n$ by the maximum likelihood estimate (MLE) $\hat{\boldsymbol{\theta}}_n$ and $\tilde{\boldsymbol{I}}_n$ by the observed Fisher information matrix

$$\hat{\boldsymbol{I}}_n = (-\frac{\partial^2}{\partial \theta_i \partial \theta_j} \log f(\boldsymbol{X}_n \mid \boldsymbol{\theta})) \mid_{\hat{\boldsymbol{\theta}}_n}$$

so that the posterior distribution of $\boldsymbol{\theta}$ is approximately $N_p(\hat{\boldsymbol{\theta}}_n, \hat{\boldsymbol{I}}_n^{-1})$.

The dispersion matrix of the approximating normal distribution may also be taken to be the expected Fisher information matrix $\boldsymbol{I}(\boldsymbol{\theta})$ evaluated at $\hat{\boldsymbol{\theta}}_n$ where $\boldsymbol{I}(\boldsymbol{\theta})$ is a matrix defined as

$$\boldsymbol{I}(\boldsymbol{\theta}) = E_{\boldsymbol{\theta}} \left(-\frac{\partial^2}{\partial \theta_i \partial \theta_j} \log f(\boldsymbol{X}_n \mid \boldsymbol{\theta}) \right).$$

Thus we have the following result.

Result. Suppose that $X_1, X_2, \ldots, X_n$ are i.i.d. observations, abbreviated as $\boldsymbol{X}_n$, having a density $f(\boldsymbol{x}_n \mid \boldsymbol{\theta})$, $\boldsymbol{\theta} \in \Theta \subset \mathcal{R}^p$. Let $\pi(\boldsymbol{\theta})$ be a prior density and $\pi(\boldsymbol{\theta} \mid \boldsymbol{X}_n)$ the posterior density as defined in (2.1). Let $\tilde{\boldsymbol{\theta}}_n$ be the posterior mode, $\hat{\boldsymbol{\theta}}_n$ the MLE and $\tilde{\boldsymbol{I}}_n$, $\hat{\boldsymbol{I}}_n$ and $\boldsymbol{I}(\boldsymbol{\theta})$ be the different forms of Fisher information matrix defined above. Then under suitable regularity conditions, for large n, the posterior distribution of $\boldsymbol{\theta}$ can be approximated by any one of the normal distributions $N_p(\tilde{\boldsymbol{\theta}}_n, \tilde{\boldsymbol{I}}_n^{-1})$ or $N_p(\hat{\boldsymbol{\theta}}_n, \hat{\boldsymbol{I}}_n^{-1})$ or $N_p(\hat{\boldsymbol{\theta}}_n, \boldsymbol{I}^{-1}(\hat{\boldsymbol{\theta}}_n))$.

In particular, under suitable regularity conditions, the posterior distribution of $\hat{\boldsymbol{I}}_n^{1/2}(\boldsymbol{\theta} - \hat{\boldsymbol{\theta}}_n)$, given $\boldsymbol{X}_n$, converges to $N_p(\boldsymbol{0}, I)$ with probability one under the true model for the data, where I denotes the identity matrix of order p. This is comparable with the result from classical statistical theory that the repeated sampling distribution of $\hat{\boldsymbol{I}}_n^{1/2}(\boldsymbol{\theta} - \hat{\boldsymbol{\theta}}_n)$ given $\boldsymbol{\theta}$ also converges to $N_p(\boldsymbol{0}, I)$.

For a comment on the accuracy of the different normal approximations stated in the above result and an example, see Berger (1985a, Sec. 4.7.8).

We formally state a theorem below giving a set of regularity conditions under which asymptotic normality of posterior distribution holds.

Posterior normality, in some form, was first observed by Laplace in 1774 and later by Bernstein (1917) and von Mises (1931). More recent contributors in this area include Le Cam (1953, 1958, 1986), Bickel and Yahav (1969), Walker (1969), Chao (1970), Borwanker et al. (1971), and Chen (1985). Ghosal (1997, 1999, 2000) considered cases where the number of parameters increases. A general approach that also works for nonregular problems is presented in Ghosh et al. (1994) and Ghosal et al. (1995).

We present below a version of a theorem that appears in Ghosh and Ramamoorthi (2003). For simplicity, we consider the case with a real parameter θ and i.i.d. observations $X_1, \ldots, X_n$.

Let $X_1, X_2, \ldots, X_n$ be i.i.d observations with a common distribution P_θ possessing a density $f(x|\theta)$ where $\theta \in \Theta$, an open subset of $\mathcal{R}$. We fix $\theta_0 \in \Theta$, which may be regarded as the "true value" of the parameter as the probability statements are all made under θ_0. Let $l(\theta, x) = \log f(x|\theta), L_n(\theta) = \sum_{i=1}^{n} l(\theta, X_i)$, the log-likelihood, and for a function h, let $h^{(i)}$ denote the ith derivative of h. We assume the following regularity conditions on the density $f(x|\theta)$.

(A1) The set $\{x : f(x|\theta) > 0\}$ is the same for all $\theta \in \Theta$.

(A2) $l(\theta, x)$ is thrice differentiable with respect to θ in a neighborhood $(\theta_0 - \delta, \theta_0 + \delta)$ of θ_0. The expectations $E_{\theta_0} l^{(1)}(\theta_0, X_1)$ and $E_{\theta_0} l^{(2)}(\theta_0, X_1)$ are both finite and

$$\sup_{\theta \in (\theta_0 - \delta, \theta_0 + \delta)} |l^{(3)}(\theta, x)| \leq M(x) \text{ and } E_{\theta_0} M(X_1) < \infty. \tag{4.2}$$

(A3) Interchange of the order of integration with respect to P_{θ_0} and differentiation at θ_0 is justified, so that

$$E_{\theta_0} l^{(1)}(\theta_0, X_1) = 0, \ \ E_{\theta_0} l^{(2)}(\theta_0, X_1) = -E_{\theta_0}(l^{(1)}(\theta_0, X_1))^2.$$

Also, the Fisher information number per unit observation $I(\theta_0) = E_{\theta_0}(l^{(1)}(\theta_0, X_1))^2$ is positive.

(A4) For any $\delta > 0$, with P_{θ_0}-probability one

$$\sup_{|\theta - \theta_0| > \delta} \frac{1}{n}(L_n(\theta) - L_n(\theta_0)) < -\epsilon$$

for some $\epsilon > 0$ and for all sufficiently large n.

Remark: Suppose there exists a strongly consistent sequence of estimators $\tilde{\theta}_n$ of θ. This means for all $\theta_0 \in \Theta$, $\tilde{\theta}_n \to \theta_0$ with P_{θ_0}-probability one. Then by the arguments given in Ghosh (1983), a strongly consistent solution $\hat{\theta}_n$ of the likelihood equation $L_n^{(1)}(\theta) = 0$ exists, i.e., there exists a sequence of statistics $\hat{\theta}_n$ such that with P_{θ_0}-probability one $\hat{\theta}_n$ satisfies the likelihood equation for sufficiently large n and $\hat{\theta}_n \to \theta_0$.

Theorem 4.2. *Suppose assumptions (A1) – (A4) hold and $\hat{\theta}_n$ is a strongly consistent solution of the likelihood equation. Then for any prior density $\pi(\theta)$ which is continuous and positive at θ_0,*

$$\lim_{n \to \infty} \int_{\mathcal{R}} |\pi_n^*(t|X_1, \ldots, X_n) - \frac{\sqrt{I(\theta_0)}}{\sqrt{2\pi}} e^{-\frac{1}{2}t^2 I(\theta_0)}| \, dt = 0 \qquad (4.3)$$

with P_{θ_0}-probability one, where $\pi_n^*(t|X_1, \ldots, X_n)$ is the posterior density of $t = \sqrt{n}(\theta - \hat{\theta}_n)$ given $X_1, \ldots, X_n$.
Also under the same assumptions, (4.3) holds with $I(\theta_0)$ replaced by $\hat{I}_n \equiv -\frac{1}{n}L_n^{(2)}(\hat{\theta}_n)$.

A sketch of proof. We only present a sketch of proof. Interested readers may obtain a detailed complete proof from this sketch.

The proof consists of essentially two steps. It is first shown that the tails of the posterior distribution are negligible. Then in the remaining part, the log-likelihood function is expanded by Taylor's theorem up to terms involving third derivative. The linear term in the expansion vanishes, the quadratic term is proportional to logarithm of a normal density, and the remainder term is negligible under assumption (4.2) on the third derivative.

Because $\pi_n(\theta|X_1, \ldots, X_n) \propto \prod_{i=1}^{n} f(X_i|\theta)\pi(\theta)$, the posterior density of $t = \sqrt{n}(\theta - \hat{\theta}_n)$ can be written as

$$\pi_n^*(t|X_1, \ldots, X_n) = C_n^{-1}\pi(\hat{\theta}_n + t/\sqrt{n})\exp[L_n(\hat{\theta}_n + t/\sqrt{n}) - L_n(\hat{\theta}_n)] \quad (4.4)$$

where $C_n = \int_{\mathcal{R}} \pi(\hat{\theta}_n + t/\sqrt{n})\exp[L_n(\hat{\theta}_n + t/\sqrt{n}) - L_n(\hat{\theta}_n)] \, dt$.

Most of the statements made below hold with P_{θ_0}-probability one but we will omit the phrase "with P_{θ_0}-probability one".
Let

$$g_n(t) = \pi(\hat{\theta}_n + t/\sqrt{n})\exp[L_n(\hat{\theta}_n + t/\sqrt{n}) - L_n(\hat{\theta}_n)] - \pi(\theta_0)e^{-\frac{1}{2}t^2 I(\theta_0)}.$$

We first note that in order to prove (4.3), it is enough to show

$$\int_{\mathcal{R}} |g_n(t)|\, dt \to 0. \tag{4.5}$$

If (4.5) holds, $C_n \to \pi(\theta_0)\sqrt{2\pi/I(\theta_0)}$ and therefore, the integral in (4.3), which is dominated by

$$C_n^{-1}\int_{\mathcal{R}} |g_n(t)|dt + \int_{\mathcal{R}} \left| C_n^{-1}\pi(\theta_0)e^{-\frac{1}{2}t^2 I(\theta_0)} - \sqrt{\frac{I(\theta_0)}{2\pi}}\, e^{-\frac{1}{2}t^2 I(\theta_0)} \right| dt$$

also goes to zero.

To show (4.5), we break $\mathcal{R}$ into two regions $A_1 = \{t : |t| > \delta_0\sqrt{n}\}$ and $A_2 = \{t : |t| < \delta_0\sqrt{n}\}$ for some suitably chosen small positive number δ_0 and show for $i = 1, 2$.

$$\int_{A_i} |g_n(t)|\, dt \to 0. \tag{4.6}$$

To show (4.6) for $i = 1$, we note that

$$\int_{A_1} |g_n(t)|\, dt$$

$$\leq \int_{A_1} \pi(\hat{\theta}_n + t/\sqrt{n}) \exp[L_n(\hat{\theta}_n + t/\sqrt{n}) - L_n(\hat{\theta}_n)]\, dt + \int_{A_1} \pi(\theta_0)e^{-\frac{1}{2}t^2 I(\theta_0)}\, dt.$$

It is easy to see that the second integral goes to zero. For the first integral, we note that by assumption (A4), for $t \in A_1$,

$$\frac{1}{n}[L_n(\hat{\theta}_n + t/\sqrt{n}) - L_n(\hat{\theta}_n)] < -\epsilon$$

for all sufficiently large n. It follows that (4.6) holds for $i = 1$.

To show (4.6) for $i = 2$, we use the dominated convergence theorem. Expanding in Taylor series and noting that $L_n^{(1)}(\hat{\theta}_n) = 0$ we have for large n,

$$L_n(\hat{\theta}_n + \frac{t}{\sqrt{n}}) - L_n(\hat{\theta}_n) = -\frac{1}{2}t^2 \hat{I}_n + R_n(t) \tag{4.7}$$

where $R_n(t) = \frac{1}{6}(t/\sqrt{n})^3 L_n^{(3)}(\theta_n')$ and θ_n' lies between $\hat{\theta}_n$ and $\hat{\theta}_n + t/\sqrt{n}$. By assumption (A2), for each t, $R_n(t) \to 0$ and $\hat{I}_n \to I(\theta_0)$ and therefore, $g_n(t) \to 0$. For suitably chosen δ_0, for any $t \in A_2$,

$$|R_n(t)| \leq \frac{1}{6}\delta_0 t^2 \frac{1}{n}\sum_{i=1}^{n} M(X_i) < \frac{1}{4}t^2 \hat{I}_n$$

for sufficiently large n so that from (4.7),

$$\exp[L_n(\hat{\theta}_n + t/\sqrt{n}) - L_n(\hat{\theta}_n)] < e^{-\frac{1}{4}t^2 \hat{I}_n} < e^{-\frac{1}{8}t^2 I(\theta_0)}.$$

Therefore, for suitably chosen small δ_0, $|g_n(t)|$ is dominated by an integrable function on the set A_2. Thus (4.6) holds for $i = 2$. This completes the proof of (4.3). The second part of the theorem follows as $\hat{I}_n \to I(\theta_0)$. $\Box$

Remark. We assume in the proof above that $\pi(\theta)$ is a proper probability density. However, Theorem 4.2 holds even if π is improper, if there is an n_0 such that the posterior distribution of θ given $(x_1, x_2, \ldots, x_{n_0})$ is proper for a.e. $(x_1, x_2, \ldots, x_{n_0})$.

The following theorem states that in the regular case with a large sample, a Bayes estimate is approximately the same as the maximum likelihood estimate $\hat{\theta}_n$. If we consider the squared error loss the Bayes estimate for θ is given by the posterior mean $\theta_n^* = \int_\Theta \theta \pi_n(\theta | X_1, \ldots, X_n) \, d\theta$.

Theorem 4.3. *In addition to the assumptions of Theorem 4.2, assume that that prior density $\pi(\theta)$ has a finite expectation. Then $\sqrt{n}(\theta_n^* - \hat{\theta}_n) \to 0$ with P_{θ_0}-probability one.*

Proof. Proceeding as in the proof of Theorem 4.2 and using the assumption of finite expectation for π, (4.3) can be strengthened to

$$\int_{\mathcal{R}} |t| \, |\pi_n^*(t | X_1, \ldots, X_n) - \frac{\sqrt{I(\theta_0)}}{\sqrt{2\pi}} e^{-\frac{1}{2}t^2 I(\theta_0)}| \, dt \to 0$$

with P_{θ_0}-probability one. This implies

$$\int_{\mathcal{R}} t \pi_n^*(t | X_1, \ldots, X_n) \, dt \to \int_{\mathcal{R}} t \frac{\sqrt{I(\theta_0)}}{\sqrt{2\pi}} e^{-\frac{1}{2}t^2 I(\theta_0)} \, dt = 0.$$

Now $\theta_n^* = E(\theta | X_1, \ldots, X_n) = E(\hat{\theta}_n + t/\sqrt{n} | X_1, \ldots, X_n)$ and therefore, $\sqrt{n}(\theta_n^* - \hat{\theta}_n) = \int_{\mathcal{R}} t \pi_n^*(t | X_1, \ldots, X_n) \, dt \to 0.$ $\Box$

Theorems 4.2 and 4.3 and their variants can be used to make inference about θ for large samples. We have seen in Chapter 2 how our inference can be based on the posterior distribution. If the sample size is sufficiently large, for a wide variety of priors we can replace the posterior distribution by the approximating normal distribution having mean $\hat{\theta}_n$ and dispersion $\hat{I}_n^{-1}$ or $(n\hat{I}_n)^{-1}$ which do not depend on the prior. Theorem 4.3 tells that in the problem of estimating a real parameter with squared error loss, the Bayes estimate is approximately the same as the MLE $\hat{\theta}_n$. Indeed, Theorem 4.3 can be extended to show that this is also true for a wide variety of loss functions. Also the moments and quantiles of the posterior distribution can be approximated by the corresponding measures of the approximating normal distribution. We consider an example at the end of Section 4.3 to illustrate the use of asymptotic posterior normality in the problems of interval estimation and testing.

4.2 Asymptotic Expansion of Posterior Distribution

Consider the setup of Theorem 4.2. Let

$$F_n(u) = \Pi_n(\{\sqrt{n}\hat{I}_n^{1/2}(\theta - \hat{\theta}_n) < u\}|X_1, \ldots, X_n)$$

be the posterior distribution function of $\sqrt{n}\hat{I}_n^{1/2}(\theta - \hat{\theta}_n)$. Then under certain regularity assumptions, $F_n(u)$ is approximately equal to $\Phi(u)$, where Φ is the standard normal distribution function. Theorem 4.2 states that under assumptions (A1)–(A4) on the density $f(x|\theta)$, for any prior density $\pi(\theta)$ which is continuous and positive at θ_0,

$$\lim_{n\to\infty} \sup_u |F_n(u) - \Phi(u)| = 0 \text{ a.s. } P_{\theta_0}. \tag{4.8}$$

Recall that this is proved essentially in two steps. It is first shown that the tails of the posterior distribution are negligible. Then in the remaining part, the log-likelihood function is expanded by Taylor's theorem up to terms involving third derivative. The linear term in the expansion vanishes, the quadratic term is proportional to logarithm of a normal density, and the remainder term is negligible under assumption (4.2) on the third derivative. Suppose now that $l(\theta, x) = \log f(x|\theta)$ is $(k + 3)$ times continuously differentiable and $\pi(\theta)$ is $(k+1)$ times differentiable at θ_0 with $\pi(\theta_0) > 0$. Then the subsequent higher order terms in the Taylor expansion provide a refinement of the posterior normality result stated in Theorem 4.2 or in (4.8) above. Under conditions similar to (4.2) for the derivatives of $l(\theta, x)$ of order $3, 4, \ldots, k+3$, and some more conditions on $f(x|\theta)$, Johnson (1970) proved the following rigorous and precise version of a refinement due to Lindley (1961).

$$\sup_u |F_n(u) - \Phi(u) - \phi(u) \sum_{j=1}^{k} \psi_j(u; X_1, \ldots, X_n)n^{-j/2}| \leq M_k n^{-(k+1)/2} \tag{4.9}$$

eventually with P_{θ_0}-probability one for some $M_k > 0$, depending on k, where $\phi(u)$ is the standard normal density and each $\psi_j(u; X_1, \ldots, X_n)$ is a polynomial in u having coefficients bounded in $X_1, \ldots, X_n$.

Under the same assumptions one can obtain a similar result involving the L_1 distance between the posterior density and an approximation.

The case $k = 0$ corresponds to that considered in Section (4.1.2) as (4.9) becomes

$$\sup_u |F_n(u) - \Phi(u)| \leq M_0 n^{-1/2}. \tag{4.10}$$

Another (uniform) version of the above result, as stated in Ghosh et al. (1982) is as follows. Let Θ_1 be a bounded open interval whose closure $\bar{\Theta}_1$ is properly contained in Θ and the prior π be positive on $\bar{\Theta}_1$. Then, as stated in Ghosh et al. (1982), for $r > 0$, (4.9) holds with P_{θ_0}-probability $1 - O(n^{-r})$, uniformly in $\theta_0 \in \Theta_1$ under certain regularity conditions (depending on r) which are stronger than those of Johnson (1970).

For a formal argument showing how the terms in the asymptotic expansion given in (4.9) are calculated, see Johnson (1970) and Ghosh et al. (1982). For example, if we want to obtain an approximation of the posterior distribution upto an error of order $o(n^{-1})$, we take $k = 2$ and proceed as follows. This is taken from Ghosh (1994).

Let $t = \sqrt{n}(\theta - \hat{\theta}_n)$ and $a_i = \frac{1}{n}\frac{d^i L_n(\theta)}{d\theta^i}|_{\hat{\theta}}, i \geq 1$, so that $a_2 = -\hat{I}_n$. The posterior density of t is given by (4.4) and by Taylor expansion

$$\pi(\hat{\theta}_n + t/\sqrt{n}) = \pi(\hat{\theta}_n)[1 + n^{-1/2}t\frac{\pi'(\hat{\theta}_n)}{\pi(\hat{\theta}_n)} + \frac{1}{2}n^{-1}t^2\frac{\pi''(\hat{\theta}_n)}{\pi(\hat{\theta}_n)}] + o(n^{-1})$$

and

$$L_n(\hat{\theta}_n + t/\sqrt{n}) - L_n(\hat{\theta}_n) = \frac{1}{2}t^2 a_2 + \frac{1}{6}n^{-1/2}t^3 a_3 + \frac{1}{24}n^{-1}t^4 a_4 + o(n^{-1}).$$

Therefore,

$$\pi(\hat{\theta}_n + t/\sqrt{n})\exp[L_n(\hat{\theta}_n + t/\sqrt{n}) - L_n(\hat{\theta}_n)]$$
$$= \pi(\hat{\theta}_n)\exp[a_2 t^2/2]$$
$$\times \left[1 + n^{-1/2}\alpha_1(t; X_1, \ldots, X_n) + n^{-1}\alpha_2(t; X_1, \ldots, X_n)\right] + o(n^{-1}),$$

where

$$\alpha_1(t; X_1, \ldots, X_n) = \frac{1}{6}t^3 a_3 + t\frac{\pi'(\hat{\theta}_n)}{\pi(\hat{\theta}_n)},$$

$$\alpha_2(t; X_1, \ldots, X_n) = \frac{1}{24}t^4 a_4 + \frac{1}{72}t^6 a_3^2 + \frac{1}{2}t^2\frac{\pi''(\hat{\theta}_n)}{\pi_n(\hat{\theta}_n)}$$
$$+ \frac{1}{6}t^4 a_3\frac{\pi'(\hat{\theta}_n)}{\pi(\hat{\theta}_n)}.$$

The normalizer C_n also has a similar expansion that can be obtained by integrating the above. The posterior density of t is then expressed as

$$\pi_n^*(t|X_1, \ldots, X_n) = (2\pi)^{-1/2}\hat{I}_n^{1/2}e^{-t^2/2}$$
$$\times \left[1 + \sum_{j=1}^{2} n^{-j/2}\gamma_j(t; X_1, \ldots, X_n)\right] + o(n^{-1}),$$

where $\gamma_1(t; , X_1, \ldots, X_n) = \frac{1}{6}t^3 a_3 + t\frac{\pi'(\hat{\theta}_n)}{\pi(\hat{\theta}_n)}$ and

$$\gamma_2(t; X_1, \ldots, X_n) = \frac{1}{24}t^4 a_4 + \frac{1}{72}t^6 a_3^2 + \frac{1}{2}t^2\frac{\pi''(\hat{\theta}_n)}{\pi(\hat{\theta}_n)} + \frac{1}{6}t^4 a_3\frac{\pi'(\hat{\theta}_n)}{\pi(\hat{\theta}_n)}$$
$$- \frac{a_4}{8a_2^2} - \frac{15}{72a_2^6}a_3^2 - \frac{1}{2a_2}\frac{\pi''(\hat{\theta}_n)}{\pi(\hat{\theta}_n)} + \frac{1}{2a_2^2}a^3\frac{\pi'(\hat{\theta}_n)}{\pi(\hat{\theta}_n)} + o(n^{-1}).$$

Transforming to $s = \hat{I}_n^{1/2} t$, we get the expansion for the posterior density of $\sqrt{n}\hat{I}_n^{1/2}(\theta - \hat{\theta}_n)$, and integrating it from $-\infty$ to u, we get the terms in (4.9). The above expansion for posterior density also gives an expansion for the posterior mean:

$$E(\theta|X_1, \ldots, X_n) = \hat{\theta}_n + n^{-1}\hat{I}_n^{-1}\left(\frac{a_3}{2} + \frac{\pi'(\hat{\theta}_n)}{\pi(\hat{\theta}_n)}\right) + o(n^{-3/2}).$$

Similar expansions can also be obtained for other moments and quantiles. For more details and discussion see Johnson (1970), Ghosh et al. (1982), and Ghosh (1994). Ghosh et al. (1982) and Ghosh (1994) also obtain expansions of Bayes estimate and Bayes risk. These expansions are rather delicate in the sense that the terms in the expansion can tend to infinity, see, e.g., the discussion in Ghosh et al. (1982).

The expansions agree with those obtained by Tierney and Kadane (1986) (see Section 4.3) up to $o(n^{-2})$. Although the Tierney-Kadane approximation is more convenient for numerical calculations, the expansions obtained in Ghosh et al. (1982) and Ghosh (1994) are more suitable for theoretical applications.

A Bayesian would want to prove an expansion like (4.9) under the marginal distribution of $X_1, \ldots, X_n$ derived from the joint distribution of X's and θ. There are certain technical difficulties in proving this from (4.9). Such a result will hold if the prior $\pi(\theta)$ is supported on a bounded interval and behaves smoothly at the boundary points in the sense that $\pi(\theta)$ and $(d^i/d\theta^i)\pi(\theta)$, $i = 1, 2, \ldots, k$ are zero at the boundary points. A rather technical proof is given in Ghosh et al. (1982). See also in this context Bickel and Ghosh (1990). For the uniform prior on a bounded interval, there can be no asymptotic expansion of the integrated Bayes risk (with squared error loss) of the form $a_0 + \frac{a_1}{n} + \frac{a_2}{n^2} + o(n^{-2})$ (Ghosh et al. (1982)).

4.2.1 Determination of Sample Size in Testing

In this subsection, we consider certain testing problems and find asymptotic approximations to the corresponding (minimum) Bayes risks. These approximations can be used to determine sample sizes required to achieve given bounds for Bayes risks.

We first consider the case with a real parameter $\theta \in \Theta$, an open interval in $\mathcal{R}$, and the problem of testing

$$H_0 : \theta \leq \theta_0 \text{ versus } H_1 : \theta > \theta_0$$

for some specified value θ_0. Let $X_1, \ldots X_n$ be i.i.d. observations with a common density $f(x|\theta)$ involving the parameter θ. Let $\pi(\theta)$ be a prior density over Θ and $\pi(\theta|\boldsymbol{x})$ be the corresponding posterior. Set

$$R_0(\boldsymbol{x}) = P(\theta > \theta_0|\boldsymbol{x}) = \int_{\theta > \theta_0} \pi(\theta|\boldsymbol{x})d\theta,$$

$$R_1(\boldsymbol{x}) = 1 - R_0(\boldsymbol{x}).$$

As mentioned in Section 2.7.2, the Bayes rule for the usual $0 - 1$ loss (see Section 2.5) is to choose H_0 if $R_0(\boldsymbol{X}) \leq R_1(\boldsymbol{X})$ or equivalently $R_1(\boldsymbol{X}) \geq \frac{1}{2}$ and to choose H_1 otherwise. The (minimum) Bayes risk is then given by

$$r(\pi) = \int_{\theta > \theta_0} P_\theta[R_1(\boldsymbol{X}) \geq 1/2]\pi(\theta)d\theta + \int_{\theta \leq \theta_0} P_\theta[R_1(\boldsymbol{X}) < 1/2]\pi(\theta)d\theta.$$

$$(4.11)$$

By Theorem 2.7 an alternative expression for the Bayes risk is given by

$$r(\pi) = E[\min\{R_0(\boldsymbol{X}), R_1(\boldsymbol{X})\}] \tag{4.12}$$

where the expectation is with respect to the marginal distribution of $\boldsymbol{X}$.

Suppose $|\theta - \theta_0| > \delta$ where δ is chosen suitably. For each such θ, $\hat{\theta}_n$ is close to θ with large probability and hence $|\hat{\theta}_n - \theta_0| > \delta$. Intuitively, for such $\hat{\theta}_n$ it will be relatively easy to choose the correct hypothesis. This suggests most of the contribution to the right hand side of (4.11) comes from θ close to θ_0, i.e., from $|\theta - \theta_0| < \delta$. A formal argument that we skip shows

$$r(\pi) = \int_{\theta_0 < \theta < \theta_0 + \delta_n} P_\theta[R_1(\boldsymbol{X}) \geq 1/2]\pi(\theta)d\theta$$

$$+ \int_{\theta_0 - \delta_n < \theta \leq \theta_0} P_\theta[R_1(\boldsymbol{X}) < 1/2]\pi(\theta)d\theta + o(n^{-1}), \tag{4.13}$$

if $\delta_n = c\sqrt{\log n}/\sqrt{n}$ with c sufficiently large. You are invited to verify this for the $N(\theta, 1)$ model in Problem 7.

An approximation to the first integral of (4.13) can be obtained as follows. By the result on normal approximation to posterior stated in the paragraph following (4.10),

$$R_1(\boldsymbol{X}) = P[\sqrt{n}\hat{I}_n^{1/2}(\theta - \hat{\theta}_n) \leq \sqrt{n}\hat{I}_n^{1/2}(\theta_0 - \hat{\theta}_n)|\boldsymbol{X}]$$

can be approximated by $\Phi[\sqrt{n}\hat{I}_n^{1/2}(\theta_0 - \hat{\theta}_n)]$. Hence

$$P_\theta[R_1(\boldsymbol{X}) \geq 1/2] \approx P_\theta[\Phi(\sqrt{n}\hat{I}_n^{1/2}(\theta_0 - \hat{\theta}_n)) \geq 1/2]$$
$$= P_\theta[\sqrt{n}\hat{I}_n^{1/2}(\hat{\theta}_n - \theta) < -\sqrt{n}\hat{I}_n^{1/2}(\theta - \theta_0)]$$
$$\approx \Phi[-\sqrt{n}I^{1/2}(\theta)(\theta - \theta_0)].$$

Indeed, using appropriate uniform versions of the results on asymptotic expansions of posterior distribution (as stated above) and sampling distribution of $\sqrt{n}\hat{I}_n^{1/2}(\hat{\theta}_n - \theta)$ given θ (see, e.g., Ghosh (1994)), one obtains

$$P_\theta[R_1(\boldsymbol{X}) \geq 1/2] = \Phi[-\sqrt{n}I^{1/2}(\theta)(\theta - \theta_0)] + o(n^{-1/2})$$

uniformly in θ belonging to bounded intervals contained in Θ. Thus

$$\int_{\theta_0 < \theta < \theta_0 + \delta_n} P_\theta[R_1(\boldsymbol{X}) \geq 1/2]\pi(\theta)\, d\theta$$

$$= \int_{\theta_0 < \theta < \theta_0 + \delta_n} \Phi[-\sqrt{n}I^{1/2}(\theta)(\theta - \theta_0)]\pi(\theta)\, d\theta + o(n^{-1/2}).$$

With similar approximation for the second integral of (4.13), we have

$$r(\pi) = \int_{\theta_0 < \theta < \theta_0 + \delta_n} \Phi[-\sqrt{n}I^{1/2}(\theta)(\theta - \theta_0)]\pi(\theta)\, d\theta$$

$$+ \int_{\theta_0 - \delta_n < \theta < \theta_0} \Phi[\sqrt{n}I^{1/2}(\theta)(\theta - \theta_0)]\pi(\theta)\, d\theta + o(n^{-1/2})$$

$$= \frac{1}{\sqrt{n}} \int_{o < t < c\sqrt{\log n}} \Phi[-tI^{1/2}(\theta_0 + t/\sqrt{n})]\pi(\theta_0 + t/\sqrt{n})\, dt$$

$$+ \frac{1}{\sqrt{n}} \int_{-c\sqrt{\log n} < t < 0} \Phi[tI^{1/2}(\theta_0 + t/\sqrt{n})]\pi(\theta_0 + t/\sqrt{n})\, dt + o(n^{-1/2}).$$

If we assume $\pi(\theta)$ and $I(\theta)$ have bounded derivatives in some neighborhoods of θ_0, the above reduces to

$$r(\pi) = \frac{\pi(\theta_0)}{\sqrt{n}} \int_0^\infty \Phi(-tI^{1/2}(\theta_0))\, dt + \frac{\pi(\theta_0)}{\sqrt{n}} \int_{-\infty}^0 \Phi(tI^{1/2}(\theta_0))\, dt + o(n^{-1/2})$$

$$= \frac{2\pi(\theta_0)C}{\sqrt{nI(\theta_0)}} + o(n^{-1/2}), \tag{4.14}$$

where $C = \int_0^\infty [1 - \Phi(u)]du \approx 0.3989423$.

From (4.14) it follows that if one wants to have Bayes risk at most equal to some specified r_0 then the required sample size n_0 with which one can achieve this (approximately) is given by

$$n_0 \geq \frac{4C^2(\pi(\theta_0))^2}{r_0^2 I(\theta_0)}. \tag{4.15}$$

In the same way we can handle, say, a two-parameter problem with parameter $\boldsymbol{\theta} = (\theta_1, \theta_2)$. Suppose θ_1 and θ_2 are comparable and the quantity of interest is $\eta = \theta_1 - \theta_2$.

The problem is to test

$$H_0 : \eta \leq \eta_0$$

for some specified η_0. Let $\pi(\boldsymbol{\theta})$ be the joint prior density of θ_1, θ_2 and $p(\eta)$ be the marginal prior density of η. Let $\hat{\boldsymbol{I}}_n$ be the observed Fisher information matrix as defined in the first part of Subsection 4.1.2. Then a normal approximation to the posterior distribution of $\boldsymbol{\theta}$ is $N(\hat{\boldsymbol{\theta}}_n, \hat{\boldsymbol{I}}_n^{-1})$, vide Subsection 4.1.2. This implies that a normal approximation to the posterior of η is given by $N(\hat{\theta}_{1n} - \hat{\theta}_{2n}, v_n)$ with

$$v_n = \hat{\boldsymbol{I}}_n^{11} + \hat{\boldsymbol{I}}_n^{22} - 2\hat{\boldsymbol{I}}_n^{12}$$

where $\hat{\boldsymbol{I}}_n^{ij}$ denotes the (i,j)th element of $\hat{\boldsymbol{I}}_n^{-1}$. Note that $(nv_n)^{-1/2} \to b(\boldsymbol{\theta}) = [I^{11}(\boldsymbol{\theta}) + I^{22}(\boldsymbol{\theta}) - 2I^{12}(\boldsymbol{\theta})]^{-1/2}$ under $\boldsymbol{\theta}$ where $I^{ij}(\boldsymbol{\theta})$ denotes the (i,j)th element of $I^{-1}(\boldsymbol{\theta})$, $I(\boldsymbol{\theta})$ being the expected Fisher information matrix per unit observation.

Let $\pi^*(\beta)$ and $\pi^*(\eta|\beta)$ be respectively the marginal prior density of $\beta = \theta_1 + \theta_2$ and conditional prior density of η given β and $a(\eta, \beta)$ be $b(\boldsymbol{\theta})$ expressed in terms of η and β. Then by arguments similar to those used above, an approximation to the Bayes risk for this problem is

$$r(\pi) \approx \frac{2}{\sqrt{n}} \int \pi^*(\eta_0|\beta) \left\{ \int_0^\infty [1 - \Phi(ta(\eta_0, \beta))]dt \right\} \pi^*(\beta)d\beta$$

$$= \frac{2C}{\sqrt{n}} \int \frac{\pi^*(\eta_0|\beta)}{a(\eta_0, \beta)} \pi^*(\beta)d\beta$$

where C is as in (4.14).

It would be a matter of taste whether one would use simulation or asymptotic approximation. In any case, each method can confirm the accuracy of the other. Advantage of asymptotics is that we get an overview quickly. In specific cases, simulation may be a more efficient alternative, and asymptotics can be used to confirm calculation.

Example 4.4. Let the observations $X_1, \ldots, X_n$ be i.i.d. $B(1, \theta), 0 < \theta < 1$, and suppose we want to test $H_0 : \theta \leq 1/2$ versus $H_1 : \theta > 1/2$.

If we consider the uniform prior $\pi(\theta) \equiv 1$ on $(0, 1)$, we have

$$R_0(\boldsymbol{X}) = R_0(T) = \frac{\Gamma(n + 2)}{\Gamma(T + 1)\Gamma(n - T + 1)} \int_{1/2}^1 \theta^T (1 - \theta)^{n-T} \, d\theta$$

which is a function of $T = \sum_{i=1}^n X_i$, and the marginal distribution of T is uniform over $\{0, 1, \ldots, n\}$. Then from (4.12) the Bayes risk is given by

$$r(\pi) = \frac{1}{n + 1} \sum_{t=0}^n \min\{R_0(t), 1 - R_0(t)\}.$$

Here $I(\theta) = [\theta(1 - \theta)]^{-1}$ and the approximation suggested in (4.14) is

$$r^*(\pi) = \frac{1}{\sqrt{n}} \int_0^\infty [1 - \Phi(u)]du.$$

Table 4.1 gives the exact values of Bayes risk $r(\pi)$ and its approximation $r^*(\pi)$ for different values of n. If one wants to have Bayes risk at most equal to $r_0 = 0.04$, from the approximate formula (4.15), the required sample size n is at least 100 while the exact expression for $r(\pi)$ yields $n \geq 99$.

The above calculations are relevant in the planning stage, when there are no data. If we have a sample of size n and want to control the posterior Bayes risk by drawing m additional observations, we can follow a similar procedure replacing the prior by the posterior from the first stage of data. Ideally, the first-stage sample would be a pilot sample of relatively small size, and the bulk of the data would come from the second stage. In this case, we may even allow an improper noninformative prior for one-sided alternatives.

Table 4.1. The Exact Values of Bayes Risk $r(\pi)$ and Its Approximation $r^*(\pi)$ for Example 4.4.

n	10	20	30	40	50	60	70	80	90	100	150	200	250
$r(\pi)$	.1230	.0881	.0722	.0627	.0561	.0513	.0475	.0445	.0419	.0398	.0325	.0282	.0252
$r^*(\pi)$	.1262	.0892	.0728	.0631	.0564	.0515	.0477	.0446	.0421	.0399	.0326	.0282	.0252

4.3 Laplace Approximation

Bayesian analysis requires evaluation of integrals of the form

$$\int g(\boldsymbol{\theta}) f(\mathbf{x}|\boldsymbol{\theta}) \pi(\boldsymbol{\theta}) \, d\boldsymbol{\theta}$$

where $f(\mathbf{x}|\boldsymbol{\theta})$ is the likelihood function, $\pi(\boldsymbol{\theta})$ is the prior density, and $g(\boldsymbol{\theta})$ is some function of $\boldsymbol{\theta}$. For example, with $g(\boldsymbol{\theta}) \equiv 1$ we have the integrated likelihood required for calculation of Bayes factor in testing or model selection. Various other characteristics of posterior and predictive distribution may also be expressed in terms of such integrals. Laplace's method (see Laplace (1774)) is a technique for approximating integrals when the integrand has a sharp maximum.

4.3.1 Laplace's Method

Let us consider an integral of the form

$$I = \int_{-\infty}^{\infty} q(\theta) e^{nh(\theta)} \, d\theta$$

where q and h are smooth functions of θ with h having a unique maximum at $\hat{\theta}$. In applications, $nh(\theta)$ may be the log-likelihood function or logarithm of the unnormalized posterior density $f(\mathbf{x}|\theta)\pi(\theta)$, and $\hat{\theta}$ may be the MLE or posterior mode. The idea is that if h has a unique sharp maximum at $\hat{\theta}$, then most contribution to the integral I comes from the integral over a small neighborhood $(\hat{\theta} - \delta, \hat{\theta} + \delta)$ of $\hat{\theta}$. We study the behavior of I as $n \to \infty$. As $n \to \infty$, we have

$$I \sim I_1 = \int_{\hat{\theta}-\delta}^{\hat{\theta}+\delta} q(\theta) e^{nh(\theta)} \, d\theta.$$

Here $I \sim I_1$ means $I/I_1 \to 1$. Laplace's method involves Taylor series expansion of q and h about $\hat{\theta}$, which gives

$$I \sim \int_{\hat{\theta}-\delta}^{\hat{\theta}+\delta} \left[q(\hat{\theta}) + (\theta - \hat{\theta})q'(\hat{\theta}) + \frac{1}{2}(\theta - \hat{\theta})^2 q''(\hat{\theta}) + \text{ smaller terms} \right]$$
$$\times \exp\left[nh(\hat{\theta}) + nh'(\hat{\theta})(\theta - \hat{\theta}) + \frac{n}{2}h''(\hat{\theta})(\theta - \hat{\theta})^2 + \text{ smaller terms} \right]$$

$$\sim e^{nh(\hat{\theta})} q(\hat{\theta}) \int_{\hat{\theta}-\delta}^{\hat{\theta}+\delta} \left[1 + (\theta - \hat{\theta}) q'(\hat{\theta})/q(\hat{\theta}) + \frac{1}{2}(\theta - \hat{\theta})^2 q''(\hat{\theta})/q(\hat{\theta}) \right]$$
$$\times \exp\left[\frac{n}{2} h''(\hat{\theta})(\theta - \hat{\theta})^2 \right] d\theta.$$

Assuming that $c = -h''(\hat{\theta})$ is positive and using a change of variable $t = \sqrt{nc}(\theta - \hat{\theta})$, we have

$$I \sim e^{nh(\hat{\theta})} q(\hat{\theta}) \frac{1}{\sqrt{nc}} \int_{-\delta\sqrt{nc}}^{\delta\sqrt{nc}} \left[1 + \frac{t}{\sqrt{nc}} q'(\hat{\theta})/q(\hat{\theta}) + \frac{t^2}{2nc} q''(\hat{\theta})/q(\hat{\theta}) \right] e^{-t^2/2} \, dt$$

$$\sim e^{nh(\hat{\theta})} \frac{\sqrt{2\pi}}{\sqrt{nc}} q(\hat{\theta}) \left[1 + \frac{q''(\hat{\theta})}{2ncq(\hat{\theta})} \right]$$

$$= e^{nh(\hat{\theta})} \frac{\sqrt{2\pi}}{\sqrt{nc}} q(\hat{\theta}) \left[1 + O(n^{-1}) \right].$$

In general, for the case with a p-dimensional parameter $\boldsymbol{\theta}$,

$$I = e^{nh(\hat{\boldsymbol{\theta}})} (2\pi)^{p/2} n^{-p/2} \det(\Delta_h(\hat{\boldsymbol{\theta}}))^{-1/2} q(\hat{\boldsymbol{\theta}})(1 + O(n^{-1})) \qquad (4.16)$$

where $\Delta_h(\boldsymbol{\theta})$ denotes the Hessian of $-h$, i.e.,

$$\Delta_h(\boldsymbol{\theta}) = \left(-\frac{\partial^2}{\partial\theta_i \partial\theta_j} h(\boldsymbol{\theta}) \right)_{p \times p}.$$

Example 4.5. (Sterling's approximation to $n!$) Note that $n!$ can be written as a gamma integral

$$n! = \Gamma(n+1) = \int_0^\infty e^{-x} x^n \, dx = \int_0^\infty e^{n(\log x - x/n)} \, dx.$$

One can use the Laplace method described above to approximate $n!$ as (Problem 9)

$$n! \sim n^{n+1/2} e^{-n} \sqrt{2\pi}.$$

The Bayesian Information Criterion (BIC)

Consider a model with likelihood $f(\mathbf{x}|\boldsymbol{\theta})$ and prior $\pi(\boldsymbol{\theta})$. Equation (4.16), with $q = \pi$ and $nh(\boldsymbol{\theta})$ equal to the log-likelihood, yields an approximation to the integrated likelihood that can be used to find an approximation to the Bayes factor defined in (2.11). Schwarz (1978) proposed a criterion, known as the BIC, based on (4.16) ignoring the terms that stay bounded as the sample size $n \to \infty$. The criterion given by

$$BIC = \log f(\boldsymbol{x}|\hat{\boldsymbol{\theta}}) - (p/2) \log n$$

serves as an approximation to the logarithm of the integrated likelihood of the model and is free from the choice of prior.

Connection Between Laplace Approximation and Posterior Normality

Posterior normality discussed in Section 4.1.2 and Laplace approximation are closely connected. The proof of posterior normality is essentially an application of Laplace approximation with a rigorous handling of the error term. We illustrate this below by re-deriving posterior normality by an application of Laplace approximation.

Let $X_1, X_2, \ldots, X_n$ be i.i.d. observations with a density $f(x|\theta)$ and $\hat{\theta}$ be the MLE of θ. We will find an approximation to the posterior distribution of $t = \sqrt{n}(\theta - \hat{\theta})$ using Laplace's method. Let $\pi(\theta)$ be the posterior density and $\Pi(\cdot|\boldsymbol{x})$ denote the posterior distribution. Then for $a > 0$,

$$\Pi(-a < t < a|\boldsymbol{x}) = \Pi(\hat{\theta} - a/\sqrt{n} < \theta < \hat{\theta} + a/\sqrt{n}|\boldsymbol{x})$$
$$= J_n/I_n$$

where

$$J_n = \int_{\hat{\theta}-a/\sqrt{n}}^{\hat{\theta}+a/\sqrt{n}} e^{nh(\theta)}\pi(\theta)\, d\theta, \ \ I_n = \int e^{nh(\theta)}\pi(\theta)\, d\theta,$$

and $h(\theta) = \frac{1}{n}L(\theta) = \frac{1}{n}\sum \log f(X_i|\theta)$.
As obtained above

$$I_n \sim e^{nh(\hat{\theta})}\pi(\hat{\theta})\sqrt{2\pi}/\sqrt{nc},$$

with $c = -h''(\hat{\theta})$ which is observed Fisher information per unit observation.

Using Laplace's method for J_n we have

$$J_n \sim e^{nh(\hat{\theta})} \int_{\hat{\theta}-a/\sqrt{n}}^{\hat{\theta}+a/\sqrt{n}} [\pi(\hat{\theta}) + (\theta - \hat{\theta})\pi'(\hat{\theta}) + \text{ smaller terms}]$$

$$\times \exp\left[-nc(\theta - \hat{\theta})^2/2\right] d\theta$$

$$\sim e^{nh(\hat{\theta})}\pi(\hat{\theta}) \int_{\hat{\theta}-a/\sqrt{n}}^{\hat{\theta}+a/\sqrt{n}} \exp\left[-nc(\theta - \hat{\theta})^2/2\right] d\theta$$

$$= e^{nh(\hat{\theta})}\pi(\hat{\theta})\frac{1}{\sqrt{n}} \int_{-a}^{a} e^{-ct^2/2}\, dt.$$

Thus, for $a > 0$,

$$\Pi(-a < t < a|\boldsymbol{x}) \sim \frac{\sqrt{c}}{\sqrt{2\pi}} \int_{-a}^{a} e^{-ct^2/2}\, dt$$

$$= P(-a < Z < a) \text{ where } Z \sim N(0, c^{-1}).$$

4.3.2 Tierney-Kadane-Kass Refinements

Suppose

$$E^{\pi}(g(\boldsymbol{\theta})|\mathbf{x}) = \frac{\int g(\boldsymbol{\theta}) f(\mathbf{x}|\boldsymbol{\theta}) \pi(\boldsymbol{\theta}) \, d\boldsymbol{\theta}}{\int f(\mathbf{x}|\boldsymbol{\theta}) \pi(\boldsymbol{\theta}) \, d\boldsymbol{\theta}} \tag{4.17}$$

is the Bayesian quantity of interest where g, f, and π are smooth functions of $\boldsymbol{\theta}$. If we express (4.17) as

$$E^{\pi}(g(\boldsymbol{\theta})|\mathbf{x}) = \frac{\int g(\boldsymbol{\theta}) e^{nh(\boldsymbol{\theta})} \, d\boldsymbol{\theta}}{\int e^{nh(\boldsymbol{\theta})} \, d\boldsymbol{\theta}}$$

with $h(\boldsymbol{\theta}) = \frac{1}{n} \log\{f(\mathbf{x}|\boldsymbol{\theta}) \pi(\boldsymbol{\theta})\}$ and apply the Laplace approximation (4.16) to both the numerator and denominator (with q equal to g and 1), we obtain a first-order approximation

$$E^{\pi}(g(\boldsymbol{\theta})|\mathbf{x}) = g(\hat{\boldsymbol{\theta}}) \left\{ 1 + O(n^{-1}) \right\}$$

(here $\hat{\boldsymbol{\theta}}$ denotes the posterior mode). This has been derived by Tierney and Kadane (1986), Kass et al. (1988), and Tierney et al. (1989).

Suppose now that g in (4.17) is positive, and let $nh(\boldsymbol{\theta}) = \log f(\mathbf{x}|\boldsymbol{\theta}) + \log \pi(\boldsymbol{\theta})$, $nh^{*}(\boldsymbol{\theta}) = nh(\boldsymbol{\theta}) + \log g(\boldsymbol{\theta}) = nh(\boldsymbol{\theta}) + G(\boldsymbol{\theta})$, say. Now apply (4.16) to both the numerator and denominator of (4.17) with q equal to 1. Then, letting $\hat{\boldsymbol{\theta}}^{*}$ denote the mode of h^{*}, $\Sigma = \Delta_h^{-1}(\hat{\boldsymbol{\theta}})$, $\Sigma^{*} = \Delta_{h^{*}}^{-1}(\hat{\boldsymbol{\theta}}^{*})$, Tierney and Kadane (1986) obtain the surprisingly accurate approximation

$$E^{\pi}(g(\boldsymbol{\theta})|\mathbf{x}) = \frac{|\Sigma^{*}|^{1/2} \exp\left(nh^{*}(\hat{\boldsymbol{\theta}}^{*})\right)}{|\Sigma|^{1/2} \exp\left(nh(\hat{\boldsymbol{\theta}})\right)} \left\{ 1 + O(n^{-2}) \right\}. \tag{4.18}$$

It is shown below how the approximation (4.18) is obtained in Tierney and Kadane (1986) for the case with a real parameter.

Let $\sigma^2 = -1/h''(\hat{\theta})$, $\sigma^{*2} = -1/h^{*''}(\hat{\theta}^{*})$. Also let $h_k = h_k(\hat{\theta})$ and $h_k^{*} = h_k^{*}(\hat{\theta}^{*})$ where $\psi_k(\theta) = (d/d\theta)^k \psi(\theta)$ for any function $\psi(\theta)$. Note that under the usual regularity conditions $\sigma, \sigma^{*}, h_k, h_k^{*}$ are all of order $O(1)$.

Consider first the denominator of (4.17), which can be written as

$$\int e^{nh(\theta)} \, d\theta = \int \exp\left[nh(\hat{\theta}) - \frac{n}{2\sigma^2}(\theta - \hat{\theta})^2 + R_n(\theta) \right]$$

$$= e^{nh(\hat{\theta})} \sqrt{2\pi} \sigma n^{-1/2} \int \exp(R_n(\theta)) \phi(\theta; \hat{\theta}, \sigma^2/n) \, d\theta$$

where $\phi(\theta; \hat{\theta}, \sigma^2/n)$ is the $N(\hat{\theta}, \sigma^2/n)$ density for θ and

$$R_n = nh(\theta) - nh(\hat{\theta}) + \frac{n}{2\sigma^2}(\theta - \hat{\theta})^2$$

$$= \frac{1}{6}(\theta - \hat{\theta})^3 nh_3 + \frac{(\theta - \hat{\theta})^4}{4!} nh_4 + \cdots$$

Using the expansion of e^x at zero and the expressions for moments of a normal distribution, we can obtain an approximation of order $O(n^{-r})$ for any $r \geq 1$.

Retaining terms upto the 6-th derivative h_6 in the expansion of R_n, Tierney and Kadane (1986) obtain

$$\int e^{nh(\theta)}\, d\theta = e^{nh(\hat\theta)}\sqrt{2\pi}\,\sigma n^{-1/2}\left(1 + \frac{a}{n} + \frac{b}{n^2} + O(n^{-3})\right) \qquad (4.19)$$

where

$$a = \frac{1}{8}\sigma^4 h_4 + \frac{5}{24}\sigma^6 h_3^2,$$

$$b = \frac{1}{48}\sigma^6 h_6 + \frac{35}{384}\sigma^8 h_4^2 + \frac{7}{48}\sigma^8 h_3 h_5 + \frac{35}{64}\sigma^{10} h_3^2 h_4 + \frac{385}{1152}\sigma^{12} h_3^4.$$

We have an exactly similar approximation for the numerator of (4.17) with σ and h_k replaced by σ^* and h_k^*. We then have

$$E^\pi(g(\boldsymbol\theta)|\mathbf{x})$$

$$= \frac{\sigma^*}{\sigma}\exp\{n(h^*(\hat\theta^*) - h(\hat\theta))\}\frac{\left(1 + \frac{a^*}{n} + \frac{b^*}{n^2} + O(n^{-3})\right)}{\left(1 + \frac{a}{n} + \frac{b}{n^2} + O(n^{-3})\right)}$$

$$= \frac{\sigma^*}{\sigma}\exp\{n(h^*(\hat\theta^*) - h(\hat\theta))\}\left(1 + \frac{a^* - a}{n} + \frac{b^* - b - a(a^* - a)}{n^2} + O(n^{-3})\right).$$

Now note that

$$\begin{aligned}
0 &= h^{*'}(\hat\theta^*)\\
&= h'(\hat\theta^*) + (1/n)G'(\hat\theta^*)\\
&\approx h'(\hat\theta) + (\hat\theta^* - \hat\theta)h''(\hat\theta) + (1/n)G'(\hat\theta) + (1/n)(\hat\theta^* - \hat\theta)G''(\hat\theta)\\
&= (\hat\theta^* - \hat\theta)(h''(\hat\theta) + (1/n)G''(\hat\theta)) + (1/n)G'(\hat\theta)
\end{aligned}$$

which implies $\hat\theta^* - \hat\theta = O(n^{-1})$. This, together with the fact that $h_k^*(\theta) = h_k(\theta) + (1/n)G_k(\theta)$, implies $a^* - a$ and $b^* - b$ are both of order $O(n^{-1})$. It then follows that

$$E^\pi(g(\boldsymbol\theta)|\mathbf{x}) = \frac{\sigma^*}{\sigma}\exp\{n(h^*(\hat\theta^*) - h(\hat\theta))\}\left(1 + O(n^{-2})\right).$$

Example 4.6. We consider the data in Table 2.1 presented in Example 2.3. This is a set of data on food poisoning and we focus on the main suspect, namely, potato salad. Separately for Crabmeat and No Crabmeat, we wish to test the null hypothesis that there is no association between potato salad and illness.

Let p_1 be the probability of being ill given that potato salad is taken and p_2 be the same given no potato salad. If X_1 denotes the number of people falling ill out of a total of n_1 people taking potato salad and X_2 denotes the same out of a total of n_2 people taking no potato salad, then X_1 and X_2 may be modeled as independent binomial variables with X_i following $B(n_i, p_i)$,

$i = 1, 2$. The test for no association between potato salad and illness is then equivalent to testing $H_0 : p_1 = p_2$.

We first carry out the test through credible intervals for $p_1 - p_2$ as described in Section 2.7.4. In order to obtain an exact Bayes test we have to choose prior densities for p_1 and p_2. We have seen in Example 2.2 that the choice of a Beta prior for a binomial proportion simplifies the calculation of posterior. If we consider a Beta (α_i, β_i) prior for p_i, $i = 1, 2$, the posterior density of $\theta \equiv p_1 - p_2$ can be obtained as

$$\pi(\theta | X_1, X_2) \propto$$
$$\int_0^1 (\theta + p_2)^{X_1 + \alpha_1 - 1} (1 - \theta - p_2)^{n_1 - X_1 + \beta_1 - 1} p_2^{X_2 + \alpha_2 - 1} (1 - p_2)^{n_2 - X_2 + \beta_2 - 1} \, dp_2$$

which can only be numerically calculated for a given θ. Because the sample size here is sufficiently large, we will, however, find an approximation to the posterior distribution using asymptotic posterior normality. This does not involve specification of the prior distributions. One can easily calculate the Fisher information matrix $\hat{I}_n$ and show that the approximate distribution of $\theta = (p_1 - p_2)$ is $N(a, b^2)$ where

$$a = \hat{p}_1 - \hat{p}_2, \ b^2 = \hat{p}_1(1 - \hat{p}_1)/n_1 + \hat{p}_2(1 - \hat{p}_2)/n_2, \ \hat{p}_1 = X_1/n_1, \ \hat{p}_2 = X_2/n_2.$$

A $100(1 - \alpha)\%$ HPD cedible interval for θ is then

$$a - bz_{\alpha/2} < \theta < a + bz_{\alpha/2}$$

where $z_{\alpha/2}$ is the $100(1 - \alpha/2)\%$ quantile of $N(0, 1)$.

For the case with crabmeat, $X_1 = 120$, $n_1 = 200$, $X_2 = 4$, $n_2 = 35$. The 99% HPD credible interval turns out to be $(0.337, 0.635)$. For the case with no crabmeat, $X_1 = 22$, $n_1 = 46$, $X_2 = 0$, $n_2 = 23$ and the 99% HPD credible interval is $(0.307, 0.650)$. In both the cases, the hypothesized value (0) of θ falls well outside the credible intervals implying strong evidence against the null hypothesis of no association.

We can calculate the significance level P by finding the $100(1 - P)\%$ credible interval that has the value 0 of null hypothesis on its boundary. More directly this will be the usual P-value corresponding with the observed χ^2 with one d.f. We consider only the case with crabmeat. The other case can be handled similarly. The logarithm of the ratio of the maximized likelihoods under H_0 and H_1 is obtained as $\log \Lambda = -15.4891$. Therefore P-value $= P(\chi_1^2 > 30.9782) \approx 0$

We now look at the same problem through the Bayes factor (BF). In order to compute the BF, we may use the Beta prior as mentioned above. However, because there is no consensus prior for this problem, we use the Schwarz BIC (Section 4.3.1) to approximate the BF. For the case with crabmeat, the BF arising from BIC is given by $BF_{01}^S = 2.8754 \times 10^{-6}$. This implies that with equal prior probabilities for H_0 and H_1, the posterior probability of H_0 is

$[1 + 1/BF_{01}^S]^{-1} = 2.8754 \times 10^{-6}$. This is very small but not as small as the P-value. In any case, both the approaches indicate strong evidence in favor of potato salad being the cause of food poisoning.

4.4 Exercises

1. With Poisson likelihood and Gamma prior for the Poisson parameter θ, show that the posterior is consistent at any $\theta_0 > 0$.
2. Let $X_1, \ldots, X_n$ be i.i.d. observations with a common density $f(x|\theta)$, $\theta \in \Theta = \{\theta_1, \theta_2, \ldots, \theta_k\}$. Consider a prior $(\pi_1, \pi_2, \ldots, \pi_k)$, with $\pi_i > 0$ for all i, $\sum \pi_i = 1$. Suppose the distribution corresponding to $f(x|\theta_i)$, $i = 1, \ldots, k$ are all distinct. Show that the posterior is consistent at each θ_i. (Hint: Express the posterior in terms of
$Z_r = (1/n) \sum_{j=1}^{n} \log(f(X_j|\theta_r)/f(X_j|\theta_i)), r = 1, \ldots k.)$
3. Show that asymptotic posterior normality (as stated in Theorem 4.2) implies posterior consistency at θ_0.
4. Verify Condition (A4) (see Theorem 4.2) for the $N(\theta, 1)$ example.
5. Obtain Laplace approximation to the integrated likelihood from (4.5).
6. Consider $N(\mu, 1)$ likelihood. Generate data of size 30 from $N(0, 1)$. Consider the following priors for μ : (i) $N(0, 2)$ (ii) $N(1, 2)$ (iii) $U(-3, 3)$. For each of these priors find $P(-0.5 < \mu < 0.5)$ and $P(-0.2 < \mu < 0.6)$ using (a) exact calculation (b) normal approximation.
Do the same thing with data generated from $N(1, 1)$.
7. Let $X_1, \ldots, X_n$ be i.i.d. $N(\theta, 1)$ and the prior distribution of θ be $N(0, \tau^2)$. Consider the problem of testing $H_0 : \theta \leq 0$ versus $H_1 : \theta > 0$.
(a) Show that the Bayes risk $r(\pi)$ given by (4.11) reduces to

$$r(\pi) = 2 \int_{\theta>0} \Phi(-\sqrt{n}\theta)\pi(\theta)d\theta$$

where $\pi(.)$ denotes the $N(0, \tau^2)$ density for θ.
(b) Verify (4.13) in this case.
8. Find numerically the exact posterior density of $\theta = p_1 - p_2$ in Example 4.6 with independent uniform priors for p_1 and p_2. Compare this with the normal approximation to the posterior.
9. Using the idea of Laplace method for approximating integrals, find the following approximation for $n!$ (see Example 4.5)

$$n! \sim n^{n+1/2} e^{-n} \sqrt{2\pi}.$$

5

Choice of Priors for Low-dimensional Parameters

Given data, a Bayesian will need a likelihood function $p(x|\theta)$ and a prior $p(\theta)$. For many standard problems, the likelihood is known either from past experience or by convention. To drive the Bayesian engine, one would still need an appropriate prior. In this chapter, we consider only low-dimensional parameters. Admittedly, low dimension is not easy to define, but we expect the dimension d to be much smaller than the sample size n to qualify as low. In most of the examples in this chapter, $d = 1$ or 2 and is rarely bigger than 5.

Ideally, one wants to choose a prior or a class of priors reflecting one's prior knowledge and belief about the unknown parameters or about different hypotheses. This is a subjective choice. If one has a class of priors, it would be necessary to study robustness of various aspects of the resulting Bayesian inference. Choice of subjective priors, usually called elicitation of priors, is still rather difficult. For some systematic methods of elicitation, see Kadane et al. (1980), Garthwaite and Dickey (1988, 1992). A recent review is Garthwaite et al. (2005).

Empirical studies have shown experience and maturity help a person in quantifying uncertainty about an event in the form of a probability. However, assigning a fully specified probability distribution to an unknown parameter is difficult even when the parameter has a physical meaning like length or breadth of some article. In such cases, it may be realistic to expect elicitation of prior mean and variance or some other prior quantities but not a full specification of the distribution. Hopefully, the situation will improve with practice, but it is hard to believe that a fully specified prior distribution will be available in all but very simple situations.

It is much more common to choose and use what are called objective priors. When very little prior information is available, objective priors are also called noninformative priors. The older terminology of noninformative priors is no longer in favor among objective Bayesians because a complete lack of information is hard to define. However, it is indeed possible to construct objective priors with low information in the sense of Bernardo's information measure or non-Euclidean geometry. These priors are not unique but, as indicated for the Bernoulli example (Example 2.2) in Chapter 2, for even a small sample size the posteriors arising from them are very close to each other. All these priors are constructed through well-defined algorithms. If some prior information is available, in some cases one can modify these algorithms.

The objective priors are typically improper but have proper posteriors. They are suitable for estimation problems and also for testing problems where both null and alternative hypotheses have the same dimension. The objective priors need to be suitably modified for sharp null hypotheses — the subject of Chapter 6.

Most of this chapter (Sections 5.1, 5.2, and 5.5) is about different principles and methods of construction of objective priors (Section 5.1) and common criticisms and answers (Section 5.2). Subjective priors appear very naturally when the decision maker judges his data to be exchangeable. We deal with this in Section 5.3. An example of elicitation of a different kind is given in Section 5.4.

5.1 Different Methods of Construction of Objective Priors

Because this section is rather long, we provide an overview here.

How can we construct objective priors under general regularity conditions? We may do one of the following things.

1. Define a uniform distribution that takes into account the geometry of the parameter space.
2. Minimize a suitable measure of information in the prior.
3. Choose a prior with some form of frequentist ideas because a prior with little information should lead to inference that is similar to frequentist inference.

To fully define these methods, we have to specify the geometry in (1), the measure of information in (2) and the frequentist ideas that are to be used in (3). This will be done in Subsections 5.1.2, 5.1.3, and 5.1.4. In Subsection 5.1.1, we discuss why the usual uniform prior $\pi(\theta) = c$ has come in for a lot of criticism. Indeed, these criticisms help one understand the motivation behind (1) and (2). It is a striking fact that both (1) and (2) lead to the Jeffreys prior, namely,

$$\pi(\theta) = [\det(I_{ij}(\theta))]^{1/2}$$

where $(I_{ij}(\theta))$ is the Fisher information matrix. In the one-dimensional case, (3) also leads to the Jeffreys prior.

We have noted in Chapter 1 that many common statistical models possess additional structure. Some are exponential families of distributions, some are location-scale families, or more generally families invariant under a group of transformations. Normals belong to both classes. For each of these special classes, there is a different choice of objective priors discussed in Subsections 5.1.5 and 5.1.6. The objective priors for exponential families come from the class of conjugate priors. In the case of location-scale families with scale parameter σ, the common objective prior is the so-called right invariant Haar measure

$$\pi_1(\mu, \sigma) = \frac{1}{\sigma}$$

and the Jeffreys prior turns out to be the left invariant Haar measure

$$\pi_2(\mu, \sigma) = \frac{1}{\sigma^2}$$

(see Subsection 5.1.7 for definitions). Jeffreys had noted this and expressed his preference for the former. As we discuss later, there are several strong reasons for preferring π_1 to π_2.

To avoid some of the problems with the Jeffreys prior, Bernardo (1979) and Berger and Bernardo (1989) had suggested an important modification of the Jeffreys prior that we take up in Subsection 5.1.10. These priors are called reference priors. In the location-scale case, the reference prior is the right invariant Haar measure. They are considerably more difficult to find than the Jeffreys prior but explicit formulas are now available for many examples, vide Berger et al. (2006). A comprehensive overview and catalogue of objective priors, up to date as of 1995, is available in Kass and Wasserman (1996). A brief introduction is Ghosh and Mukerjee (1992).

5.1.1 Uniform Distribution and Its Criticisms

The first objective prior ever to be used is the uniform distribution over a bounded interval. A common argument, based on "ignorance", seems to have been that if we know nothing about θ, why should we attach more density to one point than another? The argument given by Bayes, who was the first to use the uniform as an objective prior, is a variation on this. It is indicated in Problem 1. A second argument is that the uniform maximizes the Shannon entropy. The uniform was also used a lot by Laplace who seems to have arrived at a Bayesian point of view, independently of Bayes, but his argument seems to have been based on subjective argument that in his problems the uniform was appropriate.

The principle of ignorance has been criticized by Keynes, Fisher, and many others. Essentially, the criticism is based on an invariance argument. Let $\eta = \psi(\theta)$ be a one-to-one function of θ. If we know nothing about θ, then we know nothing about η also. So the principle of ignorance applied to η will imply our prior for η is uniform (on $\psi(\Theta)$) just as it had led to a uniform prior for θ. But this leads to a contradiction. To see this suppose ψ is differentiable and $p(\eta) = c$ on $\psi(\Theta)$. Then the prior $p^*(\theta)$ for θ is

$$p^*(\theta) = p(\eta)|\psi'(\theta)| = c|\psi'(\theta)|$$

which is not a constant in general.

This argument also leads to an invariance principle. Suppose we have an algorithm that produces noninformative priors for both θ and η, then these priors $p^*(\theta)$ and $p(\eta)$ should be connected by the equation

$$p^*(\theta) = p(\eta)|\psi'(\theta)| \tag{5.1}$$

i.e., a noninformative prior should be invariant under one-to-one differentiable transformations.

The second argument in favor of the uniform, based on Shannon entropy, is also flawed. Shannon (1948) derives a measure of entropy in the finite discrete case from certain natural axioms. His entropy is

$$H(\boldsymbol{p}) = -\sum_{i=1}^{m} p_i \log p_i$$

which is maximized by the discrete uniform, i.e., at $\boldsymbol{p} = (\frac{1}{m}, \cdots, \frac{1}{m})$. Entropy is a measure of the amount of uncertainty about the outcome of the experiment. A prior that maximizes this will maximize uncertainty, so it is a noninformative prior. Because such a prior should minimize information, we take negative of entropy as information. This usage differs from Shannon's identification of information and entropy.

Shannon's entropy is a natural measure in the discrete case and the discrete uniform appears to be the right noninformative prior. The continuous case is an entirely different matter. Shannon himself pointed out that for a density p

$$H(p) \equiv -\int (\log p(x))p(x)\, dx$$

is unsatisfactory, clearly it is not derived from axioms, it is not invariant under one-one transformations, and, as pointed out by Bernardo, it depends on the measure $\mu(x)dx$ with respect to which the density $p(x)$ is taken. Note also that the measure is not non-negative. Just take $\mu(x) = 1$ and take $p(x) = $ uniform on $[0, c]$. Then $H(p) > 0$ if and only if $c > 1$, which seems quite arbitrary.

Finally, if the density is taken with respect to $\mu(x)\, dx$, then it is easy to verify that the density is p/μ and

$$H(p) = -\int \left(\log \frac{p(x)}{\mu(x)}\right) \frac{p(x)}{\mu(x)} \mu(x)\, dx$$

is maximized at $p = \mu$, i.e., the entropy is maximum at the arbitrary μ.

For all these reasons, we do not think $H(p)$ is the right entropy to maximize. A different entropy, also due to Shannon, is explored in Subsection 5.1.3.

However, $H(p)$ serves a useful purpose when we have partial information. For details, see Subsection 5.1.12.

5.1.2 Jeffreys Prior as a Uniform Distribution

This section is based on Section 8.2.1 of Ghosh and Ramamoorthi (2003). We show in this section that if we construct a uniform distribution taking into account the topology, it automatically satisfies the invariance requirement (5.1). Moreover, this uniform distribution is the Jeffreys prior. Problem 2 shows one can construct many other priors that satisfy the invariance requirement. Of course, they are not the uniform distribution in the sense of this section. Being an invariant uniform distribution is more important than just being invariant. Suppose $\Theta = \mathcal{R}^d$ and $I(\boldsymbol{\theta}) = (I_{ij}(\boldsymbol{\theta}))$ is the $d \times d$ Fisher information matrix. We assume $I(\boldsymbol{\theta})$ is positive definite for all $\boldsymbol{\theta}$. Rao (1987) had proposed the Riemannian metric ρ related to $I(\boldsymbol{\theta})$ by

$$\rho(\boldsymbol{\theta}, \boldsymbol{\theta} + d\boldsymbol{\theta}) = \sum_{i,j} I_{i,j}(\boldsymbol{\theta})\, d\theta_i\, d\theta_j (1 + o(1)).$$

It is known, vide Cencov (1982), that this is the unique Riemannian metric that transforms suitably under one-one differentiable transformations on Θ. Notice that in general Θ does not inherit the usual Euclidean metric that goes with the (improper) uniform distribution over $\mathcal{R}^d$.

Fix a $\boldsymbol{\theta}_0$ and let $\boldsymbol{\psi}(\boldsymbol{\theta})$ be a smooth one-to-one transformation such that the information matrix

$$I^\psi = \left[E_\psi \left(\frac{\partial \log p}{\partial \psi_i} \frac{\partial \log p}{\partial \psi_j} \right) \right]$$

is the identity matrix I at $\boldsymbol{\psi_0} = \boldsymbol{\psi}(\boldsymbol{\theta_0})$. This implies the local geometry in the $\boldsymbol{\psi}$-space around $\boldsymbol{\psi_0}$ is Euclidean and hence $d\boldsymbol{\psi}$ is a suitable uniform distribution there. If we now lift this back to the $\boldsymbol{\theta}$-space by using the Jacobian of transformation and the simple fact

$$\left(\left| \frac{\partial \theta_j}{\partial \psi_i} \right| \right) (I_{i,j}(\boldsymbol{\theta})) \left(\left| \frac{\partial \theta_j}{\partial \psi_i} \right| \right)' = I^{\psi_0} = I,$$

we get the Jeffreys prior in the $\boldsymbol{\theta}$-space,

$$d\boldsymbol{\psi} = \left\{ \det \left| \frac{\partial \theta_i}{\partial \psi_j} \right| \right\}^{-1} d\boldsymbol{\theta} = \{\det[I_{i,j}(\boldsymbol{\theta})]\}^{\frac{1}{2}} d\boldsymbol{\theta}.$$

A similar method is given in Hartigan (1983, pp. 48, 49). Ghosal et al. (1997) present an alternative construction where one takes a compact subset of the parameter space and approximates this by a finite set of points in the so-called Hellinger metric

$$d(P_\theta, P_{\theta'}) = \left[\int (\sqrt{p_\theta} - \sqrt{p_{\theta'}})^2\, dx \right]^{\frac{1}{2}},$$

where p_θ and $p_{\theta'}$ are the densities of P_θ and $P_{\theta'}$. One then puts a discrete uniform distribution on the approximating finite set of points and lets the degree of approximation tend to zero. Then the corresponding discrete uniforms converge weakly to the Jeffreys distribution. The Jeffreys prior was introduced in Jeffreys (1946).

5.1.3 Jeffreys Prior as a Minimizer of Information

As in Subsection 5.1.1, let the Shannon entropy associated with a random variable or vector Z be denoted by

$$H(Z) = H(p) = -E_p(\log p(Z))$$

where p is the density (probability function) of Z. Let $\boldsymbol{X} = (X_1, X_2, \ldots, X_n)$ have density or probability function $p(\boldsymbol{x}|\boldsymbol{\theta})$ where $\boldsymbol{\theta}$ has prior density $p(\boldsymbol{\theta})$. We assume $X_1, X_2, \ldots, X_n$ are i.i.d. and conditions for asymptotic normality of posterior $p(\boldsymbol{\theta}|\boldsymbol{x})$ hold. We have argued earlier that $H(p)$ is not a good measure of entropy and $-H(p)$ not a good measure of information if p is a density. Using an idea of Lindley (1956) in the context of design of experiments, Bernardo (1979) suggested that a Kullback-Leibler divergence between prior and posterior, namely,

$$J(p(\boldsymbol{\theta}), \boldsymbol{X}) = E\left\{\log \frac{p(\boldsymbol{\theta}|\boldsymbol{X})}{p(\boldsymbol{\theta})}\right\}$$

$$= \int_\Theta \left\{\int_\mathcal{X} \left[\int_\Theta \log\left\{\frac{p(\boldsymbol{\theta}'|\boldsymbol{x})}{p(\boldsymbol{\theta}')}\right\} p(\boldsymbol{\theta}'|\boldsymbol{x})d\boldsymbol{\theta}'\right] p(\boldsymbol{x}|\boldsymbol{\theta})\, d\boldsymbol{x}\right\} p(\boldsymbol{\theta})\, d\boldsymbol{\theta} \quad (5.2)$$

is a better measure of entropy and $-J$ a better measure of information in the prior. To get a feeling for this, notice that if the prior is nearly degenerate, at say some $\boldsymbol{\theta}_0$, so will be the posterior. This would imply J is nearly zero. On the other hand, if $p(\boldsymbol{\theta})$ is rather diffuse, $p(\boldsymbol{\theta}|\boldsymbol{x})$ will differ a lot from $p(\boldsymbol{\theta})$, at least for moderate or large n, because $p(\boldsymbol{\theta}|\boldsymbol{x})$ would be quite peaked. In fact, $p(\boldsymbol{\theta}|\boldsymbol{x})$ would be approximately normal with mean $\hat{\boldsymbol{\theta}}$ and variance of the order $O(\frac{1}{n})$. The substantial difference between prior and posterior would be reflected by a large value of J. To sum up J is small when p is nearly degenerate and large when p is diffuse, i.e., J captures how diffuse is the prior. It therefore makes sense to maximize J with respect to the prior.

Bernardo suggested one should not work with the sample size n of the given data and maximize the J for this n. For one thing, this would be technically forbidding in most cases and, more importantly, the functional J is expected to be a nice function of the prior only asymptotically. We show below how asymptotic maximization is to be done. Berger et al. (1989) have justified to some extent the need to maximize asymptotically. They show that if one maximizes for fixed n, maximization may lead to a discrete prior with finitely many jumps — a far cry from a diffuse prior. We also note in passing that the measure J is a functional depending on the prior but in the given context of a particular experiment with i.i.d. observations having density $p(x|\boldsymbol{\theta})$. This is a major difference from the Shannon entropy and suggests information in a prior is a relative concept, relative to a particular experiment.

We now return to the question of asymptotic maximization. Fix an increasing sequence of compact d-dimensional rectangles K_i whose union is $\mathcal{R}^d$. For a fixed K_i, we consider only priors p_i supported on K_i, and let $n \to \infty$. We assume posterior normality holds in the Kullback-Leibler sense, i.e.,

$$\lim_{n\to\infty} E\left(\log\frac{p(\boldsymbol{\theta}|\boldsymbol{X})}{\hat{p}(\boldsymbol{\theta}|\boldsymbol{X})}\right) = \lim_{n\to\infty} \int_{K_i} E_{\boldsymbol{\theta}}\left\{\log\frac{p(\boldsymbol{\theta}|X)}{\hat{p}(\boldsymbol{\theta}|X)}\right\} p_i(\boldsymbol{\theta})\, d\boldsymbol{\theta} = 0 \qquad (5.3)$$

where $\hat{p}$ is the approximating d-dimensional normal distribution $N(\hat{\boldsymbol{\theta}}, I^{-1}(\hat{\boldsymbol{\theta}})/n)$. For sufficient conditions see Clarke and Barron (1990) and Ibragimov and Has'minskii (1981).

In view of (5.3), it is enough to consider

$$\hat{J}(p_i) = \int_{K_i}\left\{\int_{\mathcal{X}}\left[\int_{\mathcal{R}^d}\log\left\{\frac{\hat{p}_i(\boldsymbol{\theta}'|\boldsymbol{x})}{p_i(\boldsymbol{\theta}')}\right\} p_i(\boldsymbol{\theta}'|\boldsymbol{x})\, d\boldsymbol{\theta}'\right] p(\boldsymbol{x}|\boldsymbol{\theta})\, d\boldsymbol{x}\right\} p_i(\boldsymbol{\theta})\, d\boldsymbol{\theta}.$$

Using appropriate results on normal approximation to posterior distribution, it can be shown that

$$\hat{J}(p_i) = \int_{K_i}\left\{\int_{\mathcal{X}}\left[\int_{\mathcal{R}^d}\log\left\{\frac{\hat{p}_i(\boldsymbol{\theta}'|\boldsymbol{x})}{p_i(\boldsymbol{\theta}')}\right\} \hat{p}_i(\boldsymbol{\theta}'|\boldsymbol{x})\, d\boldsymbol{\theta}'\right] p(\boldsymbol{x}|\boldsymbol{\theta})\, d\boldsymbol{x}\right\} p_i(\boldsymbol{\theta})\, d\boldsymbol{\theta} + o_p(1)$$

$$= \{-\frac{d}{2}\log(2\pi) - \frac{d}{2} + \frac{d}{2}\log n\} + \int_{K_i} \log(\det I(\boldsymbol{\theta}))^{\frac{1}{2}} p_i(\boldsymbol{\theta})\, d\boldsymbol{\theta}$$

$$- \int_{K_i} (\log p_i(\boldsymbol{\theta})) p_i(\boldsymbol{\theta})\, d\boldsymbol{\theta} + o_p(1). \qquad (5.4)$$

Here we have used the well-known fact about the exponent of a multivariate normal that

$$\int -\frac{1}{2}(\boldsymbol{\theta}' - \hat{\boldsymbol{\theta}})^t (I^{-1}(\hat{\boldsymbol{\theta}})/n)^{-1}(\boldsymbol{\theta}' - \hat{\boldsymbol{\theta}})\hat{p}_i(\boldsymbol{\theta}'|\boldsymbol{x})\, d\boldsymbol{\theta}' = -\frac{d}{2}.$$

Hence by (5.3) and (5.4), we may write

$$J(p_i) = \{-\frac{d}{2}\log(2\pi) - \frac{d}{2} + \frac{d}{2}\log n\} + \int_{K_i} \log\left\{\frac{(\det(I(\boldsymbol{\theta})))^{\frac{1}{2}}}{p_i(\boldsymbol{\theta})}\right\} p_i(\boldsymbol{\theta})\, d\boldsymbol{\theta} + o_p(1).$$

Thus apart from a constant and a negligible $o_p(1)$ term, J is the functional

$$\int_{K_i} \log[c_i(\det(I(\boldsymbol{\theta})))^{1/2}/p_i(\boldsymbol{\theta})] p_i(\boldsymbol{\theta})\, d\boldsymbol{\theta} - \log c_i$$

where c_i is a normalizing constant such that $c_i[\det(I(\boldsymbol{\theta}))]^{1/2}$ is a probability density on K_i. The functional is maximized by taking

$$p_i(\boldsymbol{\theta}) = \begin{cases} c_i[\det(I(\boldsymbol{\theta}))]^{1/2} & \text{on } K_i; \\ 0 & \text{elsewhere.} \end{cases} \tag{5.5}$$

Thus for every K_i, the (normalized) Jeffreys prior $p_i(\boldsymbol{\theta})$ maximizes Bernardo's entropy. In a very weak sense, the $p_i(\boldsymbol{\theta})$ of (5.5) converge to $p(\boldsymbol{\theta}) = [\det(I(\boldsymbol{\theta}))]^{1/2}$, namely, for any two fixed Borel sets B_1 and B_2 contained in K_{i_o} for some i_o,

$$\lim_{i \to \infty} \frac{\int_{B_1} p_i(\boldsymbol{\theta})\,d\boldsymbol{\theta}}{\int_{B_2} p_i(\boldsymbol{\theta})\,d\boldsymbol{\theta}} = \frac{\int_{B_1} p(\boldsymbol{\theta})\,d\boldsymbol{\theta}}{\int_{B_2} p(\boldsymbol{\theta})\,d\boldsymbol{\theta}}. \tag{5.6}$$

The convergence based on (5.6) is very weak. Berger and Bernardo (1992, Equation 2.2.5) suggest convergence based on a metric that compares the posterior of the proper priors over compact sets B_i and the limiting improper prior (whether Jeffreys or reference or other). Examples show lack of convergence in this sense may lead to severe inadmissibility and other problems with inference based on the limiting improper prior. However, checking this kind of convergence is technically difficult in general and not attempted in this book.

We end this section with a discussion of the measure of entropy or information. In the literature, it is often associated with Shannon's missing information. Shannon (1948) introduced this measure in the context of a noisy channel. Any channel has a source that produces (say, per second) messages X with p.m.f. $p_X(x)$ and entropy

$$H(X) = -\sum_x p_X(x) \log p_X(x).$$

A channel will have an output Y (per second) with entropy

$$H(Y) = -\sum_y p_Y(y) \log p_Y(y).$$

If the channel is noiseless, then $H(Y) = H(X)$.

If the channel is noisy, Y given X is still random. Let $p(x, y)$ denote their joint p.m.f. The joint entropy is

$$H(X, Y) = -\sum_{x,y} p(x, y) \log p(x, y).$$

Following Shannon, let $p_x(y) = P\{Y = y | X = x\}$ and consider the conditional entropy of Y given X namely,

$$H_X(Y) = -\sum_{x,y} p(x, y) \log p_x(y).$$

Clearly, $H(X, Y) = H(X) + H_X(Y)$ and similarly $H(X, Y) = H(Y) + H_Y(X)$. $H_Y(X)$ is called the equivocation or average ambiguity about input X given only output Y. It is the information about input X that is received given the output Y. By Theorem 10 of Shannon (1948), it is the amount of additional information that must be supplied per unit time at the receiving end to correct the received message.

Thus $H_Y(X)$ is the missing information. So amount of information produced in the channel (per unit time) is

$$H(X) - H_Y(X)$$

which may be shown to be non-negative by Shannon's basic results

$$H(X) + H(Y) \geq H(X,Y) = H(Y) + H_Y(X).$$

In statistical problems, we take X to be θ and Y to be the observation vector $\boldsymbol{X}$. Then $H(\theta) - H_{\boldsymbol{X}}(\theta)$ is the same measure as before, namely

$$E\left(\log \frac{p(\theta|\boldsymbol{X})}{p(\theta)}\right).$$

The maximum of

$$H(X) - H_Y(X)$$

with respect to the source, i.e., with respect to $p(x)$ is what Shannon calls the capacity of the channel. Over compact rectangles, the Jeffreys prior is this maximizing distribution for the statistical channel.

It is worth pointing out that the Jeffreys prior is a special case of the reference priors of Bernardo (1979).

Another point of interest is that as $n \to \infty$, most of Bernardo's information is contained in the constant term of the asymptotic decomposition. This would suggest that for moderately large n, choice of prior is not important.

The measure of information used by Bernardo was introduced earlier in Bayesian design of experiments by Lindley (1956). There $p(\theta)$ is fixed but the observable X is not fixed, and the object is to choose a design, i.e., X, to minimize the information. Minimization is for the given sample size n, not asymptotic as in Bernardo (1979).

5.1.4 Jeffreys Prior as a Probability Matching Prior

One would expect an objective prior with low information to provide inference similar to that based on the uniform prior for θ in $N(\theta, 1)$.

In the case of $N(\theta, 1)$ with a uniform prior for θ, the posterior distribution of the pivotal quantity $\theta - \bar{X}$, given $\boldsymbol{X}$, is identical with the frequentist distribution of $\theta - \bar{X}$, given θ. In the general case we will not get exactly the same distribution but only up to $O_p(n^{-1})$. A precise definition of a probability matching prior for a single parameter is given below.

Let $X_1, X_2, ..., X_n$ be i.i.d. $p(x|\theta)$, $\theta \in \Theta \subset \mathcal{R}$. Assume regularity conditions needed for expansion of the posterior with the normal $N(\hat{\theta}, (nI(\hat{\theta}))^{-1})$

as the leading term. For $0 < \alpha < 1$, choose $\theta_\alpha(\boldsymbol{X})$ depending on the prior $p(\theta)$ such that

$$P\{\theta \leq \theta_\alpha(\boldsymbol{X})|\boldsymbol{X}\} = 1 - \alpha + O_p(n^{-1}). \tag{5.7}$$

It can be verified that $\theta_\alpha(\boldsymbol{X}) = \hat{\theta} + O_p(1/\sqrt{n})$. We say $p(\theta)$ is probability matching (to first order), if

$$P_\theta\{\theta \leq \theta_\alpha(\boldsymbol{X})\} = 1 - \alpha + O(n^{-1}) \tag{5.8}$$

(uniformly on compact sets of θ). In the normal case with $p(\theta) = $ constant,

$$\theta_\alpha(X) = \bar{X} + z_\alpha/\sqrt{n}$$

where $P\{Z > z_\alpha\} = \alpha$, $Z \sim N(0,1)$.

We have matched posterior probability and frequentist probability up to $O_p(n^{-1})$. Why one chooses this particular order may be explained as follows. For any prior $p(\theta)$ satisfying some regularity conditions, the two probabilities mentioned above agree up to $O(n^{-\frac{1}{2}})$, so $O(n^{-\frac{1}{2}})$ is too weak. On the other hand, if we strengthen $O(n^{-1})$ to, say, $O(n^{-\frac{3}{2}})$, in general no prior would be probability matching. So $O(n^{-1})$ is just right.

It is instructive to view probability matching in a slightly different but equivalent way. Instead of working with $\hat{\theta}_\alpha(\boldsymbol{X})$ one may choose to work with the approximate quantile $\hat{\theta} + z_\alpha/\sqrt{n}$ and require

$$P\{\theta < \hat{\theta} + z_\alpha/\sqrt{n}|\boldsymbol{X}\} = P_\theta\{\theta < \hat{\theta} + z_\alpha/\sqrt{n}\} + O_p(n^{-1}) \tag{5.9}$$

under θ (uniformly on compact sets of θ).

Each of these two probabilities has an expansion starting with $(1 - \alpha)$ and having terms decreasing in powers of $n^{-\frac{1}{2}}$. So for probability matching, we must have the same next term in the expansion.

In principle one would have to expand the probabilities and set the two second terms equal, leading to

$$-\frac{d(I(\theta))^{-1/2}}{d\theta} = (I(\theta))^{-1/2}\frac{1}{p(\theta)}\frac{dp(\theta)}{d\theta}. \tag{5.10}$$

The left-hand side comes from the frequentist probability, the right-hand side from the posterior probability (taking into account the limits of random quantities under θ). There are many common terms in both probabilities that cancel and hence do not need to be calculated. A convenient way of deriving this is through what is called a Bayesian route to frequentist calculations or a shrinkage argument. For details see Ghosh (1994, Chapter 9) or Ghosh and Mukerjee (1992) , or Datta and Mukerjee (2004).

If one tried to match probabilities upto $O(n^{-3/2})$, one would have to match the next terms in the expansion also. This would lead to two differential equations in the prior and in general they will not have a common solution.

Clearly, the unique solution to (5.10) is the Jeffreys prior

$$p(\theta) \propto \sqrt{I(\theta)}.$$

Equation (5.10) may not hold if $\hat{\theta}$ has a discrete lattice distribution. Suppose X has a discrete distribution. Then the case where $\hat{\theta}$ has a lattice distribution causes the biggest problem in carrying through the previous theory. But the Jeffreys prior may be approximately probability matching in some sense, Ghosh (1994), Rousseau (2000), Brown et al. (2001, 2002).

If $d > 1$, in general there is no multivariate probability matching prior (even for the continuous case), vide Ghosh and Mukerjee (1993), Datta (1996). It is proved in Datta (1996) that the Jeffreys prior continues to play an important role.

We consider the special case $d = 2$ by way of illustration. For more details, see Datta and Mukerjee (2004).

Let $\theta = (\theta_1, \theta_2)$ and suppose we want to match posterior probability of θ_1 and a corresponding frequentist probability through the following equations.

$$P\{\theta_1 < \theta_{1,\alpha}(\boldsymbol{X})|\boldsymbol{X}\} = 1 - \alpha + O_p(n^{-1}), \tag{5.11}$$

$$P\{\theta_1 < \theta_{1,\alpha}(\boldsymbol{X})|\theta_1, \theta_2\} = 1 - \alpha + O(n^{-1}). \tag{5.12}$$

Here $\theta_{1,\alpha}(\boldsymbol{X})$ is the (approximate) $100(1-\alpha)$- quantile of θ_1. If θ_2 is orthogonal to θ_1 in the sense that the off-diagonal element $I_{12}(\boldsymbol{\theta})$ of the information matrix is zero, then the probability matching prior is

$$p(\boldsymbol{\theta}) = \sqrt{I_{11}(\boldsymbol{\theta})}\psi(\theta_2)$$

where $\psi(\theta_2)$ is an arbitrary function of θ_2.

For a general multiparameter model with a one-dimensional parameter of interest θ_1 and nuisance parameters $\theta_2, \ldots, \theta_d$, the probability matching equation is given by

$$\sum_{j=1}^{d} \frac{\partial}{\partial \theta_j} \left\{ p(\theta) I^{j1} (I^{11})^{-1/2} \right\} = 0 \tag{5.13}$$

where $I^{-1}(\theta) = (I^{ij})$. This is obtained by equating the coefficient of $n^{-1/2}$ in the expansion of the left-hand side of (5.12) to zero; details are given, for example, in Datta and Mukerjee (2004).

Example 5.1. Consider the location-scale model

$$p(x|(\mu, \sigma)) = \frac{1}{\sigma} f\left(\frac{x - \mu}{\sigma}\right),$$

$-\infty < \mu < \infty$, $\sigma > 0$, where $f(\cdot)$ is a probability density. Let $\theta_1 = \mu$ and $\theta_2 = \sigma$, i.e., μ is the parameter of interest. It is easy to verify that $I^{j1} \propto \sigma^2$ for $j = 1, 2$ and hence in view of (5.13) the prior

$$p(\mu, \sigma) \propto \frac{1}{\sigma}$$

is probability matching. Similarly, one can also verify that the same prior is probability matching when σ is the parameter of interest.

Example 5.2. We now consider a bivariate normal model with means μ_1, μ_2, variances σ_1^2, σ_2^2, and correlation coefficient ρ, all the parameters being unknown. Suppose the parameter of interest is the regression coefficient $\rho \sigma_2 / \sigma_1$. We reparameterize as

$$\theta_1 = \rho \sigma_2 / \sigma_1, \quad \theta_2 = \sigma_2^2(1 - \rho^2), \quad \theta_3 = \sigma_1^2, \quad \theta_4 = \mu_1, \quad \theta_5 = \mu_2$$

which is an orthogonal parameterization in the sense that $I_{1j}(\boldsymbol{\theta}) = 0$ for $2 \leq j \leq 5$. Then $I^{1j}(\boldsymbol{\theta}) = 0$ for $2 \leq j \leq 5, I^{11}(\boldsymbol{\theta}) = I_{11}^{-1}(\boldsymbol{\theta}) = \theta_2 / \theta_3$, and the probability matching equation (5.13) reduces to

$$\frac{\partial}{\partial \theta_1} \{ p(\boldsymbol{\theta}) I_{11}^{-1/2} \} = 0,$$

i.e.,

$$\frac{\partial}{\partial \theta_1} \{ p(\boldsymbol{\theta}) (\theta_2 / \theta_3)^{1/2} \} = 0.$$

Hence the probability matching prior is given by

$$p(\boldsymbol{\theta}) = \psi(\theta_2, \ldots, \theta_5)(\theta_3 / \theta_2)^{1/2}$$

where $\psi(\theta_2, \ldots, \theta_5)$ is an arbitrary smooth function of $(\theta_2, \ldots, \theta_5)$.
One can also verify that a prior of the form

$$p^*(\mu_1, \mu_2, \sigma_1, \sigma_2, \rho) = \{ \sigma_1^r \sigma_2^s (1 - \rho^2)^t \}^{-1}$$

with reference to the original parameterization is probability matching if and only if $t = \frac{1}{2}s + 1$ (vide Datta and Mukerjee, 2004, pp. 28, 29).

5.1.5 Conjugate Priors and Mixtures

Let $X_1, \ldots, X_n$ be i.i.d. with a one-parameter exponential density

$$p(x|\theta) = \exp\{ A(\theta) + \theta \psi(x) + h(x) \}.$$

We recall from Chapter 1 that $T = \sum_1^n \psi(X_i)$ is a minimal sufficient statistic and $E_\theta(\psi(X_1)) = -A'(\theta)$. The likelihood is

$$\exp\{ nA(\theta) + \theta T \} \exp\{ \sum_1^n h(X_i) \}.$$

To construct a conjugate prior, i.e., a prior leading to posteriors of the same form, start with a so-called noninformative, possibly improper density $\mu(\theta)$.

We will choose μ to be the uniform or the Jeffreys prior density. Then define a prior density

$$p(\theta) = ce^{mA(\theta)+\theta s}\mu(\theta) \tag{5.14}$$

where c is the normalizing constant,

$$c = \left[\int_{\Theta} e^{mA(\theta)+\theta s}\mu(\theta)d\theta\right]^{-1}$$

if the integral is finite, and arbitrary if $p(\theta)$ is an improper prior. The constants m and s are hyperparameters of the prior. They have to be chosen so that the posterior is proper, i.e.,

$$\int_{\Theta} e^{(m+n)A(\theta)+\theta(s+T)}\mu(\theta)d\theta < \infty.$$

In this case, the posterior is

$$p(\theta|\boldsymbol{x}) = c'e^{(m+n)A(\theta)+\theta(s+T)}\mu(\theta), \tag{5.15}$$

i.e., the posterior is of the same form as the prior. Only the hyperparameters are different.

In other words, the family of priors $p(\theta)$ (vide (5.14)) is closed with respect to the formation of posterior. The form of the posterior allows us to interpret the hyperparameters in the prior. Assume initially the prior was μ. Take m to be a positive integer and think of a hypothetical sample of size m, with hypothetical data $x'_1, \ldots, x'_m$ such that $\sum_1^m \psi(x'_i) = s$. The prior is the same as a posterior with μ as prior and s as hypothetical data based on a sample of size m. This suggests m is a precision parameter. We expect that larger the m, the stronger is our faith in such quantities as the prior mean. The hyperparameter s/m has a simple interpretation as a prior guess about $E_\theta(\psi) = -A'(\theta)$, which is usually an important parametric function.

To prove the statement about s/m, we need to assume $\mu(\theta) = \text{constant}$, i.e., μ is the uniform distribution. We also assume all the integrals appearing below are finite.

Let $\Theta = (a, b)$, where a may be $-\infty$, b may be ∞. Integrating by parts

$$E[-A'(\theta)] = c\int_a^b (-A'(\theta))e^{mA(\theta)+\theta s}d\theta$$

$$= -c\frac{e^{mA(\theta)}}{m}\left. e^{\theta s}\right|_a^b + c\frac{s}{m}\int_a^b e^{mA(\theta)}e^{\theta s}d\theta$$

$$= \frac{s}{m} \tag{5.16}$$

if $e^{mA(\theta)+\theta s} = 0$ at $\theta = a, b$, which is often true if Θ is the natural parameter space. Diaconis and Ylvisaker (1979) have shown that (5.16) characterizes the prior.

A similar calculation with the posterior, vide Problem 3, shows the posterior mean of $-A'(\theta)$ is

$$E(-A'(\theta)|\boldsymbol{X}) = \frac{m}{m+n}\frac{s}{m} + \frac{n}{m+n}(-A'(\hat{\theta})). \tag{5.17}$$

i.e., the posterior mean is a weighted mean of the prior guess s/m and the MLE $-A'(\hat{\theta}) = T/n$ and the weights are proportional to the precision parameter m and the sample size n.

If μ is the Jeffreys distribution, the right-hand side of (5.16), i.e., s/m may be interpreted as

$$E\left(-A'(\theta)/\sqrt{-A''(\theta)}\right) / E\left(1/\sqrt{-A''(\theta)}\right).$$

i.e., s/m is a ratio of two prior guesses — a less compelling interpretation than for $\mu = $ uniform.

Somewhat trivially μ itself, whether uniform or Jeffreys, is a conjugate prior corresponding to $m = 0, s = 0$. Also, in special cases like the binomial and normal, the Jeffreys prior is a conjugate prior with $\mu = $ uniform. We do not know of any general relation connecting the Jeffreys prior and the conjugate priors with $\mu = $ uniform.

Conjugate priors, specially with $\mu = $ uniform, were very popular because the posterior is easy to calculate, the hyperparameters are easy to interpret and hence elicit and the Bayes estimate for $E_\theta(\psi(X))$ has a nice interpretation. All these facts generalize to the case of multiparameter exponential family of distribution

$$p(\boldsymbol{x}|\boldsymbol{\theta}) = \exp\{A(\boldsymbol{\theta}) + \sum_1^d \theta_i\psi_i(\boldsymbol{x}) + h(\boldsymbol{x})\}.$$

The conjugate prior now takes the form

$$p(\boldsymbol{\theta}) = \exp\{mA(\boldsymbol{\theta}) + \sum_1^d \theta_i s_i\}\mu(\boldsymbol{\theta})$$

where $\mu = $ uniform or Jeffreys, m is the precision parameter and s_i/m may be interpreted as the prior guess for $E_\theta(\psi_i(\boldsymbol{X}))$ if $\mu = $ uniform. Once again the hyperparameters are easy to elicit. Also the Bayes estimate for $E_\theta(\psi_i(\boldsymbol{X}))$ is a weighted mean of the prior guess and the MLE.

It has been known for some time that all these alternative properties can also be a problem. First of all, note that having a single precision parameter even for the multiparameter case limits the flexibility of conjugate priors; one cannot represent complex prior belief. The representation of the Bayes estimate as a weighted mean can be an embarrassment if there is serious conflict between prior guess and MLE. For example, what should one do if the prior guess is 10 and MLE is 100 or vice versa? In such cases, one should usually give greater weight to data unless the prior belief is based on reliable expert opinion, in which case greater weight would be given to prior guess. In any case, a simple weighted mean seems ridiculous. A related fact is that a conjugate prior usually has a sharp tail, whereas prior knowledge about the tail of a prior is rarely strong.

A cure for these problems is to take a mixture of conjugate priors by putting a prior on the hyperparameters. The class of mixtures is quite rich and given any prior, one can in principle construct a mixture of conjugate priors that approximates it. A general result of this sort is proved in Dalal and Hall (1980). A simple heuristic argument is given below.

Given any prior one can approximate it by a discrete probability distribution $(p_1, \ldots, p_k)$ over a finite set of points, say $(\boldsymbol{\eta}_1, \cdots, \boldsymbol{\eta}_k)$ where $\boldsymbol{\eta}_j = E_{\boldsymbol{\theta}_j}(\psi_1(\boldsymbol{X}), \cdots, \psi_d(\boldsymbol{X}))$. This may be considered as a mixture over k degenerate distributions of which the jth puts all the probability on $\boldsymbol{\eta}_j$. By choosing m sufficiently small and taking the prior guess equal to $\boldsymbol{\eta}$, one can approximate the k degenerate distributions by k conjugate priors. Finally, mix them by assigning weight p_j to the jth conjugate prior.

Of course the simplest applications would be to multimodal priors. The posterior for a mixture of conjugate priors can often be calculated numerically by MCMC (Markov chain Monte Carlo) method. See Chapter 7 for examples.

As an example of a mixture we consider the Cauchy prior used in Jeffreys test for normal mean μ with unknown variance σ^2, described in Section 2.7.2. The conjugate prior for μ given σ^2 is normal and the Cauchy prior used here is a scale mixture of normals $N(0, \tau^{-1})$ where τ is a mixing Gamma variable. This mixture has heavier tail than the normal and use of such prior means the inference is influenced more by the data than the prior. It is expected that, in general, mixtures of conjugate priors will have this property, but we have not seen any investigation in the literature.

5.1.6 Invariant Objective Priors for Location-Scale Families

Let $\theta = (\mu, \sigma), -\infty < \mu < \infty, \sigma > 0$ and

$$p(x|\theta) = \frac{1}{\sigma} f\left(\frac{x-\mu}{\sigma}\right) \tag{5.18}$$

where $f(z)$ is a probability density on $\mathcal{R}$. Let $I_{\mu,\sigma}$ be the 2×2 Fisher information matrix. Then easy calculation shows

$$I_{\mu,\sigma} = \frac{1}{\sigma^2} I_{0,1}$$

which implies the Jeffreys prior is proportional to $1/\sigma^2$. We show in Section 5.1.7 that this prior corresponds with the left invariant Haar measure and

$$p_2(\mu, \sigma) = \frac{1}{\sigma}$$

corresponds to the right invariant Haar measure. See Dawid (Encyclopedia of Statistics) for other relevant definitions of invariance and their implications. We discuss some desirable properties of p_2 in Subsection 5.1.8.

5.1.7 Left and Right Invariant Priors

We now derive objective priors for location-scale families making use of invariance. Consider the linear transformations

$$g_{a,b}x = a + bx, \quad -\infty < a < \infty, b > 0.$$

Then

$$g_{c,d}.g_{a,b}.x = c + d(a + bx) = c + ad + dbx.$$

We may express this symbolically as $g_{c,d}.g_{a,b} = g_{e,f}$ where $e = c + ad, f = db$ specify the multiplication rule. Let $G = \{g_{a,b}; -\infty < a < \infty, b > 0\}$. Then G is a group.

It is convenient to represent $g_{a,b}$ by the vector (a, b) and rewrite the multiplication rule as

$$(c, d).(a, b) = (e, f). \tag{5.19}$$

Then we may identify $\mathcal{R} \times \mathcal{R}^+$ with G and use both notations freely. We give $\mathcal{R} \times \mathcal{R}^+$ its usual topological or geometric structure. The general theory of (locally compact) groups (see, e.g., Halmos (1950, 1974) or Nachbin (1965)) shows there are two measures μ_1 and μ_2 on G such that μ_1 is left invariant, i.e., for all $g \in G$ and $A \subset G$,

$$\mu_1(gA) = \mu_1(A)$$

and μ_2 is right invariant, i.e., for all g and A,

$$\mu_2(Ag) = \mu_2(A)$$

where $gA = \{gg'; g' \in A\}, Ag = \{g'g; g' \in A\}$.

The measures μ_1 and μ_2 are said to be the left invariant and right invariant Haar measures. They are unique up to multiplicative constants. We now proceed to determine them explicitly.

Suppose we assume μ_2 has a density f_2, i.e., denoting points in $\mathcal{R} \times \mathcal{R}^+$ by (a_1, a_2)

$$\mu_2(A) = \int_A f_2(a_1, a_2)\, da_1\, da_2 \tag{5.20}$$

and assume f_2 is a continuous function. With $g = (b_1, b_2)$ and $(c_1, c_2) = (a_1, a_2).(b_1, b_2)$ with

$$c_1 = a_1 + a_2 b_1, \quad c_2 = a_2 b_2, \tag{5.21}$$

one may evaluate $\mu_2(Ag)$ in several ways, e.g.,

$$\mu_2(Ag) = \int_{Ag} h(c_1, c_2)\, dc_1\, dc_2 \tag{5.22}$$

where

$$h(c_1, c_2) = f_2(a_1, a_2)(J)^{-1} \tag{5.23}$$

with

$$J = \frac{\partial(c_1, c_2)}{\partial(a_1, a_2)} = \begin{vmatrix} 1 & b_1 \\ 0 & b_2 \end{vmatrix} = b_2.$$

Also, by definition of f_2

$$\mu_2(Ag) = \int_{Ag} f_2(c_1, c_2)\, dc_1\, dc_2. \tag{5.24}$$

Because (5.22) and (5.24) hold for all A and f_2 is continuous, we must have

$$f_2(c_1, c_2) = h(c_1, c_2), \tag{5.25}$$

$$\text{i.e.,} \quad f_2(c_1, c_2) = f_2(a_1, a_2)\frac{1}{b_2},$$

for all $(a_1, a_2), (b_1, b_2) \in \mathcal{R} \times \mathcal{R}^+$. Set $a_1 = 0$, $a_2 = 1$. Then $f_2(b_1, b_2) = f_2(0, 1)\frac{1}{b_2}$, i.e.,

$$f_2(b_1, b_2) = \text{ constant }\frac{1}{b_2}. \tag{5.26}$$

It is easy to verify that μ_2 defined by (5.20) is right invariant if f_2 is as in (5.26). One has merely to verify (5.25) and then (5.22).

Proceeding in the same way, one can show that the left invariant Haar measure has density

$$f_1(b_1, b_2) = \frac{1}{b_2^2}.$$

We have now to lift these measures to the (μ, σ)-space. To do this, we first define an isomorphic group of transformations on the parameter space. Each transformation $g_{a,b}x = a + bx$ on the sample space induces a transformation $\bar{g}_{a,b}$ defined by

$$\bar{g}_{a,b}(\mu, \sigma) = (a + b\mu, b\sigma),$$

i.e., the right-hand side gives the location and scale of the distribution of $g_{a,b}X$ where X has density (5.18). The transformation $g_{a,b} \to \bar{g}_{a,b}$ is an isomorphism, i.e.,

$$(\bar{g}_{a,b})^{-1} = \overline{(g_{a,b}^{-1})}$$

and if

$$g_{a,b} \cdot g_{c,d} = g_{e,f}$$

then

$$\bar{g}_{a,b} \cdot \bar{g}_{c,d} = \bar{g}_{e,f}.$$

In view of this, we may write $\bar{g}_{a_1,a_2}$ also as (a_1, a_2) and define the group multiplication by (5.19) or (5.21). Consequently, left and right invariant measures for $\bar{g}$ are the same as before and

$$d\mu_1(b_1, b_2) = \text{constant } \frac{1}{b_2^2} db_1 db_2,$$

$$d\mu_2(b_1, b_2) = \text{constant } \frac{1}{b_1} db_1 db_2.$$

We now lift these measures on to the (μ, σ)-space by setting up a canonical transformation

$$(\mu, \sigma) = \bar{g}_{\mu,\sigma}(0, 1)$$

that converts a single fixed point in the parameter space into an arbitrary point (μ, σ). Because $(0, 1)$ is fixed, we can think of the above relation as setting up a one-to-one transformation between (μ, σ) and $\bar{g}_{\mu,\sigma} = (\mu, \sigma)$. Because this is essentially an identity transformation from (μ, σ) into a group of transformations, given any μ^* on the space of $\bar{g}$'s we define ν on $\Theta = \mathcal{R} \times \mathcal{R}^+$ as

$$\nu(A) = \mu^*\{(\mu, \sigma); g_{\mu,\sigma} \in A\} = \mu^*(A).$$

Thus

$$d\nu_1(\mu, \sigma) = d\mu_1(\mu, \sigma) = \frac{1}{\sigma^2} d\mu \, d\sigma$$

and

$$d\nu_2(\mu, \sigma) = d\mu_2(\mu, \sigma) = \frac{1}{\sigma} d\mu \, d\sigma$$

are the left and right invariant priors for (μ, σ).

5.1.8 Properties of the Right Invariant Prior for Location-Scale Families

The right invariant prior density

$$p_r(\mu, \sigma) = \frac{1}{\sigma}$$

has many attractive properties. We list some of them below. These properties do not hold in general for the left invariant prior

$$p_l(\mu, \sigma) = \frac{1}{\sigma^2}.$$

Heath and Sudderth (1978, 1989) show that inference based on the posterior corresponding with p_r is coherent in a sense defined by them. Similar properties have been shown by Eaton and Sudderth (1998, 2004). Dawid et al. (1973) show that the posterior corresponding to p_r is free from the marginalization paradox. It is free from the marginalization paradox if the group is amenable.

Dawid (Encyclopedia of Statistics) also provides counter-examples in case the group is not amenable. Amenability is a technical condition that is also called the Hunt-Stein condition that is needed to prove theorems relating invariance to minimaxity in classical statistics, vide Bondar and Milnes (1981). Datta and Ghosh (1996) show p_r is probability matching in a certain strong sense.

A famous classical theorem due to Hunt and Stein, vide Lehmann (1986), or Kiefer (1957), implies that under certain invariance assumptions (that include amenability of the underlying group), the Bayes solution is minimax as well as best among equivariant rules, see also Berger (1985a). We consider a couple of applications. Suppose we have two location-scale families $\sigma^{-1}f_i((x-\mu)/\sigma)$, $i = 0, 1$. For example, f_0 may be standard normal i.e., $N(0, 1)$ and f_1 may be standard Cauchy, i.e.,

$$f_1(x) = \frac{1}{\pi}\frac{1}{1+x^2}.$$

The observations $X_1, X_2, \ldots, X_n$ are i.i.d. with density belonging to one of these two families. One has to decide which is true.

Consider the Bayes rule which accepts f_1 if

$$BF = \frac{\int\int \prod_{j=1}^{n}\left[\sigma^{-1}f_1(\frac{X_j-\mu}{\sigma})\right]\frac{1}{\sigma}d\mu\,d\sigma}{\int\int \prod_{j=1}^{n}\left[\sigma^{-1}f_0(\frac{X_j-\mu}{\sigma})\right]\frac{1}{\sigma}\,d\mu\,d\sigma} > c.$$

If c is chosen such that the Type 1 and Type 2 error probabilities are the same, then this is a minimax test, i.e., it minimizes the maximum error probability among all tests, where the maximum is over $i = 0, 1$ and $(\mu, \sigma) \in \mathcal{R} \times \mathcal{R}^+$.

Suppose we consider the estimation problem of a location parameter with a squared error loss. Let $X_1, X_2, \ldots, X_n$ be i.i.d. $\sim f_0(x-\theta)$. Here $p_r = p_l =$ constant. The corresponding Bayes estimate is

$$\frac{\int_{-\infty}^{\infty}\theta\prod_1^n f(X_j - \theta)\,d\theta}{\int_{-\infty}^{\infty}\prod_1^n f(X_j - \theta)\,d\theta}$$

which is both minimax and best among equivariant estimators, i.e., it minimizes $R(\theta, T(\boldsymbol{X})) = E_\theta(T(\boldsymbol{X}) - \theta)^2$ among all T satisfying

$$T(x_1 + a, \ldots, x_n + a) = T(x_1, \ldots, x_n) + a, \ a \in \mathcal{R}.$$

A similar result for scale families is explored in Problem 4.

5.1.9 General Group Families

There are interesting statistical problems that are left invariant by groups of transformations other than the location-scale transformations discussed in the preceding subsections. It is of interest then to consider invariant Haar measures for such general groups also. An example follows.

Example 5.3. Suppose $\mathbf{X} \sim N_p(\boldsymbol{\theta}, I)$. It is desired to test

$$H_0 : \boldsymbol{\theta} = \mathbf{0} \text{ versus } H_1 : \boldsymbol{\theta} \neq \mathbf{0}.$$

This testing problem is invariant under the group $\mathcal{G}_O$ of all orthogonal transformations; i.e., if H is an orthogonal matrix of order p, then $g_H \mathbf{X} = H\mathbf{X} \sim N_p(H\boldsymbol{\theta}, I)$, so that $\bar{g}_H \boldsymbol{\theta} = H\boldsymbol{\theta}$. Also, $\bar{g}_H \mathbf{0} = \mathbf{0}$. Further discussion of this example as well as treatment of invariant tests is taken up in Chapter 6. Discussion on statistical applications involving general groups can be found in sources such as Eaton (1983, 1989), Farrell (1985), and Muirhead (1982).

5.1.10 Reference Priors

In statistical problems that are left invariant by a group of nice transformations, the Jeffreys prior turns out to be the left invariant prior, vide Datta and Ghosh (1996). But for reasons outlined in Subsection 5.1.8, one would prefer the right invariant prior. In all the examples that we have seen, an interesting modification of the Jeffreys prior, introduced in Bernardo (1979) and further refined in Berger and Bernardo (1989, 1992a and 1992b), leads to the right invariant prior. These priors are called reference priors after Bernardo (1979). A reference prior is simply an objective prior constructed in a particular way, but the term reference prior could, in principle, be applied to any objective prior because any objective prior is taken as some sort of objective or conventional standard, i.e., a reference point with which one may compare subjective priors to calibrate them.

As pointed out by Bernardo (1979), suitably chosen reference priors can be appropriate in high-dimensional problems also. We discuss this in a later chapter.

Our presentation in this section is based on Ghosh (1994, Chapter 9) and Ghosh and Ramamoorthi (2003, Chapter 1).

If we consider all the parameters of equal importance, we maximize the entropy of Subsection 5.1.3. This leads to the Jeffreys prior. To avoid this, one assumes parameters as having different importance. We consider first the case of $d = 2$ parameters, namely, (θ_1, θ_2), where we have an ordering of the parameters in order of importance. Thus θ_1 is supposed to be more important than θ_2. For example, suppose we are considering a random sample from $N(\mu, \sigma^2)$. If our primary interest is in μ, we would take $\theta_1 = \mu, \theta_2 = \sigma^2$. If our primary interest is in σ^2, then $\theta_1 = \sigma^2, \theta_2 = \mu$. If our primary interest is in μ/σ, we take $\theta_1 = \mu/\sigma$ and $\theta_2 = \mu$ or σ or any other function such that (θ_1, θ_2) is a one-to-one sufficiently smooth function of (μ, σ).

For fixed θ_1, the conditional density $p(\theta_2|\theta_1)$ is one dimensional. Bernardo (1979) recommends setting this equal to the conditional Jeffreys prior

$$c(\theta_1)\sqrt{I_{22}(\boldsymbol{\theta})}.$$

Having fixed this, the marginal $p(\theta_1)$ is chosen by maximizing

$$E\left(\log \frac{p(\theta_1|X_1,\ldots,X_n)}{p(\theta_1)}\right)$$

in an asymptotic sense as explained below.

Fix an increasing sequence of closed and bounded, i.e., compact rectangles $K_{1i} \times K_{2i}$ whose union is Θ. Let $p_i(\theta_2|\theta_1)$ be the conditional Jeffreys prior, restricted to K_{2i} and $p_i(\theta_1)$ a prior supported on K_{1i}. As mentioned before,

$$p_i(\theta_2|\theta_1) = c_i(\theta_1)\sqrt{I_{22}(\boldsymbol{\theta})}$$

where $c_i(\theta_1)$ is a normalizing constant such that

$$\int_{K_{2i}} p_i(\theta_2|\theta_1)d\theta_2 = 1.$$

Then $p_i(\theta_1,\theta_2) = p_i(\theta_1)p_i(\theta_2|\theta_1)$ on $K_{1i} \times K_{2i}$ and we consider

$$\begin{aligned}
J(p_i(\theta_1),\boldsymbol{X}) &= E\left\{\log \frac{p_i(\theta_1|\boldsymbol{X})}{p_i(\theta_1)}\right\}\\
&= E\left[\log \frac{p_i(\boldsymbol{\theta}|\boldsymbol{X})}{p_i(\boldsymbol{\theta})}\right] - E\left[\log \frac{p_i(\theta_2|\theta_1,\boldsymbol{X})}{p_i(\theta_2|\theta_1)}\right]\\
&= J(p_i(\theta_1,\theta_2),\boldsymbol{X}) - \int_{K_{1i}} p_i(\theta_1)J(p_i(\theta_2|\theta_1),\boldsymbol{X})d\theta_1
\end{aligned}$$

where for fixed θ_1, $J(p_i(\theta_2|\theta_1),\boldsymbol{X})$ is the Lindley-Bernardo functional

$$J = E\left\{\log \frac{p_i(\theta_2|\theta_1,\boldsymbol{X})}{p_i(\theta_2|\theta_1)}\right\}$$

with $p_i(\theta_2|\theta_1)$ being regarded as a conditional prior for θ_2 for fixed θ_1.

Applying the asymptotic normal approximation to the first term on the right-hand side as well as the second term on the right-hand side, as in Subsection 5.1.3,

$$\begin{aligned}
&J(p_i(\theta_1),\boldsymbol{X})\\
&= K_n + \left[\int_{K_{1i}\times K_{2i}} p_i(\boldsymbol{\theta})\log\{\det(I(\boldsymbol{\theta}))\}^{\frac{1}{2}}d\boldsymbol{\theta} - \int_{K_{1i}\times K_{2i}} p_i(\boldsymbol{\theta})\log p_i(\boldsymbol{\theta})d\boldsymbol{\theta}\right]\\
&\quad - \int_{K_{1i}} p_i(\theta_1)\int_{K_{2i}} p_i(\theta_2|\theta_1)\left[\log\sqrt{I_{22}(\boldsymbol{\theta})} - \log p_i(\theta_2|\theta_1)\right]d\theta_2 d\theta_1\\
&= K_n + \int_{K_{1i}} p_i(\theta_1)\int_{K_{2i}} p_i(\theta_2|\theta_1)\log\left\{\frac{\det(I(\boldsymbol{\theta}))}{I_{22}(\boldsymbol{\theta})}\right\}^{\frac{1}{2}} d\theta_2 d\theta_1\\
&\quad - \int_{K_{1i}} p_i(\theta_1)\log p_i(\theta_1)d\theta_1\\
&= K_n + \int_{K_{1i}} p_i(\theta_1)\int_{K_{2i}} p_i(\theta_2|\theta_1)\log\left\{\frac{1}{I^{11}(\boldsymbol{\theta})}\right\}^{\frac{1}{2}} d\theta_2 d\theta_1\\
&\quad - \int_{K_{1i}} p_i(\theta_1)\log p_i(\theta_1)d\theta_1
\end{aligned}$$

$$\tag{5.27}$$

where K_n is a constant depending on n. Let $\psi_i(\theta_1)$ be the geometric mean of $(I^{11}(\boldsymbol{\theta}))^{-\frac{1}{2}}$ with respect to $p_i(\theta_2|\theta_1)$. Then (5.27) can be written as

$$K_n + \int_{K_{1i}} p_i(\theta_1) \log \frac{\psi_i(\theta_1)}{p_i(\theta_1)} d\theta_1$$

which is maximized if

$$p_i(\theta_1) = c_i' \psi_i(\theta_1) \text{ on } K_{1i}$$
$$= 0 \quad \text{outside.}$$

The product

$$p_i(\boldsymbol{\theta}) = c_i' \psi_i(\theta_1) c_i(\theta_1) [I_{22}(\boldsymbol{\theta})]^{\frac{1}{2}}$$

is the reference prior on $K_{1i} \times K_{2i}$. If we can write this as

$$p_i(\boldsymbol{\theta}) = d_i A(\theta_1, \theta_2) \text{ on } K_{1i} \times K_{2i}$$
$$= 0 \quad \text{elsewhere}$$

then the reference prior on Θ may be taken as proportional to $A(\theta_1, \theta_2)$.

Clearly, the reference prior depends on the choice of (θ_1, θ_2) and the compact sets K_{1i}, K_{2i}. The normalization on $K_{1i} \times K_{2i}$ first appeared in Berger and Bernardo (1989). If an improper objective prior is used for fixed n, one might run into paradoxes of the kind exhibited by Fraser et al. (1985). See in this connection Berger and Bernardo (1992a). Recently there has been some change in the definition of reference prior, but we understand the final results are similar (Berger et al. (2006)).

The above procedure is based on Ghosh and Mukerjee (1992) and Ghosh (1994). Algebraically it is more convenient to work with $[I(\boldsymbol{\theta})]^{-1}$ as in Berger and Bernardo (1992b). We illustrate their method for $d = 3$, the general case is handled in the same way.

First note the following two facts.

A. Suppose $\theta_1, \theta_2, \ldots, \theta_j$ follow multivariate normal with dispersion matrix Σ. Then the conditional distribution of θ_j given $\theta_1, \theta_2, \ldots, \theta_{j-1}$ is normal with variance equal to the reciprocal of the (j, j)-th element of Σ^{-1}.

B. Following the notations of Berger and Bernardo (1992b), let $S = [I(\boldsymbol{\theta})]^{-1}$ where $I(\boldsymbol{\theta})$ is the $d \times d$ Fisher information matrix and S_j be the $j \times j$ principal submatrix of S. Let $H_j = S_j^{-1}$ and h_j be the (j, j)-th element of H_j. Then by **A**, the asymptotic variance of θ_j given $\theta_1, \theta_2, \ldots, \theta_{j-1}$ is $(h_j)^{-1}/n$. To get some feeling for h_j, note that for arbitrary d, $j = d$, $h_j = I_{dd}$ and for arbitrary d, $j = 1$, $h_1 = 1/I^{11}$.

We now provide a new asymptotic formula for Lindley-Bernardo information measure, namely,

$$E\left(\log \frac{p(\theta_j|\theta_1, \ldots, \theta_{j-1}, \boldsymbol{X})}{p(\theta_j|\theta_1, \ldots, \theta_{j-1})}\right)$$

$$= E(\log(N(\hat{\theta}_j(\theta_1,\ldots,\theta_{j-1}), h_j^{-1}(\boldsymbol{\theta})/n))) - E(\log p_i(\theta_j|\theta_1,\ldots,\theta_{j-1}))$$

(where $\hat{\theta}_j(\theta_1,\ldots,\theta_{j-1})$ is the MLE for θ_j given $\theta_1,\ldots,\theta_{j-1}$ and p_i is a prior supported on a compact rectangle $K_{1i} \times K_{2i} \times \ldots \times K_{di}$.)

$$= K_n + E\left[\int_{K_{ji}} \log \frac{\psi_j(\theta_1,\ldots,\theta_j)}{p_i(\theta_j|\theta_1,\ldots,\theta_{j-1})} p(\theta_j|\theta_1,\ldots,\theta_{j-1})d\theta_j\right]$$
$$+ o_p(1) \tag{5.28}$$

where K_n is a constant depending on n and

$$\psi_j(\theta_1,\ldots,\theta_j) = \exp\left\{\int \frac{1}{2}\log h_j(\boldsymbol{\theta})p(\theta_{j+1},\ldots,\theta_d|\theta_1,\ldots,\theta_j)d\theta_{j+1}\ldots d\theta_d\right\}$$

is the geometric mean of $h_j^{1/2}(\boldsymbol{\theta})$ with respect to $p(\theta_{j+1},\ldots,\theta_d|\theta_1,\ldots,\theta_j)$.

The proof of (5.28) parallels the proof of (5.4). It follows that asymptotically (5.28) is maximized if we set

$$p(\theta_j|\theta_1,\ldots,\theta_{j-1}) = c_i'(\theta_1,\ldots,\theta_{j-1})\psi_j(\theta_1,\ldots,\theta_j) \text{ on } K_{ji}$$
$$= 0 \quad \text{elsewhere .} \tag{5.29}$$

If the dimension exceeds 2, say $d = 3$, we merely start with a compact rectangle $K_{1i} \times K_{2i} \times K_{3i}$ and set $p_i(\theta_3|\theta_1,\theta_2) = c_i(\theta_1,\theta_2)\sqrt{I_{33}}(\boldsymbol{\theta})$ on K_{3i}. Then we first determine $p_i(\theta_2|\theta_1)$ and then $p_i(\theta_1)$ by applying (5.29) twice. Thus,

$$p_i(\theta_2|\theta_1) = c_i'(\theta_1)\psi_2(\theta_1,\theta_2) \text{ on } K_{2i}$$
$$p_i(\theta_1) = c_i'\psi_1(\theta_1) \text{ on } K_{1i}.$$

One can also verify that the formulas obtained for $d = 2$ can be rederived in this way. A couple of examples follow.

Example 5.4. Let $X_1, X_2, \ldots, X_n$ be i.i.d. normal with mean θ_2 and variance θ_1, with θ_1 being the parameter of interest. Here

$$I(\boldsymbol{\theta}) = \begin{pmatrix} \frac{1}{2\theta_1^2} & 0 \\ 0 & \frac{1}{\theta_1} \end{pmatrix}, \quad I_{22}(\boldsymbol{\theta}) = \frac{1}{\theta_1}, \quad I^{11}(\boldsymbol{\theta}) = 2\theta_1^2.$$

Thus $p_i(\theta_2|\theta_1) = c_i$ on K_{2i} where $c_i^{-1} =$ volume of K_{2i}, and therefore,

$$\psi_i(\theta_1) = \exp\left\{\int_{K_{2i}} c_i \log(\frac{1}{\sqrt{2\theta_1}})d\theta_2\right\} = \frac{1}{\sqrt{2\theta_1}}.$$

We then have
$$p_i(\theta_1,\theta_2) = d_i(1/\theta_1)$$

for some constant d_i and the reference prior is taken as

$$p(\theta_1, \theta_2) \propto \frac{1}{\theta_1}.$$

This is the right invariant Haar measure unlike the Jeffreys prior which is left invariant (see Subsection 5.1.7).

Example 5.5. (Berger and Bernardo, 1992b) Let (X_1, X_2, X_3) follow a multinomial distribution with parameters $(n; \theta_1, \theta_2, \theta_3)$, i.e., (X_1, X_2, X_3) has density

$$p(x_1, x_2, x_3 | \theta_1, \theta_2, \theta_3)$$
$$= \frac{n!}{x_1! \, x_2! \, x_3! (n - x_1 - x_2 - x_3)!} \theta_1^{x_1} \theta_2^{x_2} \theta_3^{x_3} (1 - \theta_1 - \theta_2 - \theta_3)^{n - x_1 - x_2 - x_3},$$
$$x_i \geq 0, i = 1, 2, 3, \ \sum_1^3 x_i \leq n, \quad \theta_i > 0, i = 1, 2, 3, \ \sum_1^3 \theta_i < 1.$$

Here the information matrix is

$$I(\boldsymbol{\theta}) = n \, \mathrm{Diag}\{\theta_1^{-1}, \theta_2^{-1}, \theta_3^{-1}\} + n(1 - \theta_1 - \theta_2 - \theta_3)^{-1} \mathbf{1}_3,$$

where $\mathrm{Diag}\,\{a_1, a_2, a_3\}$ denotes the diagonal matrix with diagonal elements a_1, a_2, a_3 and $\mathbf{1}_3$ denotes the 3×3 matrix with all elements equal to one. Hence

$$S(\boldsymbol{\theta}) = \frac{1}{n} \mathrm{Diag}\,\{\theta_1, \theta_2, \theta_3\} - \frac{1}{n} \boldsymbol{\theta}\boldsymbol{\theta}'$$

and for $j = 1, 2, 3$

$$S_j(\boldsymbol{\theta}) = \frac{1}{n} \mathrm{Diag}\,\{\theta_1, \ldots, \theta_j\} - \frac{1}{n} \boldsymbol{\theta}_{[j]} \boldsymbol{\theta}'_{[j]}$$

with $\boldsymbol{\theta}_{[j]} = (\theta_1, \ldots, \theta_j)'$,

$$H_j(\boldsymbol{\theta}) = n \, \mathrm{Diag}\,\{\theta_1^{-1}, \ldots, \theta_j^{-1}\} + n(1 - \theta_1 - \cdots - \theta_j)^{-1} \mathbf{1}_j$$

and

$$h_j(\boldsymbol{\theta}) = n\theta_j^{-1}(1 - \theta_1 - \cdots - \theta_{j-1})(1 - \theta_1 - \cdots - \theta_j)^{-1}.$$

Note that $h_j(\boldsymbol{\theta})$ depends only on $\theta_1, \theta_2, \ldots, \theta_j$ so that

$$\psi_j(\theta_1, \ldots, \theta_j) = h_j^{1/2}(\boldsymbol{\theta}).$$

Here we need not restrict to compact rectangles as all the distributions involved have finite mass. As suggested above for the general case, the reference prior can now be obtained as

$$p(\theta_3 | \theta_1, \theta_2) = \pi^{-1}\theta_3^{-1/2}(1 - \theta_1 - \theta_2 - \theta_3)^{-1/2}, \quad 0 < \theta_3 < 1 - \theta_1 - \theta_2,$$
$$p(\theta_2 | \theta_1) = \pi^{-1}\theta_2^{-1/2}(1 - \theta_1 - \theta_2)^{-1/2}, \quad 0 < \theta_2 < 1 - \theta_1,$$
$$p(\theta_1) = \pi^{-1}\theta_1^{-1/2}(1 - \theta_1)^{-1/2}, \quad 0 < \theta_1 < 1,$$

i.e.,

$$p(\boldsymbol{\theta}) = \pi^{-3}\theta_1^{-1/2}(1-\theta_1)^{-1/2}\theta_2^{-1/2}(1-\theta_1-\theta_2)^{-1/2}\theta_3^{-1/2}(1-\theta_1-\theta_2-\theta_3)^{-1/2},$$
$$\theta_i > 0,\ i = 1,2,3,\ \textstyle\sum_1^3 \theta_i < 1.$$

As remarked by Berger and Bernardo (1992b), inferences about θ_1 based on the above prior depend only on x_1 and not on the frequencies of other cells. This is not the case with standard noninformative priors such as Jeffreys prior. See in this context Berger and Bernardo (1992b, Section 3.4).

5.1.11 Reference Priors Without Entropy Maximization

Construction of reference priors involves two interesting ideas. The first is the new measure of information in a prior obtained by comparing it with the posterior. The second idea is the step by step algorithm based on arranging the parameters in ascending order of importance. The first throws light on why an objective prior would depend on the model for the likelihood. But it is the step by step algorithm that seems to help more in removing the problems associated with the Jeffreys prior. We explore below what kind of priors would emerge if we follow only part of the Berger-Bernardo algorithm (of Berger and Bernardo (1992a)).

We illustrate with two (one-dimensional) parameters θ_1 and θ_2 of which θ_1 is supposed to be more important. This would be interpreted as meaning that the marginal prior for θ_1 is more important than the marginal prior for θ_2. Then the prior is to be written as $p(\theta_1)p(\theta_2|\theta_1)$, with

$$p(\theta_2|\theta_1) \propto \sqrt{I_{22}(\boldsymbol{\theta})}$$

and $p(\theta_1)$ is to be determined suitably.

Suppose, we determine $p(\theta_1)$ from the probability matching conditions

$$P\{\theta_1 \leq \theta_{1,\alpha}(\boldsymbol{X})|\boldsymbol{X}\} = 1 - \alpha + O_p(n^{-1}), \qquad (5.30)$$

$$\int P\{\theta_1 \leq \theta_{1,\alpha}(X)|\theta_1,\theta_2\}p(\theta_2|\theta_1)d\theta_2 = 1 - \alpha + O(n^{-1}). \qquad (5.31)$$

Here (5.30) defines the Bayesian quantile $\theta_{1,\alpha}$ of θ_1, which depends on data $\boldsymbol{X}$ and (5.31) requires that the posterior probability on the left-hand side of (5.30) matches the frequentist probability averaged out with respect to $p(\vartheta_2|\theta_1)$. Under the assumption that θ_1 and θ_2 are orthogonal, i.e., $I_{12}(\boldsymbol{\theta}) = I_{21}(\boldsymbol{\theta}) = 0$, one gets (Ghosh (1994))

$$p_i(\theta_1) = \text{ constant } \left(\int_{K_{2i}} I_{11}^{-1/2}(\boldsymbol{\theta})p(\theta_2|\theta_1)d\theta_2 \right)^{-1} \qquad (5.32)$$

where $K_{1i} \times K_{2i}$ is a sequence of increasing bounded rectangles whose union is $\Theta_1 \times \Theta_2$. This equation shows the marginal of θ_1 is a (normalized) harmonic mean of $\sqrt{I_{11}}$ which equals the harmonic mean of $1/\sqrt{I^{11}}$.

What about a choice of $p_i(\theta_1)$ equal to the geometric or arithmetic mean? The Berger-Bernardo reference prior is the geometric mean. The marginal prior is the arithmetic mean if we follow the approach of taking weak limits of suitable discrete uniforms, vide Ghosal et al. (1997). In many interesting cases involving invariance, for example, in the case of location-scale families, all three approaches lead to the right invariant Haar measures as the joint prior.

5.1.12 Objective Priors with Partial Information

Suppose we have chosen our favorite so-called noninformative prior, say p_0. How can we utilize available prior information on a few moments of $\boldsymbol{\theta}$? Let p be an arbitrary prior satisfying the following constraints based on available information

$$\int_\Theta g_j(\boldsymbol{\theta})p(\boldsymbol{\theta})\, d\boldsymbol{\theta} = A_j, \quad j = 1, 2, \ldots, k. \tag{5.33}$$

If $g_j(\boldsymbol{\theta}) = \theta_1^{j_1}\ldots\theta_p^{j_r}$, we have the usual moments of $\boldsymbol{\theta}$. We fix increasing compact rectangles K_i with union equal to Θ and among priors p_i supported on K_i and satisfying (5.33), minimize the Kullback-Leibler number

$$K(p_i, p_0) = \int_{K_i} p_i(\boldsymbol{\theta}) \log \frac{p_i(\boldsymbol{\theta})}{p_0(\boldsymbol{\theta})}\, d\boldsymbol{\theta}.$$

The minimizing prior is

$$p_i^*(\boldsymbol{\theta}) = \text{constant} \times \exp\left\{\sum_1^k \lambda_j g_j(\boldsymbol{\theta})\right\} p_0(\boldsymbol{\theta})$$

where λ_j's are hyperparameters to be chosen so as to satisfy (5.33). This can be proved by noting that for all priors p_i satisfying (5.33),

$$K(p_i, p_0) = \int_{K_i} p_i(\boldsymbol{\theta}) \log \frac{p_i^*(\theta)}{p_0(\theta)}\, d\boldsymbol{\theta} + K(p_i, p_i^*)$$

$$= \text{constant} + \sum \lambda_j A_j + K(p_i, p_i^*)$$

which is minimized at $p_i = p_i^*$.

If instead of moments we know values of some quantiles for (a one-dimensional) θ or more generally the prior probabilities a_j of some disjoint subsets B_j of Θ, then it may be assumed $\cup B_j = \Theta$ and one would use the prior given by

$$p_i^*(A) = \sum_j a_j \frac{p_0(A \cap B_j)}{p_0(B_j)}.$$

Sun and Berger (1998) have shown how reference priors can be constructed when partial information is available.

5.2 Discussion of Objective Priors

This section is based on Ghosh and Samanta (2002b). We begin by listing some of the common criticisms of objective priors. We refer to them below as "noninformative priors".

1. Noninformative priors do not exist. How can one define them?
2. Objective Bayesian analysis is ad hoc and hence no better than the ad hoc paradigms subjective Bayesian analysis tries to replace.
3. One should try to use prior information rather than waste time trying to find noninformative priors.
4. There are too many noninformative priors for a problem. Which one is to be used?
5. Noninformative priors are typically improper. Improper priors do not make sense as quantification of belief. For example, consider the uniform distribution on the real line. Let L be any large positive number. Then $P\{-L \leq \theta \leq L\}/P\{\theta \notin (-L, L)\} = 0$ for all L but for a sufficiently large L, depending on the problem, we would be pretty sure that $-L \leq \theta \leq L$.
6. If θ has uniform distribution because of lack of information, then this should also be true for any smooth one-to-one function $\eta = g(\theta)$.
7. Why should a noninformative prior depend on the model of the data?
8. What are the impacts of 7 on coherence and the likelihood principle?

We make a couple of general comments first before replying to each of these criticisms. The purpose of introducing an objective prior is to produce a posterior that depends more on the data than the objective prior. One way of checking this would be to compare the posteriors for different objective priors as in Example 2.2 of Chapter 2. The objective prior is only the means for producing the posterior. Moreover, objective Bayesian analysis agrees that it is impossible to define a noninformative prior on an unbounded parameter space because maximum entropy need not be finite. This is the reason that increasing bounded sets were used in the construction. One thinks of the objective priors as consensus priors with low information — at least in those cases where no prior information is available. In all other cases, the choice of an objective prior should depend on available prior information (Subsection 5.1.12). We now turn to the criticisms individually.

Points 1 and 2 are taken care of in the general comments. Point 3 is well taken, we do believe that elicitation of prior information is very important and any chosen prior should be consistent with what we know. A modest attempt towards this is made in Subsection 5.1.12. However, we feel it would rarely be the case that a prior would be fully elicited, only a few salient points or aspects with visible practical consequences can be ascertained, but subjected to this knowledge the construction of the prior would still be along the lines of Subsection 5.1.12 even though in general no explicit solution will exist.

As to point 4, we have already addressed this issue in the general comments. Even though there is no unique objective prior, the posteriors will usually be very similar even with a modest amount of data. Where this is not the case, one would have to undertake a robustness analysis restricted to the class of chosen objective priors. This seems eminently doable.

Even though usually objective priors are improper, we only work with them when the posterior is proper. Once again we urge the reader to go over the general comments. We would only add that many improper objective priors lead to same posteriors as coherent, proper, finitely additive priors. This is somewhat technical, but the interested reader can consult Heath and Sudderth (1978, 1989).

Point 6 is well taken care of by Jeffreys prior. Also in a given problem not all one-to-one transformations are allowed. For example, if the coordinates of $\boldsymbol{\theta}$ are in a decreasing order of importance, then we need only consider $\boldsymbol{\eta} = (\eta_1, \ldots, \eta_d)$ such that η_j is a one-to-one continuously differentiable function of θ_j. There are invariance theorems for reference and probability matching priors in such cases, Datta and Ghosh (1996).

We have discussed Point 7 earlier in the context of the entropy of Bernardo and Lindley. This measure depends on the experiment through the model of likelihood. Generally information in a prior cannot be defined except in the context of an experiment. Hence it is natural that a low-information prior will not be the same for all experiments. Because a model is a mathematical description of an experiment, a low-information prior will depend on the model.

We now turn to the last point. Coherence in the sense of Heath and Sudderth (1978) is defined in the context of a model. Hence the fact that an objective prior depends on a model will not automatically lead to incoherence. However, care will be needed. As we have noted earlier, a right Haar prior for location-scale families ensures coherent inference but in general a left Haar prior will not.

The impact on likelihood principle is more tricky. The likelihood principle in its strict sense is violated because the prior and hence the posterior depends on the experiment through the form of the likelihood function. However, for a fixed experiment, decision based on the posterior and the corresponding posterior risk depend only on the likelihood function. We pursue this a bit more below.

Inference based on objective priors does violate the stopping rule principle, which is closely related to the likelihood principle. In particular, in Example 1.2 of Carlin and Louis (1996), originally suggested by Lindley and Phillips (1976), one would get different answers according to a binomial or a negative binomial model. This example is discussed in Chapter 6.

To sum up we do seem to have good answers to most of the criticisms but have to live with some violations of the likelihood and stopping rule principles.

5.3 Exchangeability

A sequence of real valued random variables $\{X_i\}$ is exchangeable if for all n, all distinct suffixes $\{i_1, i_2, \ldots, i_n\}$ and all $B_1, B_2, \ldots, B_n \subset \mathcal{R}$,

$$P\{X_{i_1} \in B_1, X_{i_2} \in B_2, \ldots X_{i_n} \in B_n\} = P\{X_1 \in B_1, X_2 \in B_2, \ldots, X_n \in B_n\}.$$

In many cases, a statistician will be ready to assume exchangeability as a matter of subjective judgment.

Consider now the special case where each X_i assumes only the values 0 and 1. A famous theorem of de Finetti then shows that the subjective judgment of exchangeability leads to both a model for the likelihood and a prior.

Theorem 5.6. (de Finetti) *If X_i's are exchangeable and assume only values 0 and 1, then there exists a distribution Π on $(0,1)$ such that the joint distribution of $X_1, \ldots, X_n$ can be represented as*

$$P(X_1 = x_1, \ldots, X_n = x_n) = \int_0^1 \prod_{i=1}^n \theta^{x_i}(1-\theta)^{1-x_i} \, d\Pi(\theta).$$

This means X_i's can be thought of as i.i.d. $B(1, \theta)$ variables, given θ, where θ has the distribution Π. For a proof of this theorem and other results of this kind, see, for example, Bernardo and Smith (1994, Chapter 4).

The prior distribution Π can be determined in principle from the joint distribution of all the X_i's, but one would not know the joint distribution of all the X_i's. If one wants to actually elicit Π, one could ask oneself what is one's subjective predictive probability $P\{X_{i+1} = 1 | X_1, X_2, \ldots, X_i\}$. Suppose the subjective predictive probability is $(\alpha + \sum X_i)/(\alpha + \beta + i)$ where $\alpha > 0$, $\beta > 0$. Then the prior for θ is the Beta distribution with hyperparameters α and β. Nonparametric elicitations of this kind are considered in Fortini et al. (2000).

5.4 Elicitation of Hyperparameters for Prior

Although a full prior is not easy to elicit, one may be able to elicit hyperparameters in an assumed model for a prior. We discuss this problem somewhat informally in the context of two examples, a univariate normal and a bivariate normal likelihood.

Suppose $X_1, X_2, \ldots, X_n$ are i.i.d. $N(\mu, \sigma^2)$ and we assume a normal prior for μ given σ^2 and an inverse gamma prior for σ^2. How do we choose the hyperparameters? We think of a scenario where a statistician is helping a subject matter expert to articulate his judgment.

Let $p(\mu | \sigma^2)$ be normal with mean η and variance $c^2 \sigma^2$ where c is a constant. The hyperparameter η is a prior guess for the mean of X. The statistician has to make it clear that what is involved is not a guess about μ but a guess about the mean of μ. So the expert has to think of the mean μ itself as uncertain and subject to change.

Assuming that the expert can come up with a number for η, one may try to elicit a value for c in two different ways. If the expert can assign a range of variation for μ (given σ) and is pretty sure η will be in this range, one may equate this to $\eta \pm 3c\sigma$, η would be at the center of the range and distance of upper or lower limit from the center would be $3c\sigma$. To check consistency, one would like to elicit the range for several values of σ and see if one gets nearly the same c. In the second method for eliciting c, one notes that c^2 determines the amount of shrinking of $\bar{X}$ to the prior guess η in the posterior mean

$$E(\mu|\boldsymbol{X}) = \frac{\frac{n}{\sigma^2}\bar{X} + \frac{1}{c^2\sigma^2}\eta}{\frac{n}{\sigma^2} + \frac{1}{c^2\sigma^2}} = \frac{n\bar{X} + \eta/c^2}{n + 1/c^2}$$

(vide Example 2.1 of Section 2.2).

Thus if $c^2 = 1/n$, $\bar{X}$ and η have equal weights. If $c^2 = 5/n$, the weight of η diminishes to one fifth of the weight for $\bar{X}$. In most problems one would not have stronger prior belief.

We now discuss elicitation of hyperparameters for the inverse Gamma prior for σ^2 given by

$$p(\sigma^2) = \frac{1}{\Gamma(\alpha)\beta^\alpha} \frac{1}{(\sigma^2)^{\alpha+1}} e^{-1/(\beta\sigma^2)}.$$

The prior guess for σ^2 is $[\beta(\alpha-1)]^{-1}$. This is likely to be more difficult to elicit than the prior guess η about μ. The shape parameter α can be elicited by deciding how much to shrink the Bayes estimate of σ^2 towards the prior guess $[\beta(\alpha-1)]^{-1}$. Note that the Bayes estimate has the representation

$$E(\sigma^2|\boldsymbol{X}) = \frac{\alpha-1}{\alpha-1+n/2}\frac{1}{\beta(\alpha-1)} + \frac{(n-1)/2}{\alpha-1+n/2}s^2 + \frac{n(\bar{X}-\eta)^2}{(2\alpha+n-2)(1+nc^2)}.$$

where $(n-1)s^2 = \sum(X_i - \bar{X})^2$. In order to avoid dependence on $\bar{X}$ one may want to do the elicitation based on

$$E(\sigma^2|s^2) = \frac{\alpha-1}{\alpha-1+(n-1)/2}\frac{1}{\beta(\alpha-1)} + \frac{(n-1)/2}{\alpha-1+(n-1)/2}s^2.$$

The elicitation of prior for μ and σ, specially the means and variances of priors, may be based on examining related similar past data.

We turn now to i.i.d. bivariate normal data $(X_i, Y_i), i = 1, 2, \ldots, n$. There are now five parameters, (μ_X, σ_X^2), (μ_Y, σ_Y^2) and the correlation coefficient ρ. Also $E(Y|X = x) = \beta_0 + \beta_1 x$, $\mathrm{Var}(Y|X = x) = \sigma^2$, where $\sigma^2 = \sigma_Y^2(1 - \rho^2)$, $\beta_1 = \rho\sigma_Y/\sigma_X$, $\beta_0 = \mu_Y - (\rho\sigma_Y/\sigma_X)\mu_X$.

One could reparameterize in various ways. We adopt a parameterization that is appropriate when prediction of Y given X is a primary concern. We consider (μ_X, σ_X^2) as parameters for the marginal distribution of X, which may be handled as in the univariate case. We then consider the three parameters $(\sigma^2, \beta_0, \beta_1)$ of the conditional distribution of Y given $X = x$.

The joint density may be written as a product of the marginal density of X, namely $N(\mu_X, \sigma_X^2)$ and the conditional density of Y given $X = x$, namely $N(\beta_0 + \beta_1 x, \sigma^2)$. The full likelihood is

$$\prod_{i=1}^{n} \left[\frac{1}{\sqrt{2\pi}\sigma_X} \exp\left\{ -\frac{(x_i - \mu_X)^2}{2\sigma_X^2} \right\} \frac{1}{\sqrt{2\pi}\sigma} \exp\left\{ -\frac{1}{2\sigma^2}(y_i - \beta_0 - \beta_1 x_i)^2 \right\} \right].$$

It is convenient to rewrite $\beta_0 + \beta_1 x_i$ as $\gamma_0 + \gamma_1(x_i - \bar{x})$, with $\gamma_1 = \beta_1$, $\gamma_0 = \beta_0 + \gamma_1 \bar{x}$. Suppose we think of x_i's to be fixed. We concentrate on the second factor to find its conjugate prior. The conditional likelihood given the x_i's is

$$\left[\frac{1}{\sqrt{2\pi}\sigma} \right]^n \exp\{-\frac{1}{2\sigma^2} \sum (y_i - \hat{\gamma}_0 - \hat{\gamma}_1(x_i - \bar{x}))^2 - \frac{n}{2\sigma^2}(\hat{\gamma}_0 - \gamma_0)^2 - \frac{s_{xx}}{2\sigma^2}(\hat{\gamma}_1 - \gamma_1)^2\}$$

where $\hat{\gamma}_0 = \bar{y}$, $\hat{\gamma}_1 = \sum(y_i - \bar{y})(x_i - \bar{x})/s_{xx}$, and $s_{xx} = \sum(x_i - \bar{x})^2$.

Clearly, the conjugate prior for σ^2 is an inverse Gamma and the conjugate priors for γ_0 and γ_1 are independent normals whose parameters may be elicited along the same lines as those for the univariate normal except that more care is needed. The statistician could fix several values of x and invite the expert to guess the corresponding values of y. A straight line through the scatter plot will yield a prior guess on the linear relation between x and y. The slope of this line may be taken as the prior mean for β_1 and the intercept as the prior mean for β_0. These would provide prior means for γ_0, γ_1 (for the given values of x_i's in the present data). The prior variances can be determined by fixing how much shrinkage towards a prior mean is desired.

Suppose that the prior distribution of σ^2 has the density

$$p(\sigma^2) = \frac{1}{\Gamma(a)b^a} \frac{1}{(\sigma^2)^{a+1}} e^{-1/(b\sigma^2)}.$$

Given σ^2, the prior distributions for γ_0 and γ_1 are taken to be independent normals $N(\mu_0, c_0^2\sigma^2)$ and $N(\mu_1, c_1^2\sigma^2)$ respectively.

The marginal posterior distributions of γ_0 and γ_1 with these priors are Student's t with posterior means given by

$$E(\gamma_0|\boldsymbol{x}, \boldsymbol{y}) = \frac{n}{n + c_0^{-2}}\hat{\gamma}_0 + \frac{c_0^{-2}}{n + c_0^{-2}}\mu_0 \tag{5.34}$$

$$\text{and } E(\gamma_1|\boldsymbol{x}, \boldsymbol{y}) = \frac{S_{xx}}{S_{xx} + c_1^{-2}}\hat{\gamma}_1 + \frac{c_1^{-2}}{S_{xx} + c_1^{-2}}\mu_1. \tag{5.35}$$

As indicated above, these expressions may be used to elicit the values of c_0 and c_1. For elicitation of the shape parameter a of the prior distribution of σ^2 we may use similar representation of $E(\sigma^2|\boldsymbol{x}, \boldsymbol{y})$. Note that the statistics $S^2 = \sum(y_i - \hat{\gamma}_0 - \hat{\gamma}_1(x_i - \bar{x}))^2$, $\hat{\gamma}_0$ and $\hat{\gamma}_1$ are jointly sufficient for $(\sigma^2, \gamma_0, \gamma_1)$.

Table 5.1. Data on Water Flow (in 100 Cubic Feet per Second) at Two Points (Libby and Newgate) in January During 1931–43 in Kootenai River (Ezekiel and Fox, 1959)

Year	1931	32	33	34	35	36	37	38	39	40	41	42	43
Newgate(y)	19.7	18.0	26.1	44.9	26.1	19.9	15.7	27.6	24.9	23.4	23.1	31.3	23.8
Libby(x)	27.1	20.9	33.4	77.6	37.0	21.6	17.6	35.1	32.6	26.0	27.6	38.7	27.8

As in the univariate normal case considered above we do the elicitation based on

$$E(\sigma^2|S^2) = \frac{a-1}{(n/2)+a-2}\frac{1}{b(a-1)} + \frac{(n/2)-1}{(n/2)+a-2}\hat{\sigma}^2 \qquad (5.36)$$

where $\hat{\sigma}^2 = S^2/(n-2)$ is the classical estimate of σ^2. We illustrate with an example.

Example 5.7. Consider the bivariate data of Table 5.1 (Ezekiel and Fox, 1959). This relates to water flow in Kootenai River at two points, Newgate (British Columbia, Canada) and Libby (Montana, USA). A dam was being planned on the river at Newgate, B.C., where it crossed the Canadian border. The question was how the flow at Newgate could be estimated from that at Libby.

Consider the above setup for this set of bivariate data. Calculations yield $\bar{x} = 32.5385$, $S_{xx} = 2693.1510$, $\hat{\gamma}_0 = 24.9615$, $\hat{\gamma}_1 = 0.4748$ and $\hat{\sigma}^2 = 3.186$ so that the classical (least squares) regression line is

$$y = 24.9615 + 0.4748(x - \bar{x}),$$
$$\text{i.e., } y = 9.5122 + 0.4748x.$$

Suppose now that we have similar past data D for a number of years, say, the previous decade, for which the fitted regression line is given by

$$y = 10.3254 + 0.4809x$$

with an estimate of error variance (σ^2) 3.9363. We don't present this set of "historical data" here, but a scatter plot is shown in Figure 5.1. As suggested above, we may take the prior means for β_0, β_1 and σ^2 to be 10.3254, 0.4809, and 3.9363, respectively. We, however, take these to be 10.0, 0.5, and 4.0 as these are considered only as prior guesses. This gives $\mu_0 = 10 + 0.5\bar{x} = 26.2693$ and $\mu_1 = 0.5$. Given that the historical data set D was obtained in the immediate past (before 1931), we have considerable faith in our prior guess, and as indicated above, we set the weights for the prior means μ_0, μ_1, and $1/(b(a-1))$ of γ_0, γ_1, and σ^2 in (5.34), (5.35), and (5.36) equal to 1/6 so that the ratio of the weights for the prior estimate and the classical estimate is $1:5$ in each case. Thus we set

$$\frac{c_0^{-2}}{n+c_0^{-2}} = \frac{c_1^{-2}}{S_{xx}+c_1^{-2}} = \frac{a-1}{(n/2)+a-2} = \frac{1}{6}$$

which yields, with $n = 13$ and $S_{xx} = 2693.151$ for the current data, $c_0^{-2} = 2.6$, $c_1^{-2} = 538.63$, and $a = 2.1$. If, however, the data set D was older, we would attach less weight (less than $1/6$) to the prior means. Our prior guess for σ^2 is $1/(b(a - 1))$. Equating this to 4.0 we get $b = 0.2273$. Now from (5.34)–(5.36) we obtain the Bayes estimates of γ_0, γ_1 and σ^2 as 25.1795, 0.4790 and 3.3219 respectively. The estimated regression line is

$$y = 25.1795 + 0.4790(x - \bar{x}),$$

$$\text{i.e., } y = 9.5936 + 0.4790x.$$

The scatter plots for the current Kootenai data of Table 5.1 and the historical data D as well as the classical and Bayesian regression line estimates derived above are shown in Figure 5.1. The symbols "∘" for current and "∗" for historical data are used here. The continuous line stands for the Bayesian regression line based on the current data and the prior, and the broken line represents the classical regression line based on the current data. The Bayesian line seems somewhat more representative of the whole data set than the classical regression line, which is based on the current data only. If one fits a classical regression line to the whole data set, it will attach equal importance to the current and the historical data; it is a choice between all or nothing. The Bayesian method has the power to handle both current data and other available information in a flexible way.

The 95% HPD credible intervals for γ_0 and γ_1 based on the posterior t-distributions are respectively (21.4881, 28.8708) and (0.2225, 0.7355), which are comparable with the classical 95% confidence intervals — (23.8719, 26.0511) for γ_0 and (0.3991, 0.5505) for γ_1. Note that, as expected, the Bayesian intervals are more conservative than the classical ones, the Bayesian providing for the possible additional variation in the parameters. If one uses the objective prior $p(\gamma_0, \gamma_1, \sigma^2) \propto 1/\sigma^2$, the objective Bayes solutions would agree with the classical estimates and confidence intervals.

All of the above would be inapplicable if x and y have the same footing and the object is estimation of the parameters in the model rather than prediction of unobserved y's for given x's. In this case, one would write the bivariate normal likelihood

$$\prod_{i=1}^{n} \frac{1}{2\pi\sigma_X\sigma_Y\sqrt{1-\rho^2}}$$

$$\times \exp\left\{-\frac{1}{2(1-\rho^2)}\left(\frac{(x_i - \mu_X)^2}{\sigma_X^2} + \frac{(y_i - \mu_Y)^2}{\sigma_Y^2} - 2\rho\frac{(x_i - \mu_X)}{\sigma_X}\frac{(y_i - \mu_Y)}{\sigma_Y}\right)\right\}.$$

The conjugate prior is a product of a bivariate normal and an inverse-Wishart distribution. Notice that we have discussed elicitation of hyperparameters for a conjugate prior for several parameters. Instead of substituting these elicited values in the conjugate prior, we could treat the hyperparameters as having

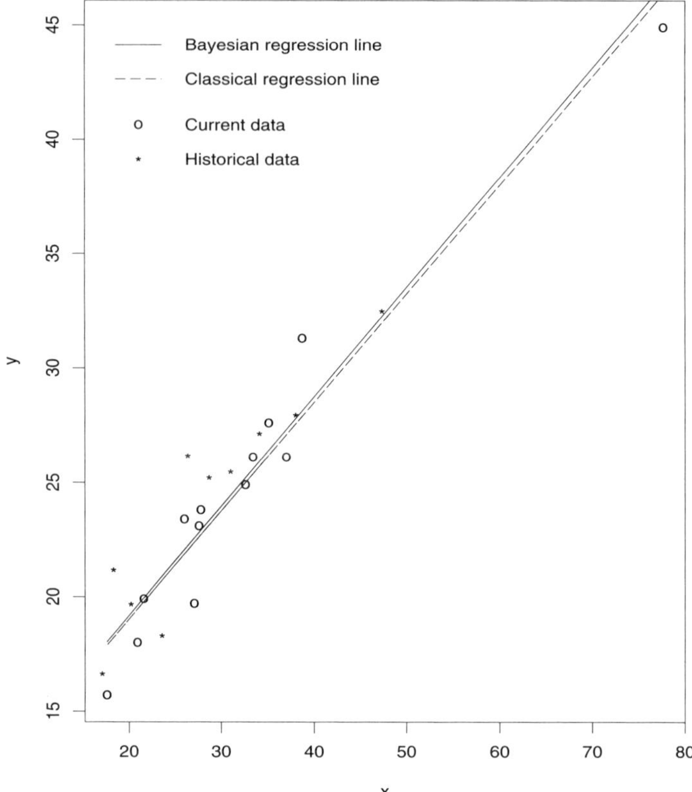

Fig. 5.1. Scatter plots and regression lines for the Kootenai River data.

a prior distribution over a set of values around the elicited numbers. The prior distribution for the hyperparameters could be a uniform distribution on the set of values around the elicited numbers. This would be a hierarchical prior. An alternative would be to use several conjugate priors with different hyperparameter values from the set around the elicited numbers and check for robustness.

Elicitation of hyperparameters of a conjugate prior for a linear model is treated in Kadane et al. (1980), Garthwaite and Dickey (1988, 1992), etc. A recent review is Garthwaite et al. (2005). Garthwaite and Dickey (1988) observe that the prior variance-covariance matrix of the regression coefficients, specially the off-diagonal elements of the matrix, are the most difficult to elicit. We have assumed the off-diagonal elements are zero, a common simplifying assumption, and determined the diagonal elements by eliciting how much shrinkage towards the prior is sought in the Bayes estimates of the means of the regression coefficients. Garthwaite and Dickey (1988) indicate an indirect way of eliciting the variance-covariance matrix.

5.5 A New Objective Bayes Methodology Using Correlation

As we have already seen, there are many approaches for deriving objective, reference priors and also for conducting default Bayesian analysis. One such approach that relies on some new developments is discussed here. Using the Pearson correlation coefficient in a rather different way, DasGupta et al. (2000) and Delampady et al. (2001) show that some of its properties can lead to substantial developments in Bayesian statistics.

Suppose X is distributed with density $f(x|\theta)$ and θ has a prior π. Let the joint probability distribution of X and θ under the prior π be denoted by P. We can then consider the Pearson correlation coefficient ρ_P between two functions $g_1(X, \theta)$ and $g_2(X, \theta)$ under this probability distribution P. An objective prior in the spirit of reference prior can then be derived by maximizing the correlation between two post-data summaries about the parameter θ, namely the posterior density and the likelihood function. Given a class Γ of priors, Delampady et al. (2001) show that the prior π that maximizes $\rho_P\{f(x|\theta), \pi(\theta|x)\}$ is the one with the least Fisher information $I(\pi) = E^\pi\{\frac{d}{d\theta}\log \pi(\theta)\}^2$ in the class Γ. Actually, Delampady et al. (2001) note that it is very difficult to work with the exact correlation coefficient $\rho_P\{f(x|\theta), \pi(\theta|x)\}$ and hence they maximize an appropriate large sample approximation by assuming that the likelihood function and the prior density are sufficiently smooth. The following example is from Delampady et al. (2001).

Example 5.8. Consider a location parameter θ with $|\theta| \leq 1$. Assume that f and π are sufficiently smooth. Then the prior density which achieves the minimum Fisher information in the class of priors compactly supported on $[-1, 1]$ is what is desired. Bickel (1981) and Huber (1974) show that this prior is

$$\pi(\theta) = \begin{cases} \cos^2(\pi\theta/2), & \text{if } |\theta| \leq 1; \\ 0, & \text{otherwise.} \end{cases}$$

Thus, the Bickel prior is the default prior under the correlation criterion. The variational result that obtains this prior as the one achieving the minimum Fisher information was rediscovered by Ghosh and Bhattacharya in 1983 (see Ghosh (1994); a proof of this interesting result is also given there).

In addition to the above problem, it is also possible to identify a robust estimate of θ using this approach. Suppose π is a reference prior, and Γ is a class of plausible priors. π may or may not belong to Γ. Let δ_ν be the Bayes estimate of θ with respect to prior $\nu \in \Gamma$. To choose an optimal Bayes

Table 5.2. Values of $\delta_\pi(X)$ and $\delta_\nu(X)$ for Various $X = x$

x	0	.5	1	1.5	2	3	5	8	10	15
$\delta_\pi(x)$	0	.349	.687	1.002	1.284	1.735	2.197	2.216	2.065	1.645
$\delta_\nu(x)$	0	.348	.683	.993	1.267	1.685	1.976	1.60	1.15	.481

estimate, maximize (over $\nu \in \Gamma$) the correlation coefficient $\rho_P(\theta, \delta_\nu)$ between θ and δ_ν. Note that ρ_P is calculated under P, and in this joint distribution θ follows the reference prior π. Again, by maximizing an appropriate large sample approximation $\hat\rho_P(\theta, \delta_\nu)$ (assuming that the likelihood function and the prior density are sufficiently smooth), Delampady et al. (2001) obtain the following theorem.

Theorem 5.9. *The estimate $\delta_\nu(X)$ maximizing $\hat\rho_P(\theta, \delta_\nu)$ is Bayes with respect to the prior density*

$$\nu(\theta) = c\pi(\theta) \exp\left\{ -\frac{1}{2}\, \tau^2(\theta - \mu)^2 \right\}, \tag{5.37}$$

where μ, τ^2 are arbitrary and c is a normalizing constant.

The interesting aspect of this reference prior ν is evident from the fact that it is simply a product of the initial reference prior π and a Gaussian factor. This may be interpreted as ν being the posterior density when one begins with a flat prior π and it is revised with an observation θ from the Gaussian distribution to pull in its tails. Consider the following example again from Delampady et al. (2001).

Example 5.10. Consider the reference prior density $\pi(\theta) \propto (1 + \theta^2/3)^{-2}$, density of the Student's t_3 prior, a flat prior. Suppose that the family Γ contains only symmetric priors and so $\nu(\theta)$ is of the form $\nu(\theta) = c(1 + \theta^2/3)^{-2} \exp\left\{-\theta^2/(2\tau^2)\right\}$. Let $X \sim \text{Cauchy}(\theta, \sigma)$, with known σ and having density

$$f(x|\theta) = \frac{1}{\sigma\pi} \left\{ 1 + \left(\frac{x - \theta}{\sigma} \right)^2 \right\}^{-1}.$$

Some selected values are reported in Table 5.2 for $\sigma = 0.2$. For small and moderate values of x, δ_π and δ_ν behave similarly, whereas for large values, δ_ν results in much more shrinkage than δ_π. This is only expected because the penultimate ν has normal tails, whereas the reference π has flat tails.

5.6 Exercises

1. Let $X \sim B(n, p)$. Choose a prior on p such that the marginal distribution of X is uniform on $\{0, 1, \ldots, n\}$.

2. (Schervish (1995, p.121)). Let h be a function of (x, θ) that is differentiable in θ. Define a prior

$$p^*(\theta) = [\mathrm{Var}_\theta \left((\partial/\partial\theta) h(X, \theta) \right)]^{1/2}.$$

   (a) Show that $p^*(\theta)$ satisfies the invariance condition (5.1).
   (b) Choose a suitable h such that $p^*(\theta)$ is the Jeffreys prior.
3. Prove (5.16) and generalize it to the multiparameter case.
4. (Lehmann and Casela (1998)) For a scale family, show that there is an equivariant estimate of σ^k that minimizes $E(T - \sigma^k)^2/\sigma^{2k}$. Display the estimate as the ratio of two integrals and interpret as a Bayes estimate.
5. Consider a multinomial distribution.
   (a) Show that the Dirichlet distribution is a conjugate prior.
   (b) Identify the precision parameter for the Dirichlet prior distribution.
   (c) Let the precision parameter go to zero and identify the limiting prior and posterior. Suggest why the limiting posterior, but not the limiting prior, is used in objective Bayesian analysis.
6. Find the Jeffreys prior for the multinomial model.
7. Find the Jeffreys prior for the multivariate normal model with unknown mean vector and dispersion matrix.
8. (a) Examine why the Jeffreys prior may not be appropriate if the parameter is not identifiable over the full parameter space.
   (b) Show with an example that the Jeffreys prior may not have a proper posterior. (Hint. Try the following mixture: $X = 0$ with probability $1/2$ and is $N(\mu, 1)$ with probability $1/2$.)
   (c) Suggest a heuristic reason as to why the posterior is often proper if we use a Jeffreys prior.
9. Bernardo has recently proposed the use of $\min\{K(f_0, f_1), K(f_1, f_0)\}$, instead of $K(f_0, f_1)$, as the criterion to maximize at each stage of reference prior. Examine the consequence of this change for the examples of reference priors discussed in this chapter.
10. Given (μ, σ^2), let $X_1, \ldots, X_n$ be i.i.d. $N(\mu, \sigma^2)$ and consider the prior $\pi(\mu, \sigma^2) \propto 1/\sigma^2$. Verify that

$$\begin{aligned}
&P\{\bar{X} - t_{\alpha/2, n-1} s/\sqrt{n} \leq \mu \leq \bar{X} + t_{\alpha/2, n-1} s/\sqrt{n} | \mu, \sigma^2\} \\
&= P\{\bar{X} - t_{\alpha/2, n-1} s/\sqrt{n} \leq \mu \leq \bar{X} + t_{\alpha/2, n-1} s/\sqrt{n} | X_1, \ldots, X_n\} \\
&= 1 - \alpha,
\end{aligned}$$

   where $\bar{X}$ is the sample mean, $s^2 = \sum(X_i - \bar{X})^2/(n-1)$, and $t_{\alpha/2}$ is the upper $\alpha/2$ point of t_{n-1}, $0 < \alpha < 1$.
11. Given $0 < \theta < 1$, let $X_1, \ldots, X_n$ be i.i.d. $B(1, \theta)$. Consider the Jeffreys prior for θ. Find by simulation the frequentist coverage of θ by the two-tailed 95% credible interval for $\theta = \frac{1}{8}, \frac{1}{4}, \frac{1}{2}, \frac{3}{4}, \frac{7}{8}$. Do the same for the usual frequentist interval $\hat{\theta} \pm z_{0.025}\sqrt{\hat{\theta}(1-\hat{\theta})/n}$ where $\hat{\theta} = \sum X_i/n$.
12. Derive (5.32) from an appropriate probability matching equation.

6

Hypothesis Testing and Model Selection

For Bayesians, model selection and model criticism are extremely important inference problems. Sometimes these tend to become much more complicated than estimation problems. In this chapter, some of these issues will be discussed in detail. However, all models and hypotheses considered here are low-dimensional because high-dimensional models need a different approach. The Bayesian solutions will be compared and contrasted with the corresponding procedures of classical statistics whenever appropriate. Some of the discussion in this chapter is technical and it will not be used in the rest of the book. Those sections that are very technical (or otherwise can be omitted at first reading) are indicated appropriately. These include Sections 6.3.4, 6.4, 6.5, and 6.7. In Sections 6.2 and 6.3, we compare frequentist and Bayesian approaches to hypothesis testing. We do the same in an asymptotic framework in Section 6.4. Recently developed methodologies such as the Bayesian P-value and some non-subjective Bayes factors are discussed in Sections 6.5 and 6.7.

6.1 Preliminaries

First, let us recall some notation from Chapter 2 and also let us introduce some specific notation for the discussion that follows.

Suppose X having density $f(x|\boldsymbol{\theta})$ is observed, with $\boldsymbol{\theta}$ being an unknown element of the parameter space Θ. Suppose that we are interested in comparing two models M_0 and M_1, which are given by

$$M_0 : X \text{ has density } f(x|\boldsymbol{\theta}) \text{ where } \boldsymbol{\theta} \in \Theta_0;$$
$$M_1 : X \text{ has density } f(x|\boldsymbol{\theta}) \text{ where } \boldsymbol{\theta} \in \Theta_1. \tag{6.1}$$

For $i = 0, 1$ let $g_i(\boldsymbol{\theta})$ be the prior density of $\boldsymbol{\theta}$, conditional on M_i being the true model. Then, to compare models M_0 and M_1 on the basis of a random sample $\mathbf{x} = (x_1, \ldots, x_n)$ one would use the Bayes factor

$$B_{01}(\mathbf{x}) = \frac{m_0(\mathbf{x})}{m_1(\mathbf{x})}, \tag{6.2}$$

where

$$m_i(\mathbf{x}) = \int_{\Theta_i} f(\mathbf{x}|\boldsymbol{\theta})g_i(\boldsymbol{\theta})\,d\boldsymbol{\theta}, \quad i = 0, 1. \tag{6.3}$$

We also use the notation BF_{01} for the Bayes factor. Recall from Chapter 2 that the Bayes factor is the ratio of posterior odds ratio of the hypotheses to the corresponding prior odds ratio. Therefore, if the prior probabilities of the hypotheses, $\pi_0 = P^\pi(M_0) = P^\pi(\Theta_0)$ and $\pi_1 = P^\pi(M_1) = P^\pi(\Theta_1) = 1 - \pi_0$ are specified, then as in (2.17),

$$P(M_0|x) = \left\{ 1 + \frac{1 - \pi_0}{\pi_0} B_{01}^{-1}(x) \right\}^{-1}. \tag{6.4}$$

Thus, if conditional prior densities g_0 and g_1 can be specified, one should simply use the Bayes factor B_{01} for model selection. If, further π_0 is also specified, the posterior odds ratio of M_0 to M_1 can also be utilized. However, these computations may not always be easy to perform, even when the required prior ingredients are fully specified. A possible solution is the use of BIC as an approximation to a Bayes factor. We study this in Subsection 6.1.1. The situation can get much worse when the task of specifying these prior inputs itself becomes a difficult problem as in the following problem.

Example 6.1. Consider the problem that is usually called nonparametric regression. Independent responses y_i are observed along with covariates x_i, $i = 1, \ldots, n$. The model of interest is

$$y_i = g(x_i) + \epsilon_i, i = 1, \ldots, n, \tag{6.5}$$

where ϵ_i are i.i.d. $N(0, \sigma^2)$ errors with unknown error variance σ^2. The function g is called the regression function. In linear regression, g is *a priori* assumed to be linear in a set of finite regression coefficients. In general, g can be assumed to be fully unknown also. Now, if model selection involves choosing g from two different fully nonparametric classes of regression functions, this becomes a very difficult problem. Computation of Bayes factor or posterior odds ratio is then a formidable task. Various simplifications including reducing g to be semi-parametric have been studied. In such cases, some of these problems can be handled.

Consider a different model checking problem now, that of testing for normality. This is a very common problem encountered in frequentist inference, because much of the inferential methodology is based on the normality assumption. Simple or multiple linear regression, ANOVA, and many other techniques routinely use this assumption. In its simplest form, the problem can be stated as checking whether a given random sample $X_1, X_2, \ldots, X_n$ arose from a population having the normal distribution. In the setup given above in (6.1), we may write it as

$$M_0 : X \text{ is } N(\mu, \sigma^2) \text{ with arbitrary } \mu \text{ and } \sigma^2 > 0;$$

$$M_1 : X \text{ does not have the normal distribution.} \qquad (6.6)$$

However, this looks quite different from (6.1) above, because M_1 does not constitute a parametric alternative. Hence it is not clear how to use Bayes factors or posterior odds ratios here for model checking. The difficulty with this model checking problem is clear: one is only interested in M_0 and not in M_1.

This problem is addressed in Section 6.3 of Gelman et al. (1995). See also Section 9.9. We use the posterior predictive distribution of replicated future data to assess whether the predictions show systematic differences. In practice, replicated data will not be available, so cross-validation of some form has to be used, as discussed in Section 9.9. Gelman et al. (1995) have not used cross-validation and their P-values have come in for some criticism (see Section 6.5).

The object of model checking is not to decide whether the model is true or false but to check whether the model provides plausible approximation to the data. It is clear that we have to use posterior predictive values and Bayesian P-values of some sort, but consensus on details does not seem to have emerged yet. It remains an important problem.

6.1.1 BIC Revisited

Under appropriate regularity conditions on f, g_0, and g_1, the Bayes factor given in (6.2) can be approximated using the Laplace approximation or the *saddle point* approximation. Let us change notation and express (6.3) as follows:

$$m_i(\mathbf{x}) = \int f(\mathbf{x}|\boldsymbol{\theta}_i) g_i(\boldsymbol{\theta}_i) \, d\boldsymbol{\theta}_i, i = 0, 1. \qquad (6.7)$$

where $\boldsymbol{\theta}_i$ is the p_i-dimensional vector of parameters under M_i, assumed to be independent of n (the dimension of the observation vector $\mathbf{x}$). Let $\widetilde{\boldsymbol{\theta}}_i$ be the posterior mode of $\boldsymbol{\theta}_i$, $i = 0, 1$. Assume $\widetilde{\boldsymbol{\theta}}_i$ is an interior point of Θ_i. Then, expanding the logarithm of the integrand in (6.7) around $\widetilde{\boldsymbol{\theta}}_i$ using a second-order Taylor series approximation, we obtain

$$\log\left(f(\mathbf{x}|\boldsymbol{\theta}_i) g_i(\boldsymbol{\theta}_i)\right) \approx \log\left(f(\mathbf{x}|\widetilde{\boldsymbol{\theta}}_i) g_i(\widetilde{\boldsymbol{\theta}}_i)\right) - \frac{1}{2}\left(\boldsymbol{\theta}_i - \widetilde{\boldsymbol{\theta}}_i\right)' H_{\widetilde{\boldsymbol{\theta}}_i}\left(\boldsymbol{\theta}_i - \widetilde{\boldsymbol{\theta}}_i\right),$$

where $H_{\widetilde{\boldsymbol{\theta}}_i}$ is the corresponding negative Hessian. Applying this approximation to (6.7) yields,

$$m_i(\mathbf{x}) \approx f(\mathbf{x}|\widetilde{\boldsymbol{\theta}}_i)g_i(\widetilde{\boldsymbol{\theta}}_i) \int \exp\left\{-\frac{1}{2}\left(\boldsymbol{\theta}_i - \widetilde{\boldsymbol{\theta}}_i\right)' H_{\widetilde{\boldsymbol{\theta}}_i}\left(\boldsymbol{\theta}_i - \widetilde{\boldsymbol{\theta}}_i\right)\right\} d\boldsymbol{\theta}_i$$

$$= f(\mathbf{x}|\widetilde{\boldsymbol{\theta}}_i)g_i(\widetilde{\boldsymbol{\theta}}_i)(2\pi)^{p_i/2}|H_{\widetilde{\boldsymbol{\theta}}_i}^{-1}|^{1/2}. \tag{6.8}$$

$2\log B_{01}$ is a commonly used evidential measure to compare the support provided by the data $\mathbf{x}$ for M_0 relative to M_1. Under the above approximation we have,

$$2\log(B_{01}) \approx 2\log\left(\frac{f(\mathbf{x}|\widetilde{\boldsymbol{\theta}}_0)}{f(\mathbf{x}|\widetilde{\boldsymbol{\theta}}_1)}\right) + 2\log\left(\frac{g_0(\widetilde{\boldsymbol{\theta}}_0)}{g_1(\widetilde{\boldsymbol{\theta}}_1)}\right)$$
$$+(p_0 - p_1)\log(2\pi) + \log\left(\frac{|H_{\widetilde{\boldsymbol{\theta}}_0}^{-1}|}{|H_{\widetilde{\boldsymbol{\theta}}_1}^{-1}|}\right).$$

A variation of this approximation is also commonly used, where instead of the posterior mode $\widetilde{\boldsymbol{\theta}}_i$, the maximum likelihood estimate $\widehat{\boldsymbol{\theta}}_i$ is employed. Then, instead of (6.8), one obtains

$$m_i(\mathbf{x}) \approx f(\mathbf{x}|\widehat{\boldsymbol{\theta}}_i)g_i(\widehat{\boldsymbol{\theta}}_i)(2\pi)^{p_i/2}|H_{\widehat{\boldsymbol{\theta}}_i}^{-1}|^{1/2}. \tag{6.9}$$

Here $H_{\widehat{\boldsymbol{\theta}}_i}$ is the observed Fisher information matrix evaluated at the maximum likelihood estimator. If the observations are i.i.d. we have that $H_{\widehat{\boldsymbol{\theta}}_i} = nH_{1,\widehat{\boldsymbol{\theta}}_i}$, where $H_{1,\widehat{\boldsymbol{\theta}}_i}$ is the observed Fisher information matrix obtained from a single observation. In this case,

$$m_i(\mathbf{x}) \approx f(\mathbf{x}|\widehat{\boldsymbol{\theta}}_i)g_i(\widehat{\boldsymbol{\theta}}_i)(2\pi)^{p_i/2}n^{-p_i/2}|H_{1,\widehat{\boldsymbol{\theta}}_i}^{-1}|^{1/2},$$

and hence

$$2\log(B_{01}) \approx 2\log\left(\frac{f(\mathbf{x}|\widehat{\boldsymbol{\theta}}_0)}{f(\mathbf{x}|\widehat{\boldsymbol{\theta}}_1)}\right) + 2\log\left(\frac{g_0(\widehat{\boldsymbol{\theta}}_0)}{g_1(\widehat{\boldsymbol{\theta}}_1)}\right)$$
$$-(p_0 - p_1)\log\frac{n}{2\pi} + \log\left(\frac{|H_{1,\widehat{\boldsymbol{\theta}}_0}^{-1}|}{|H_{1,\widehat{\boldsymbol{\theta}}_1}^{-1}|}\right). \tag{6.10}$$

An approximation to (6.10) correct to $O(1)$ is

$$2\log(B_{01}) \approx 2\log\left(\frac{f(\mathbf{x}|\widehat{\boldsymbol{\theta}}_0)}{f(\mathbf{x}|\widehat{\boldsymbol{\theta}}_1)}\right) - (p_0 - p_1)\log n. \tag{6.11}$$

This is the approximate Bayes factor based on the Bayesian information criterion (BIC) due to Schwarz (1978). The term $(p_0 - p_1)\log n$ can be considered a penalty for using a more complex model.

A related criterion is

$$2 \log \left(\frac{f(\mathbf{x}|\widehat{\boldsymbol{\theta}}_0)}{f(\mathbf{x}|\widehat{\boldsymbol{\theta}}_1)} \right) - 2(p_0 - p_1) \tag{6.12}$$

which is based on the Akaike information criterion (AIC), namely,

$$AIC = 2 \log f(\mathbf{x}|\widehat{\boldsymbol{\theta}}) - 2p$$

for a model $f(\mathbf{x}|\boldsymbol{\theta})$. The penalty for using a complex model is not as drastic as that in BIC.

A Bayesian interpretation of AIC for high-dimensional prediction problems is presented in Chapter 9. Problem 16 of Chapter 9 invites you to explore if AIC is suitable for low-dimensional testing problems.

6.2 P-value and Posterior Probability of H_0 as Measures of Evidence Against the Null

One particular tool from classical statistics that is very widely used in applied sciences for model checking or hypothesis testing is the P-value. It also happens to be one of the concepts that is highly misunderstood and misused. The basic idea behind R.A. Fisher's (see Fisher (1973)) original (1925) definition of P-value given below did have a great deal of appeal: It is the probability under a (simple) null hypothesis of obtaining a value of a test statistic that is at least as extreme as that observed in the sample data.

Suppose that it is desired to test

$$H_0 : \theta = \theta_0 \text{ versus } H_1 : \theta \neq \theta_0, \tag{6.13}$$

and that a classical significance test is available and is based on a test statistic $T(X)$, large values of which are deemed to provide evidence against the null hypothesis. If data $X = x$ is observed, with corresponding $t = T(x)$, the P-value then is

$$\alpha = P_{\theta_0}\left(T(X) \geq T(x) \right).$$

Example 6.2. Suppose we observe $X_1, \ldots, X_n$ i.i.d. from $N(\theta, \sigma^2)$, where σ^2 is known. Then $\bar{X}$ is sufficient for θ and it has the $N(\theta, \sigma^2/n)$ distribution. Noting that $T = T(\bar{X}) = |\sqrt{n}\left(\bar{X} - \theta_0 \right)/\sigma|$, is a natural test statistic to test (6.13), one obtains the usual P-value as $\alpha = 2[1 - \Phi(t)]$, where $t = |\sqrt{n}\left(\bar{x} - \theta_0 \right)/\sigma|$ and Φ is the standard normal cumulative distribution function.

Fisher meant P-value to be used informally as a measure of degree of surprise in the data relative to H_0. This use of P-value as a post-experimental or conditional measure of statistical evidence seems to have some intuitive justification. From a Bayesian point of view, various objections have been raised by Edwards et al. (1963), Berger and Sellke (1987), and Berger and Delampady (1987), against use of P-values as measures of evidence against H_0. A recent review is Ghosh et al. (2005).

To a Bayesian the posterior probability of H_0 summarizes the evidence against H_0. In many of the common cases of testing, the P-value is smaller than the posterior probability by an order of magnitude. The reason for this is that the P-value ignores the likelihood of the data under the alternative and takes into account not only the observed deviation of the data from the null hypothesis as measured by the test statistic but also more significant deviations. In view of these facts, one may wish to see if P-values can be calibrated in terms of bounds for posterior probabilities over natural classes of priors. It appears that calibration takes the form of a search for an alternative measure of evidence based on posterior that may be acceptable to a non-Bayesian. In this connection, note that there is an interesting discussion of the admissibility of P-value as a measure of evidence in Hwang et al. (1992).

6.3 Bounds on Bayes Factors and Posterior Probabilities

6.3.1 Introduction

We begin with an example where P-values and the posterior probabilities are very different.

Example 6.3. We observe $\bar{X} \sim N(\theta, \sigma^2/n)$, with known σ^2. Upon using $T = |\sqrt{n}\left(\bar{X} - \theta_0\right)/\sigma|$ as the test statistic to test (6.13), recall that the P-value comes out to be $\alpha = 2[1 - \Phi(t)]$, where $t = |\sqrt{n}\left(\bar{x} - \theta_0\right)/\sigma|$ and Φ is the standard normal cumulative distribution function. On the set $\{\theta \neq \theta_0\}$, let θ have the density (g_1) of $N(\mu, \tau^2)$. Then, we have,

$$B_{01} = \sqrt{1 + \rho^{-2}} \exp\left\{ -\frac{1}{2}\left[\frac{(t - \rho\eta)^2}{(1 + \rho^2)} - \eta^2 \right]\right\},$$

where $\rho = \sigma/(\sqrt{n}\tau)$ and $\eta = (\theta_0 - \mu)/\tau$. Now, if we choose $\mu = \theta_0$, $\tau = \sigma$ and $\pi_0 = 1/2$, we get,

$$B_{01} = \sqrt{1 + \rho^{-2}} \exp\left\{ -\frac{1}{2}\left[\frac{t^2}{(1 + \rho^2)} \right]\right\}.$$

For various values of t and n, the different measures of evidence, $\alpha =$ P-value, $B =$ Bayes factor, and $P = P(H_0|x)$ are displayed in Table 6.1 as shown in Berger and Delampady (1987). It may be noted that the posterior probability of H_0 varies between 4 and 50 times the corresponding P-value which is an indication of how different these two measures of evidence can be.

Table 6.1. Normal Example: Measures of Evidence

		n										
		1		5		10		20		50		100	
t	α	B	P	B	P	B	P	B	P	B	P	B	P
1.645	.10	.72	.42	.79	.44	.89	.47	1.27	.56	1.86	.65	2.57	.72
1.960	.05	.54	.35	.49	.33	.59	.37	.72	.42	1.08	.52	1.50	.60
2.576	.01	.27	.21	.15	.13	.16	.14	.19	.16	.28	.22	.37	.27
3.291	.001	.10	.09	.03	.03	.02	.02	.03	.03	.03	.03	.05	.05

6.3.2 Choice of Classes of Priors

Clearly, there are irreconcilable differences between the classical P-value and the corresponding Bayesian measures of evidence in the above example. However, one may argue that the differences are perhaps due to the choice of π_0 or g_1 that cannot claim to be really 'objective.' The choice of $\pi_0 = 1/2$ may not be crucial because the Bayes factor, B, which does not need this, seems to be providing the same conclusion, but the choice of g_1 does have substantial effect. To counter this argument, let us consider lower bounds on B and P over wide classes of prior densities. What is surprising is that even these lower bounds that are based on priors 'least favorable' to H_0 are typically an order of magnitude larger than the corresponding P-values for precise null hypotheses. The other motivation for looking at bounds over classes of priors is that they correspond with robust Bayesian answers that are more compelling when an objective choice for a single prior does not exist. Thus, in the case of precise null hypotheses, if G is the class of all plausible conditional prior densities g_1 under H_0, we are then lead to the consideration of the following bounds.

$$\underline{B}(G, x) = \inf_{g \in G} B_{01} = \frac{f(x|\theta_0)}{\sup_{g \in G} m_g(x)}, \tag{6.14}$$

where $m_g(x) = \int_{\theta \neq \theta_0} f(x|\theta) g(\theta) \, d\theta$, and

$$\underline{P}(H_0|G, x) = \inf_{g \in G} P(H_0|x) = \left[1 + \frac{1 - \pi_0}{\pi_0} \underline{B}(G, x)^{-1} \right]^{-1}. \tag{6.15}$$

This brings us back to the question of choice of the class G as in Chapter 3, where the robust Bayesian approach has been discussed. As explained there, robustness considerations force us to consider classes that are neither too large nor too small. Choosing the class $G_A = \{$all densities$\}$ certainly is very extreme because it allows densities that are severely biased towards H_1. Quite often, the class $G_{NC} = \{$all natural conjugate densities with mean $\theta_0\}$ is an interesting class to consider. However, this turns out to be inadequate for robustness considerations. The following class

$$G_{US} = \{\text{all densities symmetric about } \theta_0 \text{ and non-increasing in } |\theta - \theta_0|\} \tag{6.16}$$

which strikes a balance between these two extremes seems to be a good choice. Because we are comparing various measures of evidence, it is informative to examine the lower bounds for each of these classes. In particular, we can gather the magnitudes of the differences between these measures across the classes. To simplify proofs of some of the results given below, we restate a result indicated in Section 3.8.1.

Lemma 6.4. *Suppose C_T is a set of prior probability measures on $\mathcal{R}^p$ given by $C_T = \{\nu_t : t \in T\}$, $T \subset \mathcal{R}^d$, and let $\mathcal{C}$ be the convex hull of C_T. Then*

$$\sup_{\pi \in \mathcal{C}} \int_\Theta f(x|\theta)\, d\pi(\theta) = \sup_{t \in T} \int_\Theta f(x|\theta)\, d\nu_t(\theta). \tag{6.17}$$

Proof. Because $\mathcal{C} \supset C_T$, LHS $\geq$ RHS in (6.17). However, as $\int f(x|\theta)\, d\pi(\theta) = \int f(x|\theta) \int_T \nu_t(d\theta)\mu(dt)$, for some probability measure μ on T, using Fubini's theorem,

$$\int_\Theta f(x|\theta)\, d\pi(\theta) = \int_\Theta f(x|\theta) \int_T d\nu_t(\theta)\, d\mu(t)$$

$$= \int_T \left(\int_\Theta f(x|\theta)\, d\nu_t(\theta) \right) d\mu(t)$$

$$\leq \sup_{t \in T} \int_\Theta f(x|\theta)\, d\nu_t(\theta).$$

Therefore,

$$\sup_{\pi \in \mathcal{C}} \int_\Theta f(x|\theta)\, d\pi(\theta) \leq \sup_{t \in T} \int_\Theta f(x|\theta)\, d\nu_t(\theta),$$

yielding the other inequality also. $\square$

The following results are from Berger and Sellke (1987) and Edwards et al. (1963).

Theorem 6.5. *Let $\hat{\theta}(x)$ be the maximum likelihood estimate of θ for the observed value of x. Then*

$$\underline{B}(G_A, x) = \frac{f(x|\theta_0)}{f(x|\hat{\theta}(x))}, \tag{6.18}$$

$$\underline{P}(H_0|G_A, x) = \left[1 + \frac{1 - \pi_0}{\pi_0} \underline{B}(G_A, x)^{-1} \right]^{-1}. \tag{6.19}$$

In view of Lemma 6.4, the proof of this result is quite elementary, once it is noted that the extreme points of G_A are point masses.

Theorem 6.6. *Let $\mathcal{U}_S$ be the class of all uniform distributions symmetric about θ_0. Then*

$$\underline{B}(G_{US}, x) = \underline{B}(\mathcal{U}_S, x), \tag{6.20}$$

$$\underline{P}(H_0|G_{US}, x) = \underline{P}(H_0|\mathcal{U}_S, x). \tag{6.21}$$

Proof. Simply note that any unimodal symmetric distribution is a mixture of symmetric uniforms, and apply Lemma 6.4 again.   □

Because $\underline{B}(\mathcal{U}_S, x) = f(x|\theta_0)/\sup_{g \in \mathcal{U}_S} m_g(x)$, computation of

$$\sup_{g \in \mathcal{U}_S} m_g(x) = \sup_{g \in \mathcal{U}_S} \int f(x|\theta)g(\theta)\, d\theta$$

is required to employ Theorem 6.6. Also, it may be noted that as far as robustness is considered, using the class G_{US} of all symmetric unimodal priors is the same as using the class $\mathcal{U}_S$ of all symmetric uniform priors. It is perhaps reasonable to assume that many of these uniform priors are somewhat biased against H_0, and hence we should consider unimodal symmetric prior distributions that are smoother. One possibility is scale mixtures of normal distributions having mean θ_0. This class is substantially larger than just the class of normals centered at θ_0; it includes Cauchy, all Student's t and so on. To obtain the lower bounds, however, it is enough to consider

$$G_{Nor} = \{ \text{ all normal distributions with mean } \theta_0 \},$$

in view of Lemma 6.4.

Example 6.7. Let us continue with Example 6.3. We have the following results from Berger and Sellke (1987) and Edwards et al. (1963).
(i) $\underline{B}(G_A, x) = \exp(-\frac{t^2}{2})$, because the MLE of θ is $\bar{x}$; hence
(ii) $\underline{P}(H_0|G_A, x) = \left[1 + \frac{1-\pi_0}{\pi_0} \exp(\frac{t^2}{2})\right]^{-1}$.
(iii) If $t \leq 1$, $\underline{B}(G_{US}, x) = 1$, and $\underline{P}(H_0|G_{US}, x) = \pi_0$. This is because in this case, the unimodal symmetric distribution that maximizes $m_g(x)$ is the degenerate distribution that puts all its mass at θ_0.
(iv) If $t > 1$, the $g \in G_{US}$ that maximizes $m_g(x)$ is non-degenerate and from Theorem 6.6 and Example 3.4,

$$\underline{B}(G_{US}, x) = \frac{\phi(t)}{\sup_{u>0} \frac{1}{2u}\{\Phi(u - t) - \Phi(-(u + t))\}}.$$

(v) If $t \leq 1$, $\underline{B}(G_{Nor}, x) = 1$, and $\underline{P}(H_0|G_{Nor}, x) = \pi_0$. If $t > 1$,

$$\underline{B}(G_{Nor}, x) = t \exp(-\frac{(t^2 - 1)}{2}).$$

For various values of t, the different measures of evidence, $\alpha = $ P-value, $\underline{B} = $ lower bound on Bayes factor, and $\underline{P} = $ lower bound on $P(H_0|x)$ are displayed in Table 6.2. π_0 has been chosen to be 0.5.

What we note is that the differences between P-values and the corresponding Bayesian measures of evidence remain irreconcilable even when the lower bounds on such measures are considered. In other words, even the least possible Bayes factor and posterior probability of H_0 are substantially larger than the corresponding P-value. This is so, even for the choice G_A, which is rather astonishing (see Edwards et al. (1963)).

Table 6.2. Normal Example: Lower Bounds on Measures of Evidence

t	α	G_A		G_{US}		G_{Nor}	
		$\underline{B}$	$\underline{P}$	$\underline{B}$	$\underline{P}$	$\underline{B}$	$\underline{P}$
1.645	.10	.258	.205	.639	.390	.701	.412
1.960	.05	.146	.128	.408	.290	.473	.321
2.576	.01	.036	.035	.122	.109	.153	.133
3.291	.001	.0044	.0044	.018	.018	.024	.0235

6.3.3 Multiparameter Problems

It is not the case that the discrepancies between P-values and lower bounds on Bayes factor or posterior probability of H_0 are present only for tests of precise null hypotheses in single parameter problems. This phenomenon is much more prevalent. We shall present below some simple multiparameter problems where similar discrepancies have been discovered. The following result on testing a p-variate normal mean vector is from Delampady (1986).

Example 6.8. Suppose $\mathbf{X} \sim N_p(\boldsymbol{\theta}, I)$, where $\mathbf{X} = (X_1, X_2, \ldots, X_p)$ and $\boldsymbol{\theta} = (\theta_1, \theta_2, \ldots, \theta_p)$. It is desired to test

$$H_0 : \boldsymbol{\theta} = \boldsymbol{\theta}^0 \text{ versus } H_1 : \boldsymbol{\theta} \neq \boldsymbol{\theta}^0,$$

where $\boldsymbol{\theta}^0 = (\theta_1^0, \theta_2^0, \ldots, \theta_p^0)$ is a specified vector. The classical test statistic is

$$T(\mathbf{X}) = ||\mathbf{X} - \boldsymbol{\theta}^0||^2,$$

which has a χ_p^2 distribution under H_0. Thus the P-value of the data $\mathbf{x}$ is

$$\alpha = P(\chi_p^2 \geq T(\mathbf{x})).$$

Consider the class G_{USP} of unimodal spherically symmetric (about $\boldsymbol{\theta}^0$) prior distributions for $\boldsymbol{\theta}$, the natural generalization G_{US}. This will consist of densities $g(\boldsymbol{\theta})$ of the form $g(\boldsymbol{\theta}) = h((\boldsymbol{\theta} - \boldsymbol{\theta}^0)'(\boldsymbol{\theta} - \boldsymbol{\theta}^0))$, where h is non-increasing. Noting that any unimodal spherically symmetric distribution is a mixture of uniforms on symmetric spheres, and applying Lemma 6.4, we obtain

$$\sup_{g \in G_{USP}} m_g(\mathbf{x}) = \sup_{k>0} \frac{1}{V(k)} \int_{||\boldsymbol{\theta}-\boldsymbol{\theta}^0|| \leq k} f(\mathbf{x}|\boldsymbol{\theta}) \, d\boldsymbol{\theta},$$

where $V(k)$ is the volume of a sphere of radius k, and $f(\mathbf{x}|\boldsymbol{\theta})$ is the $N_p(\boldsymbol{\theta}, I)$ density. Therefore, we have that,

$$\underline{B}(G_{USP}, x) = \frac{\exp(-\frac{1}{2}||\mathbf{x} - \boldsymbol{\theta}^0||^2)}{\sup_{k>0} \frac{1}{V(k)} \int_{||\boldsymbol{\theta}-\boldsymbol{\theta}^0|| \leq k} \exp(-\frac{1}{2}||\mathbf{x} - \boldsymbol{\theta}||^2) \, d\boldsymbol{\theta}}.$$

Using this result, numerical values were computed for different dimensions, p and different P-values, α. In Table 6.3 we present these values where

Table 6.3. Multivariate Normal Example: Lower Bounds on Measures of Evidence

p	$\alpha = .001$		$\alpha = .01$		$\alpha = .05$		$\alpha = .10$	
	B	P	B	P	B	P	B	P
1	.018	.018	.122	.109	.409	.290	.639	.390
2	.014	.014	.098	.089	.348	.258	.570	.363
3	.012	.012	.090	.083	.326	.246	.540	.351
4	.011	.011	.085	.078	.314	.239	.523	.344
5	.010	.010	.082	.076	.307	.235	.513	.339
10	.009	.009	.078	.072	.293	.226	.491	.329
15	.009	.009	.075	.070	.288	.223	.483	.326
20	.009	.009	.074	.069	.284	.221	.478	.324
30	.009	.009	.074	.069	.281	.219	.473	.321
40	.009	.009	.073	.068	.279	.218	.471	.320
∞	.009	.009	.073	.068	.279	.218	.468	.319

$\underline{B}$ denotes $\underline{B}(G_{USP}, x)$ and $\underline{P}$ denotes $\underline{P}(H_0|G_{USP}, x)$ for $\pi_0 = 0.5$. As can be readily seen, the lower bounds remain substantially larger than the corresponding P-values in all dimensions.

Note that spherical symmetry is not the only generalization of symmetry from one dimension to higher dimensions. Very different answers can be obtained if, for example, elliptical symmetry is used instead. Suppose we consider densities of the form $g(\boldsymbol{\theta}) = \sqrt{|Q|}h((\boldsymbol{\theta} - \boldsymbol{\theta}^0)'Q(\boldsymbol{\theta} - \boldsymbol{\theta}^0))$, where Q is an arbitrary positive definite matrix and h is non-increasing. Then the following result, which is informally stated in Delampady and Berger (1990), obtains. For the sake of simplicity, let us take $\boldsymbol{\theta}^0 = \mathbf{0}$.

Theorem 6.9. *Let $f(\mathbf{x}|\boldsymbol{\theta})$ be a multivariate, multiparameter density. Consider the class of elliptically symmetric unimodal prior densities*

$$G_{UES} = \left\{ g : g(\boldsymbol{\theta}) = |Q|^{\frac{1}{2}} h(\boldsymbol{\theta}'Q\boldsymbol{\theta}), h \text{ non-increasing, } Q \text{ positive definite} \right\}. \tag{6.22}$$

Then

$$\sup_{g \in G_{UES}} m_g(\mathbf{x}) = \sup_{Q>0} \left\{ \sup_{k>0} \frac{1}{V(k)} \int_{||\mathbf{u}|| \leq k} f(\mathbf{x}|Q^{-\frac{1}{2}}\mathbf{u}) \, d\mathbf{u} \right\}, \tag{6.23}$$

where $V(k)$ is the volume of a sphere of radius k, and $Q > 0$ denotes that Q is positive definite.

Proof. Note that

$$\sup_{g \in G_{UES}} m_g(\mathbf{x}) = \sup_{g \in G_{UES}} \int f(\mathbf{x}|\boldsymbol{\theta})g(\boldsymbol{\theta}) \, d\boldsymbol{\theta}$$

$$= \sup_{h,Q} \int f(\mathbf{x}|\boldsymbol{\theta})h(\boldsymbol{\theta}'Q\boldsymbol{\theta})|Q|^{\frac{1}{2}} \, d\boldsymbol{\theta}$$

$$= \sup_{Q>0} \left\{ \sup_{h} \int f(\mathbf{x}|Q^{-\frac{1}{2}}\mathbf{u})h(\mathbf{u}'\mathbf{u})\,d\mathbf{u} \right\} \qquad (6.24)$$

$$= \sup_{Q>0} \left\{ \sup_{k>0} \frac{1}{V(k)} \int_{||\mathbf{u}||\leq k} f(\mathbf{x}|Q^{-\frac{1}{2}}\mathbf{u})\,d\mathbf{u} \right\},$$

because the maximization of the inside integral over non-increasing h in (6.24) is the same as maximization of that integral over the class of unimodal spherically symmetric densities, and hence Lemma 6.4 applies.   □

Consider the above result in the context of Example 6.8. The lower bounds on the Bayes factor as well as the posterior probability of the null hypothesis will be substantially lower if we use the class G_{UES} rather than G_{USP}. This is immediate from (6.23), because the lower bounds over G_{USP} correspond with the maximum in (6.23) with $Q = I$. The result also questions the suitability of G_{UES} for these lower bounds in view of the fact that the lower bounds will correspond with prior densities that are extremely biased towards H_1.

Many other esoteric classes of prior densities have also been considered by some for deriving lower bounds. In particular, generalization of symmetry from the single-parameter case to the multiparameter case has been examined. DasGupta and Delampady (1990) consider several subclasses of the symmetric star-unimodal densities. Some of these are mixtures of uniform distributions on $\mathcal{L}_p$ (for $p = 1, 2, \infty$), class of distributions with components that are independent symmetric unimodal distributions and a certain subclass of one-symmetric distributions. Note that mixtures of uniform distributions on $\mathcal{L}_2$ balls are simply unimodal spherically symmetric distributions, whereas mixtures of uniform distributions on $\mathcal{L}_1$ balls contain distributions whose components are i.i.d. exponential distributions. Uniform distributions on hypercubes form a subclass of mixtures of uniforms on $\mathcal{L}_\infty$ balls. Also considered there is the larger subclass consisting of distributions whose components are identical symmetric unimodal distributions. Another class of one-symmetric distributions considered there is of interest because it contains distributions whose components are i.i.d. Cauchy. Even though studies such as these are important from robustness considerations, we feel that they do not necessarily add to our understanding of possible interpretation of P-values from a robust Bayesian point of view. However, interested readers will find that Dharmadhikari and Joag-Dev (1988) is a good source for multivariate unimodality, and Fang et al. (1990) is a good reference for multivariate symmetry for material related to the classes mentioned above.

We have noted earlier that computation of Bayes factor and posterior probability is difficult when parametric alternatives are not available. Many frequentist statisticians claim that P-values are valuable when there are no alternatives explicitly specified, as is common with tests of fit. We consider this issue here for a particularly common test of fit, the chi-squared test of goodness of fit. It will be observed that alternatives do exist implicitly, and hence Bayes factors and posterior probabilities can indeed be computed. The following results from Delampady and Berger (1990) once again point out the discrepancies between P-values and Bayesian measures of evidence.

Example 6.10. Let $\mathbf{n} = (n_1, n_2, \ldots, n_k)$ be a sample of fixed size $N = \sum_{i=1}^{k} n_i$ from a k-cell multinomial distribution with unknown cell probabilities $\mathbf{p} = (p_1, p_2, \ldots, p_k)$ and density (mass) function

$$f(\mathbf{n}|\mathbf{p}) = \frac{N!}{\prod_{i=1}^{k} n_i!} \prod_{i=1}^{k} p_i^{n_i}.$$

Consider testing

$$H_0 : \mathbf{p} = \mathbf{p}^0 \text{ versus } H_1 : \mathbf{p} \neq \mathbf{p}^0,$$

where $\mathbf{p}^0 = (p_1^0, p_2^0, \ldots, p_k^0)$ is a specified interior point of the k-dimensional simplex. Instead of focusing on the exact multinomial setup, the most popular approach is to use the chi-squared approximation. Here the test statistic of interest is

$$T_N = \sum_{i=1}^{k} \frac{(n_i - Np_i^0)^2}{Np_i^0}$$

which has the asymptotic distribution (as $N \to \infty$) of χ_{k-1}^2 under H_0. To compare P-values so obtained with the corresponding robust Bayesian measures of evidence, the following are two natural classes of prior distributions to consider.

(i) The conjugate class G_C of Dirichlet priors with density

$$g(\mathbf{p}) = \frac{\Gamma(\sum_{i=1}^{k} \alpha_i)}{\prod_{i=1}^{k} \Gamma(\alpha_i)} \prod_{i=1}^{k} p_i^{\alpha_i - 1},$$

where $\alpha_i > 0$ satisfy

$$\frac{1}{\sum_{i=1}^{k} \alpha_i} (\alpha_1, \alpha_2, \ldots, \alpha_k)' = E^g(\mathbf{p}) = \mathbf{p}^0.$$

(ii) Consider the following transform of $(p_1, p_2, \ldots, p_{k-1})'$:

$$\mathbf{u} = \mathbf{u}(\mathbf{p}) = \left(\frac{p_1 - p_1^0}{p_1}, \frac{p_2 - p_2^0}{p_2}, \ldots, \frac{p_{k-1} - p_{k-1}^0}{p_{k-1}} \right)'$$
$$+ \left(\frac{p_k^0 - p_k}{\sqrt{p_k} + p_k} \right) \left(\sqrt{p_1}, \sqrt{p_2}, \ldots, \sqrt{p_{k-1}} \right)'.$$

The justification (see Delampady and Berger (1990)) for using such a transform is that its range is $\mathcal{R}^{k-1}$ unlike that of $\mathbf{p}$ and its likelihood function is more symmetric and closer to a multivariate normal. Now let

$$G_{USP}^* = \{\text{unimodal } g^*(\mathbf{u}) \text{ that are spherically symmetric about } \mathbf{0}\},$$

and consider the class of prior densities g obtained by transforming back to the original parameter:

$$G_{TUSP} = \left\{ g(\mathbf{p}) = g^*(\mathbf{u}(\mathbf{p}))|\frac{\partial \mathbf{u}(\mathbf{p})}{\partial \mathbf{p}}| \right\}.$$

Delampady and Berger (1990) show that as $N \to \infty$, the lower bounds on Bayes factors over G_C and G_{TUS} converge to those corresponding with the multivariate normal testing problem (chi-squared test) in Example 6.8, thus proving that irreconcilability of P-values and Bayesian measures of evidence is present in goodness of fit problems as well.

Additional discussion of the multinomial testing problem with mixture of conjugate priors can be found in Good (1965, 1967, 1975). Edwards et al. (1963) discuss the possibility of finding lower bounds on Bayes factors over the conjugate class of priors for the binomial problem. Extensive discussion of the binomial problem and further references can be found in Berger and Delampady (1987).

6.3.4 Invariant Tests[1]

A natural generalization of the symmetry assumption (on the prior distribution) is invariance under a group of transformations. Such a generalization and many examples can be found in Delampady (1989a). A couple of those examples will be discussed below to show the flavor of the results. The general results that utilize sophisticated mathematical arguments will be skipped, and instead interested readers are referred to the source indicated above. For a good discussion on invariance of statistical decision rules, see Berger (1985a).

Recall that the random observable $\mathbf{X}$ takes values in a space $\mathcal{X}$ and has density (mass) function $f(\mathbf{x}|\boldsymbol{\theta})$. The unknown parameter is $\boldsymbol{\theta} \in \Theta \subseteq \mathcal{R}^n$, for some positive integer n. It is desired to test $H_0 : \boldsymbol{\theta} \in \Theta_0$ versus $H_1 : \boldsymbol{\theta} \in \Theta_1$. We assume the following in addition.

(i) There is a group $\mathcal{G}$ (of transformations) acting on $\mathcal{X}$ that induces a group $\bar{\mathcal{G}}$ (of transformations acting) on Θ. These two groups are isomorphic (see Section 5.1.7) and elements of $\mathcal{G}$ will be denoted by g, those of $\bar{\mathcal{G}}$ by $\bar{g}$.

(ii) $f(g\mathbf{x}|\bar{g}\boldsymbol{\theta}) = f(\mathbf{x}|\boldsymbol{\theta})k(g)$ for a suitable continuous map k (from $\mathcal{G}$ to $(0, \infty)$).

(iii) $\bar{g}\Theta_0 = \Theta_0$, $\bar{g}\Theta_1 = \Theta_1$, $\bar{g}\Theta = \Theta$.

In this context, the following concept of a maximal invariant is needed.

Definition. When a group G of transformations acts on a space $\mathcal{X}$, a function $T(\mathbf{x})$ on $\mathcal{X}$ is said to be invariant (with respect to G) if $T(g(\mathbf{x})) = T(\mathbf{x})$ for all $\mathbf{x} \in \mathcal{X}$ and $g \in G$. A function $T(x)$ is maximal invariant (with respect to G) if it is invariant and further

$$T(\mathbf{x}_1) = T(\mathbf{x}_2) \text{ implies } \mathbf{x}_1 = g(\mathbf{x}_2) \text{ for some } g \in G.$$

---

[1] Section 6.3.4 may be omitted at first reading.

This means that G divides $\mathcal{X}$ into orbits where invariant functions are constant. A maximal invariant assigns different values to different orbits.

Now from (i), we have that the action of $\mathcal{G}$ and $\bar{\mathcal{G}}$ induce maximal invariants $t(\mathbf{X})$ on $\mathcal{X}$ and $\eta(\boldsymbol{\theta})$ on Θ, respectively.

Remark 6.11. The family of densities $f(\mathbf{x}|\boldsymbol{\theta})$ is said to be invariant under $\mathcal{G}$ if (ii) is satisfied. The testing problem $H_0 : \theta \in \Theta_0$ versus $H_1 : \theta \in \Theta_1$ is said to be invariant under $\mathcal{G}$ if in addition (iii) is also satisfied.

Example 6.12. Consider Example 6.8 again and suppose $\mathbf{X} \sim N_p(\boldsymbol{\theta}, I)$. It is desired to test
$$H_0 : \boldsymbol{\theta} = \mathbf{0} \text{ versus } H_1 : \boldsymbol{\theta} \neq \mathbf{0}.$$

This testing problem is invariant under the group $\mathcal{G}_O$ of all orthogonal transformations; i.e., if H is an orthogonal matrix of order p, then $g_H \mathbf{X} = H\mathbf{X} \sim N_p(H\boldsymbol{\theta}, I)$, so that $\bar{g}_H \boldsymbol{\theta} = H\boldsymbol{\theta}$. Further,

$$f(\mathbf{x}|\boldsymbol{\theta}) = (2\pi)^{-p/2} \exp(-\frac{1}{2}(\mathbf{x} - \boldsymbol{\theta})'(\mathbf{x} - \boldsymbol{\theta})), \text{ and}$$

$$f(g_H \mathbf{x}|\bar{g}_H \boldsymbol{\theta}) = (2\pi)^{-p/2} \exp(-\frac{1}{2}(H\mathbf{x} - H\boldsymbol{\theta})'(H\mathbf{x} - H\boldsymbol{\theta}))$$
$$= (2\pi)^{-p/2} \exp(-\frac{1}{2}(\mathbf{x} - \boldsymbol{\theta})'(\mathbf{x} - \boldsymbol{\theta}))$$
$$= f(\mathbf{x}|\boldsymbol{\theta}),$$

so that (ii) is satisfied. Also, $\bar{g}_H \mathbf{0} = \mathbf{0}$, and (iii) too is satisfied.

Example 6.13. Let $X_1, X_2, \cdots, X_n$ be a random sample from $N\left(\theta, \sigma^2\right)$ with both θ and σ unknown. The problem is to test the hypothesis $H_0 : \theta = 0$ against $H_1 : \theta \neq 0$. A sufficient statistic for (θ, σ) is $\mathbf{x} = (\bar{X}, S)$, $\bar{X} = \sum_1^n X_i/n$ and $S = \left[\sum_1^n (X_i - \bar{X})^2/n\right]^{1/2}$. Then

$$f(\mathbf{x}|\theta, \sigma) = K\sigma^{-n} S^{n-2} \exp(-n\left[(\bar{X} - \theta)^2 + S^2\right]/(2\sigma^2)),$$

where K is a constant. Also,

$$\mathcal{X} = \{(\bar{x}, s) : \bar{x} \in \mathcal{R}, s > 0\}, \text{ and } \Theta = \{(\theta, \sigma) : \theta \in \mathcal{R}, \sigma > 0\}.$$

The problem is invariant under the group $G = \{g_c = c : c > 0\}$, where the action of g_c is given by $g_c(x) = c(\bar{x}, s) = (c\bar{x}, cs)$. Note that $f(g_c x|\theta, \sigma) = c^{-2} f(x|\theta, \sigma)$.

A number of technical conditions in addition to the assumptions (i)–(iii) yield a very useful representation for the density of the maximal invariant statistic $t(\mathbf{X})$. Note that this density, $q(t(\mathbf{x})|\eta(\boldsymbol{\theta}))$, depends on the parameter $\boldsymbol{\theta}$ only through the maximal invariant in the parameter space, $\eta(\boldsymbol{\theta})$.

The technique involved in the derivation of these results uses an averaging over a relevant group. The general method of this kind of averaging is due to Stein (1956), but because there are a number of mathematical problems to overcome, various different approaches were discovered as can be seen in Wijsman (1967, 1985, 1986), Andersson (1982), Andersson et al. (1983), and Farrell (1985). For further details, see Eaton (1989), Kariya and Sinha (1989), and Wijsman (1990). The specific conditions and proofs of these results can be found in the above references. In particular, the groups considered here are *amenable groups* as presented in detail in Bondar and Milnes (1981). See also Delampady (1989a). The orthogonal group, and the group of location-scale transformations are amenable. The multiplicative group of non-singular $p \times p$ linear transformations is not.

Let us return to the issue of comparison of P-values and lower bounds on Bayes factors and posterior probabilities (of hypotheses) in this setup. We note that it is necessary to reduce the problem by using invariance for any meaningful comparison because the classical test statistic and hence the computation of P-value are already based on this. Therefore, the natural class of priors to be used for this comparison is the class G_I of G-invariant priors; i.e., those priors π that satisfy

(iv) $\pi(A) = \pi(A\bar{g})$.

Theorem 6.14. *If $\mathcal{G}$ is a group of transformations satisfying certain regularity conditions (see Delampady (1989a)),*

$$\underline{B}(G_I, x) = \frac{\inf_{\eta_1 \in \Theta_0/\mathcal{G}} q\left(t(\mathbf{x})|\eta_1\right)}{\sup_{\eta_2 \in \Theta_1/\mathcal{G}} q\left(t(\mathbf{x})|\eta_2\right)}, \tag{6.25}$$

where $\Theta/\mathcal{G}$ denotes the space of maximal invariants on the parameter space.

Corollary 6.15. *If $\Theta_0/\mathcal{G} = \{0\}$, then under the same conditions as in Theorem 6.14,*

$$\underline{B}(G_I, x) = \frac{q\left(t(\mathbf{x})|0\right)}{\{\sup_{\eta \in \Theta/\mathcal{G}} q\left(t(\mathbf{x})|\eta\right)\}}.$$

Example 6.16. (Example 6.12, continued.) Consider the class of all priors that are invariant under orthogonal transformations, and note that this class is simply the class of all spherically symmetric distributions. Now, application of Corollary 6.15 yields,

$$\underline{B}(G_I, x) = \frac{q\left(t(\mathbf{x})|0\right)}{q\left(t(\mathbf{x})|\hat{\eta}\right)},$$

where $q(t|\eta)$ is the density of a noncentral χ^2 random variable with p degrees of freedom and non-centrality parameter η, and $\hat{\eta}$ is the maximum likelihood estimate of η from data $t(\mathbf{x})$. For selected values of p the lower bounds, $\underline{B}$ and $\underline{P}$ (for $\pi_0 = 0.5$) are tabulated against their P-values in Table 6.4.

Table 6.4. Invariant Test for Normal Means

	$\alpha = 0.01$		$\alpha = 0.05$	
p	$\underline{B}$	$\underline{P}$	$\underline{B}$	$\underline{P}$
3	.0749	.0697	.2913	.2256
4	.0745	.0693	.2903	.2250
15	.0737	.0686	.2842	.2213
20	.0734	.0684	.2821	.2200

Notice that the lower bounds on the posterior probabilities of the null hypothesis are anywhere from 4 to 7 times as large as the corresponding P-values, indicating that there is a vast discrepancy between P-values and posterior probabilities. This is the same phenomenon as what was seen in Table 6.3. What is, however, interesting is that the class of priors considered here is larger and contains the one considered there, but the magnitude of the discrepancy is about the same.

Example 6.17. (Example 6.13, continued.) In the normal example with unknown variance, we have the maximal invariants $t(\mathbf{x}) = \overline{x}/s$ and $\eta(\theta, \sigma) = \theta/\sigma$. If we define,

$$G_I = \{\pi : d\pi(\theta, \sigma) = h_1(\eta)d\eta\frac{d\sigma}{\sigma}, h_1 \text{ is any density for } \eta\},$$

we obtain,

$$\underline{B}(G_I, x) = \frac{q\left(t(\mathbf{x})|0\right)}{q\left(t(\mathbf{x})|\hat{\eta}\right)},$$

where $q(t|\eta)$ is the density of a noncentral Student's t random variable with $n-1$ degrees of freedom, and non-centrality parameter η, and $\hat{\eta}$ is the maximum likelihood estimate of η. The fact that all the necessary conditions (which are needed to apply the relevant results) are satisfied is shown in Andersson (1982) and Wijsman (1967). For selected values of the lower bounds are tabulated along with the P-values in Table 6.5.

For small values of n, the lower bounds in Table 6.5 are comparable with the corresponding P-values, whereas as n gets large the differences between these lower bounds and the P-values get larger. See also in this connection Section 6.4.

There is substantial literature on Bayesian testing of a point null. Among these are Jeffreys (1957, 1961), Good (1950, 1958, 1965, 1967, 1983, 1985, 1986), Lindley (1957, 1961, 1965, 1977), Raiffa and Schlaiffer (1961), Edwards et al. (1963), Hildreth (1963), Smith (1965), Zellner (1971, 1984), Dickey (1971, 1973, 1974, 1980), Lempers (1971), Rubin (1971), Leamer (1978), Smith and Spiegelhalter (1980), Zellner and Siow (1980), and Diamond and Forrester (1983). Related work can also be found in Pratt (1965), DeGroot (1973), Dempster (1973), Dickey (1977), Bernardo (1980), Hill (1982), Shafer (1982), and Berger (1986).

Table 6.5. Test for Normal Mean, Variance Unknown

	$\alpha = 0.01$		$\alpha = 0.05$		$\alpha = 0.10$	
n	$\underline{B}$	$\underline{P}$	$\underline{B}$	$\underline{P}$	$\underline{B}$	$\underline{P}$
2	.0117	.0116	.0506	.0482	.0939	.0858
8	.0137	.0135	.0941	.0860	.2114	.1745
12	.0212	.0208	.1245	.1107	.2309	.1876
16	.0290	.0282	.1301	.1151	.2375	.1919
32	.0327	.0317	.1380	.1213	.2478	.1986

Invariance and Minimaxity

Our focus has been on deriving bounds on Bayes factors for invariant testing problems. There is, however, a large literature on other aspects of invariant tests. For example, if the group under consideration satisfies the technical condition of amenability and hence the Hunt-Stein theorem is valid, then the minimax invariant test is minimax among all tests. We do not discuss these results here. For details on this and other related results we would like to refer the interested readers to Berger (1985a), Kiefer (1957, 1966), and Lehmann (1986).

6.3.5 Interval Null Hypotheses and One-sided Tests

Closely related to a sharp null hypothesis $H_0 : \theta = \theta_0$ is an interval null hypothesis $H_0 : |\theta - \theta_0| \leq \epsilon$. Dickey (1976) and Berger and Delampady (1987) show that the conflict between P-values and posterior probabilities remains if ϵ is sufficiently small. The precise order of magnitude of small ϵ depends on the sample size n.

One may also ask similar questions of possible conflict between P-values and posterior probabilities for one-sided null, say, $H_0 : \theta \leq \theta_0$ versus $H_1 : \theta > \theta_0$. In the case of $\theta = $ mean of a normal, and the usual uniform prior, direct calculation shows the P-value equals posterior probability. On the other hand, Casella and Berger (1987) show in general the two are not the same and the P-value may be smaller or greater depending on the family of densities in the model. Incidentally, the ambiguity of an improper prior discussed in Section 6.7 does not apply to one-sided nulls. In this case the Bayes factor remains invariant if the improper prior is multiplied by an arbitray constant.

6.4 Role of the Choice of an Asymptotic Framework[2]

This section is based on Ghosh et al. (2005). Suppose $X_1, \ldots, X_n$ are i.i.d. $N(\theta, \sigma^2)$, σ^2 known, and consider the problem of testing $H_0 : \theta = \theta_0$ versus

---

[2] Section 6.4 may be omitted at first reading.

$H_1 : \theta \neq \theta_0$. If instead of taking a lower bound as in the previous sections, we take a fixed prior density $g_1(\theta)$ under H_1 but let n go to ∞, then the conflict between P-values and posterior probabilities is further enhanced. Historically this phenomenon was noted earlier than the conflict with the lower bound, vide Jeffreys (1961) and Lindley (1957).

Let g_1 be a uniform prior density over some interval $(\theta_0 - a, \theta_0 + a)$ containing θ_0. The posterior probability of H_0 given $\boldsymbol{X} = (X_1, \ldots, X_n)$ is

$$P(H_0|\boldsymbol{X}) = \pi_0 \exp[-n(\bar{X} - \theta_0)^2/(2\sigma^2)]/K,$$

where π_0 is the specified prior probability of H_0 and

$$K = \pi_0 \exp[-n(\bar{X} - \theta_0)^2/(2\sigma^2)] + \frac{1 - \pi_0}{2a} \int_{\theta_0 - a}^{\theta_0 + a} \exp[-n(\bar{X} - \theta)^2/(2\sigma^2)]d\theta.$$

Suppose $\boldsymbol{X}$ is such that $\bar{X} = \theta_0 + z_\alpha \sigma/\sqrt{n}$ where z_α is the $100(1 - \alpha)\%$ quantile of $N(0,1)$. Then $\bar{X}$ is significant at level α. Also, for sufficiently large n, $\bar{X}$ is well within $(\theta_0 - a, \theta_0 + a)$ because $\bar{X} - \theta_0$ tends to zero as n increases. This leads to

$$\int_{\theta_0 - a}^{\theta_0 + a} \exp[-n(\bar{X} - \theta)^2/(2\sigma^2)]d\theta \cong \sigma\sqrt{(2\pi/n)}$$

and hence

$$P(H_0|\boldsymbol{X}) \cong \pi_0 \exp(-z_\alpha^2/2)/[\pi_0 \exp(-z_\alpha^2/2) + \frac{(1 - \pi_0)}{2a}\sigma\sqrt{(2\pi/n)}].$$

Thus $P(H_0|\boldsymbol{X}) \to 1$ as $n \to \infty$ whereas the P-value is equal to α for all n. This is known as the Jeffreys-Lindley paradox. It may be noted that the same phenomenon would arise with any flat enough prior in place of uniform.

Indeed, P-values cannot be compared across sample sizes or across experiments, see Lindley (1957), Ghosh et al. (2005). Even a frequentist tends to agree that the conventional values of the significance level α like $\alpha = 0.05$ or 0.01 are too large for large sample sizes. This point is further discussed below.

The Jeffreys-Lindley paradox shows that for inference about θ, P-values and Bayes factors may provide contradictory evidence and hence can lead to opposite decisions. Once again, as mentioned in Section 6.3, the evidence against H_0 contained in P-values seems unrealistically high. We argue in this section that part of this conflict arises from the fact that different types of asymptotics are being used for the Bayes factors and the P-values. We begin with a quick review of the two relevant asymptotic frameworks in classical statistics for testing a sharp null hypothesis.

The standard asymptotics of classical statistics is based on what are called Pitman alternatives, namely, $\theta_n = \theta_0 + d/\sqrt{n}$ at a distance of $O(1/\sqrt{n})$ from the null. The Pitman alternatives are also called contiguous in the very general asymptotic theory developed by Le Cam (vide Roussas (1972), Le Cam and

Yang (2000), Hájek and Sidák (1967)). The log-likelihood ratio of a contiguous alternative with respect to the null is stochastically bounded as $n \to \infty$. On the other hand, for a fixed alternative, the log-likelihood ratio tends to $-\infty$ (under the null) or ∞ (under the fixed alternative). If the probability of Type 1 error is $0 < \alpha < 1$, then the behavior of the likelihood ratio has the following implication. The probability of Type 2 error will converge to $0 < \beta < 1$ under a contiguous alternative θ_n and to zero if θ is a fixed alternative. This means the fixed alternatives are relatively easy to detect. So in this framework it is assumed that the alternatives of importance are the contiguous alternatives. Let us call this theory Pitman type asymptotics.

There are several other frameworks in classical statistics of which Bahadur's (Bahadur, 1971; Serfling, 1980, pp. 332–341) has been studied most. We focus on Bahadur's approach. In Bahadur's theory, the alternatives of importance are fixed and do not depend on n. Given a test statistic, Bahadur evaluates its performance at a fixed alternative by the limit (in probability or a.s.) of $\frac{1}{n}(\log \text{P-value})$ when the alternative is true.

Which of these two asymptotics is appropriate in a given situation should depend on which alternatives are important, fixed alternatives or Pitman alternatives $\theta_0 + d/\sqrt{n}$ that approach the null hypothesis at a certain rate. This in turn depends on how the sample size n is chosen. If n is chosen to ensure a Type 2 error bounded away from 0 and 1 (like α), then Pitman alternatives seem appropriate. If n is chosen to be quite large, depending on available resources but not on alternatives, then Bahadur's approach would be reasonable.

6.4.1 Comparison of Decisions via P-values and Bayes Factors in Bahadur's Asymptotics

In this subsection, we essentially follow Bahadur's approach for both P-values and Bayes factors. A Pitman type asymptotics is used for both in the next subsection. We first show that if the P-value is sufficiently small, as small as it is typically in Bahadur's theory, B_{01} will tend to zero, calling for rejection of H_0, i.e., the evidence in the P-value points in the same direction as that in the Bayes factor or posterior probability, removing the sense of paradox in the result of Jeffreys and Lindley. One could, therefore, argue that the P-values or the significance level α assumed in the Jeffreys-Lindley example are not small enough. The asymptotic framework chosen is not appropriate when contiguous alternatives are not singled out as alternatives of importance.

We now verify the claim about the limit of B_{01}. Without loss of generality, take $\theta_0 = 0, \sigma^2 = 1$. First note that

$$\log B_{01} = -\frac{n}{2}\bar{X}^2 + \frac{1}{2}\log n + R_n, \tag{6.26}$$

where

$$R_n = -\log \pi(\bar{X}|H_1) - \frac{1}{2}\log(2\pi) + o(1)$$

provided the prior $g_1(\theta)$ is a continuous function of θ and is positive at all θ. If we omit R_n from the right-hand side of (6.26), we have Schwarz's (1978) approximation to the Bayes factor via BIC (Section 4.3).

The logarithm of P-value (p) corresponding to observed $\bar{X}$ is

$$\log p = \log 2[1 - \Phi(\sqrt{n} \mid \bar{X} \mid)] = -\frac{n}{2}\bar{X}^2(1 + o(1))$$

by standard approximation to a normal tail (vide Feller (1973, p. 175) or Bahadur (1971, p. 1)). Thus $\frac{1}{n}\log p \to -\theta^2/2$ and by (6.26), $\log B_{01} \to -\infty$. This result is true as long as $|\bar{X}| > c(\log n/n)^{1/2}, c > \sqrt{2}$. Such deviations are called moderate deviations, vide Rubin and Sethuraman (1965). Of course, even for such P-values, $p \sim (B_{01}/n)$ so that P-values are smaller by an order of magnitude. The conflict in measuring evidence remains but the decisions are the same.

Ghosh et al. (2005) also pursue the comparison of the three measures of evidence based on the likelihood ratio, the P-value based on the likelihood ratio test, and the Bayes factor B_{01} under general regularity conditions.

6.4.2 Pitman Alternative and Rescaled Priors

We consider once again the problem of testing $H_0 : \theta = 0$ versus $H_1 : \theta \neq 0$ on the basis of a random sample from $N(\theta, 1)$. Suppose that the Pitman alternatives are the most important ones and the prior $g_1(\theta)$ under H_1 puts most of the mass on Pitman alternatives. One such prior is $N(0, \delta/n)$. Then

$$B_{01} = \sqrt{\delta + 1}\exp\left[-\frac{n}{2}\left(\frac{\delta}{\delta + 1}\right)\bar{X}^2\right].$$

If the P-value is close to zero, $\sqrt{n}|\bar{X}|$ is large and therefore, B_{01} is also close to zero, i.e., for these priors there is no paradox. The two measures are of the same order but the result of Berger and Sellke (1987) for symmetric unimodal priors still implies that P-value is smaller than the Bayes factor.

6.5 Bayesian P-value[3]

Even though valid Bayesian quantities such as Bayes factor and posterior probability of hypotheses are in principle the correct tools to measure the evidence for or against hypotheses, they are quite often, and especially in many practical situations, very difficult to compute. This is because either the alternatives are only very vaguely specified, vide (6.6), or very complicated. Also, in some cases one may not wish to compare two or more models but check how a model fits the data. Bayesian P-values have been proposed to deal with such problems.

---

[3] Section 6.5 may be omitted at first reading.

Let M_0 be a target model, and departure from this model be of interest. If, under this model, X has density $f(x|\eta)$, $\eta \in \mathcal{E}$, then for a Bayesian with prior π on η, $m_\pi(x) = \int_{\mathcal{E}} f(x|\eta)\pi(\eta)\,d\eta$, the prior predictive distribution is the actual predictive distribution of X. Therefore, if a model departure statistic $T(X)$ is available, then one can define the *prior predictive* P-value (or tail area under the predictive distribution) as

$$p = P^{m_\pi}\left(T(X) \geq T(x_{obs})|M_0\right),$$

where x_{obs} is the observed value of X (see Box (1980)). Although it is true that this is a valid Bayesian quantity for model checking and it is useful in situations such as the ones described in Exercise 13 or Exercise 14, it does face the criticism that it may be influenced to a large extent by the prior π as can be seen in the following example.

Example 6.18. Let $X_1, X_2, \cdots, X_n$ be a random sample from $N\left(\theta, \sigma^2\right)$ with both θ and σ^2 unknown. It is of interest to detect discrepancy in the mean of the model with the target model being $M_0 : \theta = 0$. Note that $T = \sqrt{n}\bar{X}$ (actually its absolute value) is the natural model departure statistic for checking this.

(a) Case 1. It is felt *a priori* that σ^2 is known, or equivalently, we choose the prior on σ^2, which puts all its mass at some known constant σ_0^2. Then under M_0, there are no unknown parameters and hence the prior predictive P-value is simply $2\left(1 - \Phi(\sqrt{n}|\bar{x}_{obs}|/\sigma_0)\right)$, where $\bar{x}_{obs}$ is the observed value of $\bar{X}$. This can highly overestimate the evidence against M_0 if σ_0^2 happens to underestimate the actual model variance.

(b) Case 2. Consider the usual non-informative prior on σ^2: $\pi(\sigma^2) \propto 1/\sigma^2$. Then,

$$m_\pi(\mathbf{x}) = \int_0^\infty f_{\mathbf{X}}(\mathbf{x}|\sigma^2)\pi(\sigma^2)\,d\sigma^2$$

$$\propto \int_0^\infty \exp(-\frac{1}{2\sigma^2}\sum_{i=1}^n x_i^2)(\sigma^2)^{-n/2}\,\frac{d\sigma^2}{\sigma^2}$$

$$\propto \left(\sum_{i=1}^n x_i^2\right)^{-n/2},$$

which is an improper density, thus completely disallowing computation of the prior predictive P-value.

(c) Case 3. Consider an inverse Gamma prior $IG(\nu, \beta)$ with the following density for σ^2: $\pi(\sigma^2|\nu, \beta) = \frac{\beta^\nu}{\Gamma(\nu)}(\sigma^2)^{-(\nu+1)}\exp(-\frac{\beta}{\sigma^2})$ for $\sigma^2 > 0$, where ν and β are specified positive constants. Because $T|\sigma^2 \sim N(0, \sigma^2)$, under this prior the predictive density of T is then,

$$m_\pi(t) = \int_0^\infty f_T(t|\sigma^2)\pi(\sigma^2|\nu, \beta)\,d\sigma^2$$

Table 6.6. Normal Example: Prior Predictive P-values

ν	.5	.5	.5	1	1	1	2	2	2	5	5	5
β	.5	1	2	.5	1	2	.5	1	2	.5	1	2
p	.300	.398	.506	.109	.189	.300	.017	.050	.122	.0001	.001	.011

$$\propto \int_0^\infty \exp(-\frac{1}{\sigma^2}(\beta + \frac{t^2}{2}))(\sigma^2)^{-(\nu+1+1/2)}\, d\sigma^2$$

$$\propto (2\beta + t^2)^{-(2\nu+1)/2}.$$

If 2ν is an integer, under this predictive distribution,

$$\frac{T}{\sqrt{\beta/\nu}} \sim t_{2\nu}.$$

Thus we obtain,

$$p = P^{m_\pi}\left(|\bar{X}| \geq |\bar{x}_{obs}| | M_0\right)$$

$$= P^{m_\pi}\left(|\frac{T}{\sqrt{\beta/\nu}}| \geq \frac{\sqrt{n}|\bar{x}_{obs}|}{\sqrt{\beta/\nu}} | M_0\right)$$

$$= 2\left(1 - F_{2\nu}(\frac{\sqrt{n}|\bar{x}_{obs}|}{\sqrt{\beta/\nu}})\right),$$

where $F_{2\nu}$ is the c.d.f. of $t_{2\nu}$. For $\sqrt{n}\bar{x}_{obs} = 1.96$ and various values of ν and β, the corresponding values of the prior predictive P-values are displayed in Table 6.6.

Further, note that $p \to 1$ as $\beta \to \infty$ for any fixed $\nu > 0$. Thus it is clear that the prior predictive P-value, in this example, depends crucially on the values of ν and β.

What can be readily seen in this example is that if the prior π used is a poor choice, even an excellent model can come under suspicion upon employing the prior predictive P-value. Further, as indicated above, non-informative priors that are improper (thus making m_π improper too) will not allow computing such a tail area, a further undesirable feature. To rectify these problems, Gutman (1967), Rubin (1984), Meng (1994), and Gelman et al. (1996) suggest modifications by replacing π in m_π by $\pi(\eta|x_{obs})$:

$$m^*(x|x_{obs}) = \int_{\mathcal{E}} f(x|\eta)\pi(\eta|x_{obs})\, d\eta, \text{ and}$$

$$p^* = P^{m^*(\cdot|x_{obs})}\left(T(X) \geq T(x_{obs})\right).$$

This is called the *posterior predictive* P-value. This removes some of the difficulties cited above. However, this version of Bayesian P-value has come under severe criticism also. Bayarri and Berger (1998a) note that these modified quantities are not really Bayesian. To see this, they observe that there is "double use" of data in the above modifications: first to convert (a possibly improper) $\pi(\eta)$ into a proper $\pi(\eta|x_{obs})$, and then again in computing the tail area of $T(X)$ corresponding with $T(x_{obs})$. Furthermore, for large sample sizes, the posterior distribution of η will essentially concentrate at $\hat{\eta}$, the MLE of η, so that $m^*(x|x_{obs})$ will essentially equal $f(x|\hat{\eta})$, a non-Bayesian object. In other words, the criticism is that, for large sample sizes the posterior predictive P-value will not achieve anything more than rediscovering the classical approach. Let us consider Example 6.18 again.

Example 6.19. (Example 6.18, continued.) Let us consider the non-informative prior $\pi(\sigma^2) \propto 1/\sigma^2$ again. Then, as before, because $T|\sigma^2 \sim N(0, \sigma^2)$, and

$$\pi(\sigma^2|\mathbf{x}_{obs}) \propto \exp(-\frac{1}{2\sigma^2}\sum_{i=1}^{n} x_i^2)(\sigma^2)^{-\frac{n}{2}+1}$$

$$\propto \exp(-\frac{n}{2\sigma^2}(\bar{x}_{obs}^2 + s_{obs}^2))(\sigma^2)^{-\frac{n}{2}+1},$$

the posterior predictive distribution of T is

$$m_\pi(t|\mathbf{x}_{obs}) = \int_0^\infty f_T(t|\sigma^2)\pi(\sigma^2|\mathbf{x}_{obs})\, d\sigma^2$$

$$\propto \int_0^\infty (\sigma^2)^{-1/2}\exp(-\frac{t^2}{2\sigma^2})(\sigma^2)^{-n/2}\exp(-\frac{n}{2\sigma^2}(\bar{x}_{obs}^2 + s_{obs}^2))\frac{d\sigma^2}{\sigma^2}$$

$$\propto \int_0^\infty \exp(-v\{n(\bar{x}_{obs}^2 + s_{obs}^2) + t^2\})v^{(n+1)/2}\frac{dv}{v}$$

$$\propto \left(1 + \frac{1}{n}\frac{t^2}{\bar{x}_{obs}^2 + s_{obs}^2}\right)^{-(n+1)/2}.$$

Therefore, we see that, under the posterior predictive distribution,

$$\frac{T}{\sqrt{\bar{x}_{obs}^2 + s_{obs}^2}} \sim t_n.$$

Thus we obtain the posterior predictive P-value to be

$$p = P^{m_\pi(\cdot|\mathbf{x}_{obs})}\left(|\bar{X}| \geq |\bar{x}_{obs}||M_0\right)$$

$$= P^{m_\pi(\cdot|\mathbf{x}_{obs})}\left(|\frac{T}{\sqrt{\bar{x}_{obs}^2 + s_{obs}^2}}| \geq \frac{\sqrt{n}|\bar{x}_{obs}|}{\sqrt{\bar{x}_{obs}^2 + s_{obs}^2}}|M_0\right)$$

$$= 2\left(1 - F_n(\frac{\sqrt{n}|\bar{x}_{obs}|}{\sqrt{\bar{x}_{obs}^2 + s_{obs}^2}})\right),$$

where F_n is the distribution function of t_n. This definition of a Bayesian P-value doesn't seem satisfactory. Let $|\bar{x}_{obs}| \to \infty$. Note that then $p \to 2\left(1 - F_n(\sqrt{n})\right)$. Actually, p decreases to this lower bound as $|\bar{x}_{obs}| \to \infty$.

$$\textbf{Table 6.7.} \text{ Values of } p_n = 2\left(1 - F_n(\sqrt{n})\right)$$

n	1	2	3	4	5	6	7	8	9	10
p_n	.500	.293	.182	.116	.076	.050	.033	.022	.015	.010

Values of this lower bound for different n are shown in Table 6.7. Note that these values have no serious relationship with the observations and hence cannot be really used for model checking. Bayarri and Berger (1998a) attribute this behavior to the 'double' use of the data, namely, the use of $\bar{x}$ in computing both the posterior distribution and the tail area probability of the posterior predictive distribution.

In an effort to combine the desirable features of the prior predictive P-value and the posterior predictive P-value and eliminate the undesirable features, Bayarri and Berger (see Bayarri and Berger (1998a)) introduce the *conditional predictive* P-value. This quantity is based on the prior predictive distribution m_π but is more heavily influenced by the model than the prior. Further, non-informative priors can be used, and there is no double use of the data. The steps are as follows: An appropriate statistic $U(X)$, not involving the model departure statistic $T(X)$, is identified, the conditional predictive distribution $m(t|u)$ is derived, and the conditional predictive P-value is defined as

$$p_c = P^{m(.|u_{obs})}\left(T(X) \geq T(x_{obs})\right),$$

where $u_{obs} = U(x_{obs})$. The following example is from Bayarri and Berger (1998a).

Example 6.20. (Example 6.18, continued.) $T = \sqrt{n}\bar{X}$ is the model departure statistic for checking discrepancy of the mean in the normal model. Let $U(X) = s^2 = \frac{1}{n}\sum_{i=1}^n (X_i - \bar{X})^2$. Note that $nU/\sigma^2 \sim \sigma^2\chi^2_{n-1}$. Consider $\pi(\sigma^2) \propto 1/\sigma^2$ again. Then $\pi(\sigma^2|U = s^2) \propto (\sigma^2)^{(n-1)/2+1}\exp(-ns^2/(2\sigma^2))$ is the density of inverse Gamma, and hence the conditional predictive density of T given s^2_{obs} is

$$m_\pi(t|s^2_{obs}) = \int_0^\infty f_T(t|\sigma^2)\pi(\sigma^2|s^2_{obs})\,d\sigma^2$$

$$\propto \int_0^\infty (\sigma^2)^{-1/2}\exp(-\frac{t^2}{2\sigma^2})(\sigma^2)^{-(n-1)/2}\exp(-\frac{n}{2\sigma^2}s^2_{obs})\frac{d\sigma^2}{\sigma^2}$$

$$\propto \int_0^\infty \exp(-v\{ns^2_{obs} + t^2\})v^{n/2}\frac{dv}{v}$$

$$\propto \left(1 + \frac{1}{n}\frac{t^2}{s^2_{obs}}\right)^{-n/2}.$$

Thus, under the conditional predictive distribution,

$$\sqrt{\frac{n-1}{n}}\,\frac{T}{s_{obs}} \sim t_{n-1},$$

and hence we obtain the conditional predictive P-value to be

$$p_c = P^{m(\cdot|s_{obs}^2)}\left(|\bar{X}| \geq |\bar{x}_{obs}||M_0\right)$$
$$= P^{m(\cdot|s_{obs}^2)}\left(\sqrt{n-1}\frac{|\bar{X}|}{s_{obs}} \geq \sqrt{n-1}\frac{|\bar{x}|}{s_{obs}}|M_0\right)$$
$$= 2\left(1 - F_{n-1}\left(\frac{\sqrt{n-1}|\bar{x}_{obs}|}{s_{obs}}\right)\right).$$

We have thus found a Bayesian interpretation for the classical P-value from the usual t-test. It is worth noting that s_{obs}^2 was used to produce the posterior distribution to eliminate σ^2, and that $\bar{x}_{obs}$ was then used to compute the tail area probability. It is also to be noted that in this example, it was easy to find $U(\mathbf{X})$, which eliminates σ^2 upon conditioning, and that the conditional predictive distribution is a standard one. In general, however, even though this procedure seems satisfactory from a Bayesian point of view, there are problems related to identifying suitable $U(X)$ and also computing tail areas from (quite often intractable) $m(t|u_{obs})$.

Another possibility is the partial posterior predictive P-value (see Bayarri and Berger (1998a) again) defined as follows:

$$p^* = P^{m^*(\cdot)}\left(T(X) \geq T(x_{obs})\right),$$

where the predictive density m^* is obtained using a partial posterior density π^* that does not use the information contained in $t_{obs} = T(x_{obs})$ and is given by

$$m^*(t) = \int_{\mathcal{E}} f_T(t|\eta)\pi^*(\eta)\,d\eta,$$

with the partial posterior π^* defined as

$$\pi^*(\eta) \propto f_{X|T}(x_{obs}|t_{obs},\eta)\pi(\eta)$$
$$\propto \frac{f_X(x_{obs}|\eta)}{f_T(t_{obs}|\eta)}\pi(\eta).$$

Consider Example 6.18 again with $\pi(\sigma^2) \propto 1/\sigma^2$. Note that because X_i are i.i.d. $N(0,\sigma^2)$ under M_0,

$$f_{\mathbf{X}}(\mathbf{x}_{obs}|\sigma^2) \propto (\sigma^2)^{-n/2}\exp(-\frac{n}{2\sigma^2}(\bar{x}_{obs}^2 + s_{obs}^2))$$
$$\propto f_{\bar{X}}(\bar{x}_{obs}|\sigma^2)(\sigma^2)^{-(n-1)/2}\exp(-\frac{n}{2\sigma^2}s_{obs}^2),$$

so that

$$f_{\mathbf{X}|\bar{X}}(\mathbf{x}_{obs}|\bar{x}_{obs}, \sigma^2) \propto (\sigma^2)^{-(n-1)/2} \exp(-\frac{n}{2\sigma^2} s_{obs}^2).$$

Therefore,

$$\pi^*(\sigma^2) \propto (\sigma^2)^{-(n-1)/2+1} \exp(-\frac{n}{2\sigma^2} s_{obs}^2),$$

and is the same as $\pi(\sigma^2|s_{obs}^2)$, which was used to obtain the conditional predictive density earlier. Thus, in this example, the partial predictive P-value is the same as the conditional predictive P-value. Because this alternative version p^* does not require the choice of the statistic U, it appears this method may be used for any suitable goodness-of-fit test statistic T. However, we have not seen such work.

6.6 Robust Bayesian Outlier Detection

Because a Bayes factor is a weighted likelihood ratio, it can also be used for checking whether an observation should be considered an outlier with respect to a certain target model relative to an alternative model. One such approach is as follows. Recall the model selection set-up as given in (6.1). X having density $f(x|\theta)$ is observed, and it is of interest to compare two models M_0 and M_1 given by

$$M_0 : X \text{ has density } f(x|\theta) \text{ where } \theta \in \Theta_0;$$
$$M_1 : X \text{ has density } f(x|\theta) \text{ where } \theta \in \Theta_1.$$

For $i = 1, 2$, $g_i(\theta)$ is the prior density of θ, conditional on M_i being the true model. To compare M_0 and M_1 on the basis of a random sample $\mathbf{x} = (x_1, \ldots, x_n)$ the Bayes factor is given by

$$B_{01}(\mathbf{x}) = \frac{m_0(\mathbf{x})}{m_1(\mathbf{x})},$$

where $m_i(\mathbf{x}) = \int_{\Theta_i} f(\mathbf{x}|\theta)g_i(\theta)\,d\theta$ for $i = 1, 2$. To measure the effect on the Bayes factor of observation x_d one could use the quantity

$$k_d = \log\left(\frac{B(\mathbf{x})}{B(\mathbf{x}_{-d})}\right), \tag{6.27}$$

where $B(\mathbf{x}_{-d})$ is the Bayes factor excluding observation x_d. If $k_d < 0$, then when observation x_d is deleted there is an increase of evidence for M_0. Consequently, observation x_d itself favors model M_1. The extent to which x_d favors M_1 determines whether it can be considered an outlier under model M_0. Similarly, a positive value for k_d implies that x_d favors M_0. Pettit and Young (1990) discuss how k_d can be effectively used to detect outliers when the prior is non-informative. The same analysis can be done with informative priors also. This assumes that the conditional prior densities g_0 and g_1 can be fully specified. However, we take the robust Bayesian point of view that only certain broad features of these densities, such as symmetry and unimodality, can be specified, and hence we can only state that g_0 or g_1 belongs to a certain class of densities as determined by the features specified. Because k_d, derived from the Bayes factor, is the Bayesian quantity of inferential interest here, upper and lower bounds on k_d over classes of prior densities are required.

We shall illustrate this approach with a precise null hypothesis. Then we have the problem of comparing

$$M_0 : \theta = \theta_0 \text{ versus } M_1 : \theta \neq \theta_0$$

using a random sample from a population with density $f(x|\theta)$. Under M_1, suppose θ has the prior density g, $g \in \Gamma$. The Bayes factors with all the observations and without the dth observation, respectively, are

$$B_g(\mathbf{x}) = \frac{f(\mathbf{x}|\theta_0)}{\int_{\theta \neq \theta_0} f(\mathbf{x}|\theta)g(\theta)\,d\theta},$$

$$B_g(\mathbf{x}_{-d}) = \frac{f(\mathbf{x}_{-d}|\theta_0)}{\int_{\theta \neq \theta_0} f(\mathbf{x}_{-d}|\theta)g(\theta)\,d\theta}.$$

Because $f(\mathbf{x}|\theta) = f(x_d|\theta)f(\mathbf{x}_{-d}|\theta)$, we get

$$k_{d,g} = \log \left[\frac{f(\mathbf{x}|\theta_0)}{f(\mathbf{x}_{-d}|\theta_0)} \frac{\int_{\theta \neq \theta_0} f(\mathbf{x}_{-d}|\theta)g(\theta)\,d\theta}{\int_{\theta \neq \theta_0} f(\mathbf{x}|\theta)g(\theta)\,d\theta} \right].$$

$$= \log f(x_d|\theta_0) - \log \left[\frac{\int_{\theta \neq \theta_0} f(\mathbf{x}|\theta)g(\theta)\,d\theta}{\int_{\theta \neq \theta_0} f(\mathbf{x}_{-d}|\theta)g(\theta)\,d\theta} \right]. \tag{6.28}$$

Now note that to find the extreme values of $k_{d,g}$, it is enough to find the extreme values of

$$h_{d,g} = \frac{\int_{\theta \neq \theta_0} f(\mathbf{x}|\theta)g(\theta)\,d\theta}{\int_{\theta \neq \theta_0} f(\underline{x}_{-d}|\theta)g(\theta)\,d\theta} \tag{6.29}$$

over the set Γ. Further, this optimization problem can be rewritten as follows:

$$\sup_{g \in G} h_{d,g} = \sup_{g \in G} \frac{\int_{\theta \neq \theta_0} f(x_d|\theta)f(\underline{x}_{-d}|\theta)g(\theta)\,d\theta}{\int_{\theta \neq \theta_0} f(\underline{x}_{-d}|\theta)g(\theta)\,d\theta}$$

$$= \sup_{g^* \in G^*} \int_{\theta \neq \theta_0} f(x_d|\theta)g^*(\theta)\,d\theta, \tag{6.30}$$

$$\inf_{g \in G} h_{d,g} = \inf_{g \in G} \frac{\int_{\theta \neq \theta_0} f(x_d|\theta)f(\underline{x}_{-d}|\theta)g(\theta)\,d\theta}{\int_{\theta \neq \theta_0} f(\underline{x}_{-d}|\theta)g(\theta)\,d\theta}$$

$$= \inf_{g^* \in G^*} \int_{\theta \neq \theta_0} f(x_d|\theta)g^*(\theta)\,d\theta, \tag{6.31}$$

where

$$G^* = \left\{ g^* : g^*(\theta) = \frac{g(\theta)f(\underline{x}_{-d}|\theta)}{\int_{u \neq \theta_0} g(u)f(\underline{x}_{-d}|u)\,du}, g \in G \right\}.$$

Equations (6.30) and (6.31) indicate how optimization of ratio of integrals can be transformed to optimization of integrals themselves. Consider the case where Γ is the class A of all prior densities on the set $\{\theta : \theta \neq \theta_0\}$. Then we have the following result.

Theorem 6.21. *If $f(\underline{x}_{-d}|\theta) > 0$ for each $\theta \neq \theta_0$,*

$$\sup_{g \in A} h_{d,g} = \sup_{\theta \neq \theta_0} f(x_d|\theta), \quad and \tag{6.32}$$

$$\inf_{g \in A} h_{d,g} = \inf_{\theta \neq \theta_0} f(x_d|\theta). \tag{6.33}$$

Proof. From (6.30) and (6.31) above,

$$\sup_{g \in A} h_{d,g} = \sup_{g^* \in A^*} \int_{\theta \neq \theta_0} f(x_d|\theta)g^*(\theta)\,d\theta,$$

where

$$A^* = \left\{ g^* : g^*(\theta) = \frac{g(\theta)f(\underline{x}_{-d}|\theta)}{\int_{u \neq \theta_0} g(u)f(\underline{x}_{-d}|u)\,du}, g \in A \right\}.$$

Now note that extreme points of A^* are point masses. Proof for the infimum is similar.

The corresponding extreme values of k_d are

$$\sup_{g \in A} k_{d,g} = \log f(x_d|\theta_0) - \log \inf_{\theta \neq \theta_0} f(x_d|\theta), \tag{6.34}$$

$$\inf_{g \in A} k_{d,g} = \log f(x_d|\theta_0) - \log \sup_{\theta \neq \theta_0} f(x_d|\theta). \tag{6.35}$$

Example 6.22. Suppose we have a sample of size n from $N(\theta, \sigma^2)$ with known σ^2. Then, from (6.34) and (6.35),

$$\sup_{g \in A} k_{d,g} = \frac{1}{2\sigma^2} \sup_{\theta \neq \theta_0} \left[(x_d - \theta)^2 - (x_d - \theta_0)^2 \right]$$

$$= \infty, \quad and$$

$$\inf_{g \in A} k_{d,g} = \frac{1}{2\sigma^2} \inf_{\theta \neq \theta_0} \left[(x_d - \theta)^2 - (x_d - \theta_0)^2 \right]$$

$$= -\frac{(x_d - \theta_0)^2}{2\sigma^2}.$$

It can be readily seen from the above bounds on k_d that no observation, however large in magnitude, will be considered an outlier here. This is because A is too large a class of prior densities.

Instead, consider the class G of all $N(\theta_0, \tau^2)$ priors with $\tau^2 > \tau_0^2 > 0$. Note that τ^2 close to 0 will make M_1 indistinguishable from M_0, and hence it is necessary to consider τ^2 bounded away from 0. Then for $g \in G$

$$h_{d,g} = \frac{\int_{\theta \neq \theta_0} f(x_d|\theta) f(\underline{x}_{-d}|\theta) g(\theta)\, d\theta}{\int_{\theta \neq \theta_0} f(\underline{x}_{-d}|\theta) g(\theta)\, d\theta}$$

$$= \int_{\theta \neq \theta_0} f(x_d|\theta) g^*(\theta)\, d\theta,$$

where g^* is the density of $N(m, \delta^2)$ with

$$m = m(\underline{x}_{-d}, \tau^2) = \frac{(n-1)\tau^2}{(n-1)\tau^2 + \sigma^2}\overline{x}_{-d} + \frac{\sigma^2}{(n-1)\tau^2 + \sigma^2}\theta_0,$$

$$\delta^2 = \delta^2(\underline{x}_{-d}, \tau^2) = \frac{\tau^2\sigma^2}{(n-1)\tau^2 + \sigma^2}.$$

Note, therefore, that $h_{d,g} = h_{d,g}(x_d)$ is just the density of $N(m, \sigma^2 + \delta^2)$ evaluated at x_d. Thus,

$$h_{d,g} = (2\pi\sigma^2)^{-1/2}\left(1 + \frac{\tau^2}{(n-1)\tau^2 + \sigma^2}\right)^{-1/2}$$

$$\times \exp\left(-\frac{\left(x_d - \frac{(n-1)\tau^2}{(n-1)\tau^2+\sigma^2}\overline{x}_{-d} - \frac{\sigma^2}{(n-1)\tau^2+\sigma^2}\theta_0\right)^2}{2\sigma^2\left(1 + \frac{\tau^2}{(n-1)\tau^2+\sigma^2}\right)}\right).$$

For each x_d, one just needs to graphically examine the extremes of the expression above as a function of τ^2 to determine if that particular observation should be considered an outlier. Delampady (1999) discusses these results and also results for some larger nonparametric classes of prior densities.

6.7 Nonsubjective Bayes Factors[4]

Consider two models M_0 and M_1 for data $\boldsymbol{X}$ with density $f_i(\boldsymbol{x}|\boldsymbol{\theta}_i)$ under model M_i, $\boldsymbol{\theta}_i$ being an unknown parameter of dimension $p_i, i = 0, 1$. Given prior specifications $g_i(\boldsymbol{\theta}_i)$ for parameter $\boldsymbol{\theta}_i$, the Bayes factor of M_1 to M_0 is obtained as

$$B_{10} = \frac{m_1(\boldsymbol{x})}{m_0(\boldsymbol{x})} = \frac{\int f_1(\boldsymbol{x}|\boldsymbol{\theta}_1)g_1(\boldsymbol{\theta}_1)d\boldsymbol{\theta}_1}{\int f_0(\boldsymbol{x}|\boldsymbol{\theta}_0)g_0(\boldsymbol{\theta}_0)d\boldsymbol{\theta}_0}. \tag{6.36}$$

Here $m_i(\boldsymbol{x})$ is the marginal density of $\boldsymbol{X}$ under $M_i, i = 0, 1$. When subjective specification of prior distributions is not possible, which is frequently the case, one would look for automatic method that uses standard noninformative priors.

---

[4] Section 6.7 may be omitted at first reading.

There are, however, difficulties with (6.36) for noninformative priors that are typically improper. If g_i are improper, these are defined only up to arbitrary multiplicative constants c_i; $c_i g_i$ has as much validity as g_i. This implies that $(c_1/c_0)B_{10}$ has as much validity as B_{10}. Thus the Bayes factor is determined only up to an arbitrary multiplicative constant. This indeterminacy, noted by Jeffreys (1961), has been the main motivation of new objective methods. We shall confine mainly to the nested case where f_0 and f_1 are of the same functional form and $f_0(\boldsymbol{x}|\boldsymbol{\theta}_0)$ is the same as $f_1(\boldsymbol{x}|\boldsymbol{\theta}_1)$ with some of the co-ordinates of $\boldsymbol{\theta}_1$ specified. However, the methods described below can also be used for non-nested models.

It may be mentioned here that use of diffuse (flat) proper prior does not provide a good solution to the problem. Also, truncation of noninformative priors leads to a large penalty for the more complex model. An example follows.

Example 6.23. (Testing normal mean with known variance.) Suppose we observe $\boldsymbol{X} = (X_1, \ldots, X_n)$. Under M_0, X_i are i.i.d. $N(0,1)$ and under M_1, X_i are i.i.d. $N(\theta, 1)$, $\theta \in R$ is the unknown mean. With the uniform noninformative prior $g_1^N(\theta) \equiv c$ for θ under M_1, the Bayes factor of M_1 to M_0 is given by

$$B_{10}^N = \sqrt{2\pi}cn^{-1/2}\exp[n\bar{X}^2/2].$$

If one uses a uniform prior over $-K \leq \theta \leq K$, then for large K, the new Bayes factor B_{10}^K is approximately $1/(2Kc)$ times B_{10}^N. Thus for large K, one is heavily biased against M_1. This is reminiscent of the phenomenon observed by Lindley (1957). A similar conclusion is obtained if one uses a diffuse proper prior such as a normal prior $N(0, \tau^2)$, with variance τ^2 large. The corresponding Bayes factor is

$$B_{10}^{\mathrm{norm}} = (n\tau^2 + 1)^{-1/2}\exp\left[\frac{1}{2}\frac{n\tau^2}{n\tau^2 + 1}n\bar{X}^2\right]$$

which is approximately $(n\tau^2)^{-1/2}\exp[n\bar{X}^2/2]$ for large values of $n\tau^2$ and hence can be made arbitrarily small by choosing a large value of τ^2. Also B_{10}^{norm} is highly non-robust with respect to the choice of τ^2, and this non-robustness plays the same role as indeterminacy. The expressions for B_{10}^N and B_{10}^{norm} clearly indicate similar behavior of these Bayes factors and the similar roles of $\sqrt{2\pi}c$ and $(\tau^2 + 1/n)^{-1/2}$.

A solution to the above problem with improper priors is to use part of the data as a *training sample.* The data are divided into two parts, $\boldsymbol{X} = (\boldsymbol{X}_1, \boldsymbol{X}_2)$. The first part $\boldsymbol{X}_1$ is used as a training sample to obtain proper posterior distributions for the parameters (given $\boldsymbol{X}_1$) starting from the noninformative priors

$$g_i(\boldsymbol{\theta}_i|\boldsymbol{X}_1) = \frac{f_i(\boldsymbol{X}_1|\boldsymbol{\theta}_i)g_i(\boldsymbol{\theta}_i)}{\int f_i(\boldsymbol{X}_1|\boldsymbol{\theta}_i)g_i(\boldsymbol{\theta}_i)\,d\boldsymbol{\theta}_i}, \quad i = 0, 1.$$

These proper posteriors are then used as priors to compute the Bayes factor with the remainder of the data ($\boldsymbol{X}_2$). This conditional Bayes factor, conditioned on $\boldsymbol{X}_1$, can be expressed as

$$
\begin{aligned}
B_{10}(\boldsymbol{X}_1) &= \frac{\int f_1(\boldsymbol{X}_2|\boldsymbol{\theta}_1)g_1(\boldsymbol{\theta}_1|\boldsymbol{X}_1)d\boldsymbol{\theta}_1}{\int f_0(\boldsymbol{X}_2|\boldsymbol{\theta}_0)g_0(\boldsymbol{\theta}_0|\boldsymbol{X}_1)d\boldsymbol{\theta}_0} \\
&= \frac{m_1(\boldsymbol{X})\int f_0(\boldsymbol{X}_1|\boldsymbol{\theta}_0)g_0(\boldsymbol{\theta}_0)d\boldsymbol{\theta}_0}{m_0(\boldsymbol{X})\int f_1(\boldsymbol{X}_1|\boldsymbol{\theta}_1)g_1(\boldsymbol{\theta}_1)d\boldsymbol{\theta}_1} \\
&= B_{10}\frac{m_0(\boldsymbol{X}_1)}{m_1(\boldsymbol{X}_1)}
\end{aligned}
\tag{6.37}
$$

where $m_i(\boldsymbol{X}_1)$ is the marginal density of $\boldsymbol{X}_1$ under $M_i, i = 0, 1$. Note that if the priors $c_i g_i$, $i = 0, 1$, are used to compute $B_{10}(\boldsymbol{X}_1)$, the arbitrary constant multiplier c_1/c_0 of B_{10} is cancelled by (c_0/c_1) of $m_0(\boldsymbol{X}_1)/m_1(\boldsymbol{X}_1)$ so that the indeterminacy of the Bayes factor is removed in (6.37).

A part of the data, $\boldsymbol{X}_1$, may be used as a training sample as described above if the corresponding posteriors $g_i(\theta_i|\boldsymbol{X}_1)$, $i = 0, 1$ are proper or, equivalently, the marginal densities $m_i(\boldsymbol{X}_1)$ of $\boldsymbol{X}_1$ under $M_i, i = 0, 1$ are finite. One would naturally use minimal amount of data as such a training sample leaving most part of the data for model comparison. As in Berger and Pericchi (1996a), a training sample $\boldsymbol{X}_1$ may be called proper if $0 < m_i(\boldsymbol{X}_1) < \infty$, $i = 0, 1$ and *minimal* if it is proper and no subset of it is proper.

Example 6.24. (Testing normal mean with known variance.) Consider the setup of Example 6.23 and the uniform noninformative prior $g_1(\theta) \equiv 1$ for θ under M_1. The minimal training samples are subsamples of size 1 with $m_0(X_i) = (1/\sqrt{2\pi})e^{-X_i^2/2}$ and $m_1(X_i) = 1$.

Example 6.25. (Testing normal mean with variance unknown.) Let $\boldsymbol{X} = (X_1, \ldots, X_n)$.
$M_0 : X_1, \ldots, X_n$ are i.i.d. $N(0, \sigma_0^2)$,
$M_1 : X_1, \ldots, X_n$ are i.i.d. $N(\mu, \sigma_1^2)$.
Consider the noninformative priors $g_0(\sigma_0) = 1/\sigma_0$ under M_0 and $g_1(\mu, \sigma_1) = 1/\sigma_1$. Here $m_1(X_i) = \infty$ for a single observation X_i and a minimal training sample consists of two distinct observations X_i, X_j and for such a training sample (X_i, X_j),

$$
m_0(X_i, X_j) = \frac{1}{2\pi(X_i^2 + X_j^2)} \quad \text{and} \quad m_1(X_i, X_j) = \frac{1}{2|X_i - X_j|}.
\tag{6.38}
$$

6.7.1 The Intrinsic Bayes Factor

As described above, a solution to the problem with improper priors is obtained using a conditional Bayes factor $B_{10}(\boldsymbol{X}_1)$, conditioned on a training sample $\boldsymbol{X}_1$. However, this conditional Bayes factor depends on the choice of the training sample $\boldsymbol{X}_1$. Let $\boldsymbol{X}(l)$, $l = 1, 2, \ldots L$ be the list of all possible minimal training samples. Berger and Pericchi (1996a) suggest considering all these minimal training samples and taking average of the corresponding L conditional Bayes factors $B_{10}(\boldsymbol{X}(l))$'s to obtain what is called the intrinsic Bayes factor (IBF). For example, taking an arithmetic average leads to the arithmetric intrinsic Bayes factor (AIBF)

$$AIBF_{10} = B_{10} \frac{1}{L} \sum_{l=1}^{L} \frac{m_0(\boldsymbol{X}(l))}{m_1(\boldsymbol{X}(l))} \tag{6.39}$$

and the geometric average gives the geometric intrinsic Bayes factor (GIBF)

$$GIBF_{10} = B_{10} \left(\prod \frac{m_0(\boldsymbol{X}(l))}{m_1(\boldsymbol{X}(l))} \right)^{1/2}, \tag{6.40}$$

the sum and product in (6.39) and (6.40) being taken over the L possible training samples $\boldsymbol{X}(l), l = 1, \ldots, L$.

Berger and Pericchi (1996a) also suggest using trimmed averages or the median (complete trimming) of the conditional Bayes factors when taking an average of all of them does not seem reasonable (e.g., when the conditional Bayes factors vary much). AIBF and GIBF have good properties but are affected by outliers. If the sample size is very small, using a part of the sample as a training sample may be impractical, and Berger and Pericchi (1996a) recommend using expected intrinsic Bayes factors that replace the averages in (6.39) and (6.40) by their expectations, evaluated at the MLE under the more complex model M_1. For more details, see Berger and Pericchi (1996a). Situations in which the IBF reduces simply to the Bayes factor B_{10} with respect to the noninformative priors are given in Berger et al. (1998). The AIBF is justified by the possibility of its correspondence with actual Bayes factors with respect to "reasonable" priors at least asymptotically. Berger and Pericchi (1996a, 2001) have argued that these priors, known as "intrinsic" priors, may be considered to be natural "default" priors for the testing problems. The intrinsic priors are discussed here in Subsection 6.7.3.

6.7.2 The Fractional Bayes Factor

O'Hagan (1994, 1995) proposes a solution using a fractional part of the full likelihood in place of using parts of the sample as training samples and averaging over them. The resulting "partial" Bayes factor, called the *fractional Bayes factor* (FBF), is given by

$$FBF_{10} = \frac{m_1(\boldsymbol{X}, b)}{m_0(\boldsymbol{X}, b)}$$

where b is a fraction and

$$m_i(\boldsymbol{X}, b) = \frac{\int f_i(\boldsymbol{X}|\boldsymbol{\theta}_i)g_i(\boldsymbol{\theta}_i)\,d\boldsymbol{\theta}_i}{\int [f_i(\boldsymbol{X}|\boldsymbol{\theta}_i)]^b g_i(\boldsymbol{\theta}_i)d\boldsymbol{\theta}_i}.$$

Note that FBF_{10} can also be written as

$$FBF_{10} = B_{10}\frac{m_0^b(\boldsymbol{X})}{m_1^b(\boldsymbol{X})}$$

where

$$m_i^b(\boldsymbol{X}) = \int [f_i(\boldsymbol{X}|\boldsymbol{\theta}_i)]^b g_i(\boldsymbol{\theta}_i)d\boldsymbol{\theta}_i, \; i = 0, 1. \tag{6.41}$$

To make FBF comparable with the IBF, one may take $b = m/n$ where m is the size of a minimal training sample as defined above and n is the total sample size. O'Hagan also recommends other choices of b, e.g., $b = \sqrt{n}/n$ or $\log n/n$.

We now illustrate through a number of examples.

Example 6.26. (Testing normal mean with known variance.) Consider the setup of Example 6.23. The Bayes factor with the noninformative prior $g_1(\theta) \equiv 1$ was obtained as

$$B_{10} = \sqrt{2\pi}n^{-1/2}\exp[n\bar{X}^2/2] = \sqrt{2\pi}n^{-1/2}\lambda_{10}$$

where λ_{10} is the likelihood ratio statistic. Bayes factor conditioned on X_i is

$$B_{10}(X_i) = B_{10}\,m_0(X_i)/m_1(X_i) = B_{10}(1/\sqrt{2\pi})\exp(-X_i^2/2).$$

Thus

$$AIBF_{10} = n^{-1}\sum_{i=1}^{n} B_{10}(X_i) = n^{-3/2}\exp(n\bar{X}^2/2)\sum_{i=1}^{n}\exp(-X_i^2/2),$$

$$GIBF_{10} = n^{-1/2}\exp[n\bar{X}^2/2 - (1/2n)\sum X_i^2].$$

The median IBF (MIBF) is obtained as the median of the set of values $B_{10}(X_i), \; i = 1, 2, \cdots, n$.

The FBF with a fraction $o < b < 1$ is

$$FBF_{10} = b^{1/2}\exp[n(1-b)\bar{X}^2/2]$$
$$= n^{-1/2}\exp[(n-1)\bar{X}^2/2], \;\; \text{if } b = 1/n.$$

Example 6.27. (Testing normal mean with variance unknown.) Consider the setup of Example 6.25. For the standard noninformative priors considered in this example, we have

$$B_{10} = \sqrt{\frac{\pi}{n}} \times \frac{\Gamma(\frac{n-1}{2})}{\Gamma(\frac{n}{2})} \times \frac{(\sum X_i^2)^{n/2}}{[\sum(X_i - \bar{X})^2]^{\frac{n-1}{2}}}$$

$$AIBF_{10} = B_{10} \times \frac{1}{\pi\binom{n}{2}} \sum_{1 \le i < j \le n} \frac{|X_i - X_j|}{(X_i^2 + X_j^2)}$$

$$FBF_{10} = \frac{\Gamma(\frac{n-1}{2})}{\sqrt{\pi}\,\Gamma(\frac{n}{2})} \times \left[\frac{\sum_{i=1}^{n} X_i^2}{\sum_{i=1}^{n}(X_i - \bar{X})^2} \right]^{\frac{n}{2}-1}, \quad \text{with } b = \frac{2}{n}.$$

Example 6.28. (normal linear model.) This example is from Berger and Pericchi (1996b, 2001). Berger and Pericchi determined the IBF for linear models for both the nested and non-nested case. We consider here only the nested case. Suppose for the data $\boldsymbol{Y}(n \times 1)$ we consider the linear models

$$M_i : \boldsymbol{Y} = \boldsymbol{X}_i \boldsymbol{\beta}_i + \boldsymbol{\epsilon}_i, \ \boldsymbol{\epsilon}_i \sim N_n(\boldsymbol{0}, \sigma_i^2 \boldsymbol{I}_n), \ \ i = 0, 1$$

where $\boldsymbol{\beta}_i = (\beta_{i1}, \beta_{i2}, \cdots, \beta_{ip_i})'$ and σ_i^2 are unknown parameters, and $\boldsymbol{X}_i$ is an $n \times p_i$ known design matrix of rank $p_i < n$. Consider priors of the form

$$g_i(\beta_i, \sigma_i) = \sigma_i^{-(1+q_i)}, \ q_i > -1.$$

Here $q_i = 0$ gives the reference prior of Berger and Bernardo (1992a), and $q_i = p_i$ corresponds with the Jeffreys prior. For the nested case, when M_0 is nested in M_1, Berger and Pericchi (1996b) consider a *modified Jeffreys* prior for which $q_0 = 0$ and $q_1 = p_1 - p_0$. The integrated likelihoods $m_i(\boldsymbol{Y})$ with these priors can be obtained as

$$m_i(\boldsymbol{Y}) = C 2^{q_i/2} \pi^{p_i/2} \Gamma((n - p_i + q_i)/2) \left| \boldsymbol{X}_i' \boldsymbol{X}_i \right|^{-1/2} R_i^{-(n-p_i+q_i)/2}$$

where C is a constant not depending on i, and R_i is the residual sum of squares under M_i, $i = 0, 1$. The Bayes factor B_{10} with the modified Jeffreys prior is then given by

$$B_{10} = (2\pi)^{(p_1-p_0)/2} \frac{|\boldsymbol{X}_0' \boldsymbol{X}_0|^{1/2}}{|\boldsymbol{X}_1' \boldsymbol{X}_1|^{1/2}} \left(\frac{R_0}{R_1} \right)^{(n-p_0)/2}. \tag{6.42}$$

Also, one can see that a minimal training sample $\boldsymbol{Y}(l)$ in this case is a sample of size $m = p_1 + 1$ such that for the corresponding design matrices $\boldsymbol{X}_i(l)$ (under M_i), $\boldsymbol{X}_i'(l)\boldsymbol{X}_i(l)$, $i = 0, 1$ are nonsingular. The ratio $m_0(\boldsymbol{Y}(l))/m_1(\boldsymbol{Y}(l))$ can be obtained from the expression of B_{10} by inverting it and replacing n, $\boldsymbol{X}_0$, $\boldsymbol{X}_1$, R_0, and R_1 by m, $\boldsymbol{X}_0(l)$, $\boldsymbol{X}_1(l)$, $R_0(l)$, and $R_1(l)$, respectively, where $R_i(l)$ is the residual sum of squares corresponding to $(\boldsymbol{Y}(l))$ under M_i, $i = 0, 1$. Thus the conditional Bayes factor $B_{10}(\boldsymbol{Y}(l))$, conditioned on $\boldsymbol{Y}(l)$ is given by

$$B_{10}(\boldsymbol{Y}(l))$$
$$= \frac{|\boldsymbol{X}_0' \boldsymbol{X}_0|^{1/2}}{|\boldsymbol{X}_1' \boldsymbol{X}_1|^{1/2}} \left(\frac{R_0}{R_1} \right)^{(n-p_0)/2} \frac{|\boldsymbol{X}_1'(l)\boldsymbol{X}_1(l)|^{1/2}}{|\boldsymbol{X}_0'(l)\boldsymbol{X}_0(l)|^{1/2}} \left(\frac{R_1(l)}{R_0(l)} \right)^{(p_1-p_0+1)/2}.$$

One may now find an average of these conditional Bayes factors to find an IBF. For example, an arithmetic mean of $B_{10}(Y(l))$'s for all possible minimal training samples $Y(l)$'s gives the AIBF, and a median gives the median IBF.

In case of fractional Bayes factor, one obtains that (see, for example, Berger and Pericchi, 2001, page 152), with $m_i^b(\boldsymbol{X})$ as defined in (6.41),

$$\frac{m_0^b(\boldsymbol{X})}{m_1^b(\boldsymbol{X})} = \left(\frac{b}{2\pi}\right)^{(p_1-p_0)/2} \frac{|\boldsymbol{X}_1'\boldsymbol{X}_1|^{1/2}}{|\boldsymbol{X}_0'\boldsymbol{X}_0|^{1/2}} \left(\frac{R_1}{R_0}\right)^{(m-p_0)/2}$$

with $b = m/n$ and hence

$$FBF_{10} = b^{(p_1-p_0)/2}(R_0/R_1)^{(n-m)/2}.$$

See also O'Hagan (1995) in this context.

For more examples, see Berger and Pericchi (1996a, 1996b, 2001) and O'Hagan (1995).

Several other methods have been proposed as solutions to the problem with noninformative priors. Smith and Spiegelhalter (1980) and Spiegelhalter and Smith (1982) propose the imaginary minimal sample device; see also Ghosh and Samanta (2002b) for a generalization. Berger and Pericchi (2001) present comparison of four methods including the IBF and FBF with illustration through a number of examples. Ghosh and Samanta (2002b) discuss a unified derivation of some of the methods that shows that in some qualitative and conceptual sense, these methods are close to each other.

6.7.3 Intrinsic Priors

Given a default Bayes factor such as the IBF or FBF, a natural question is whether it corresponds with an actual Bayes factor based on some priors at least approximately. If such priors exist, they are called intrinsic priors. A default Bayes factor such as IBF can then be calculated as an actual Bayes factor using the intrinsic prior, and one need not consider all possible training samples and average over them. A "reasonable" intrinsic prior that corresponds to a naturally developed good default Bayes factor may be considered with be a natural default prior for the given testing or model selection problem. On the other hand, a particular default Bayesian method may be evaluated on the basis of the corresponding intrinsic prior depending on how "reasonable" the intrinsic prior is. Berger and Pericchi (1996a) describe how one can obtain intrinsic priors using an asymptotic argument. We begin with an example.

Example 6.29. (Example 6.26, continued.) Suppose that for some proper prior $\pi(\theta)$ under model M_1,

$$BF_{10}^\pi \cong AIBF_{10} \tag{6.43}$$

where BF_{10}^{π} denotes the Bayes factor based on a prior $\pi(\theta)$ under M_1. Using Laplace approximation (Section 4.3) to the integrated likelihood under M_1, we have

$$BF_{10}^{\pi} \cong \frac{f_1(\boldsymbol{X}|\hat{\theta})}{f_0(\boldsymbol{X}|\theta=0)} n^{-1/2} \pi(\hat{\theta})\sqrt{2\pi}(\det \hat{I})^{-1/2}$$

where $\hat{\theta}$ is the MLE of θ under M_1, and $\hat{I}$ is the observed Fisher information number. Thus using the expression for the AIBF in this example, and noting that $\hat{I} = 1$, (6.43) can be expressed as

$$\pi(\hat{\theta}) \cong (1/\sqrt{2\pi})\frac{1}{n}\sum_{i=1}^{n}\exp(-X_i^2/2).$$

As the RHS converges to $(1/\sqrt{2\pi})E_\theta(e^{-X_1^2/2}) = (1/\sqrt{2\pi})(1/\sqrt{2})e^{-\theta^2/4}$ with probability one under any θ, this suggests the intrinsic prior

$$\pi(\theta) = \frac{1}{\sqrt{2\pi}\sqrt{2}}\exp(-\theta^2/4)$$

which is a $N(0,2)$ density. One can easily verify that

$$BF_{10}^{\pi}/AIBF_{10} \to 1$$

with probability one under any θ, i.e., the AIBF is approximately the same as the Bayes factor with an $N(0,2)$ prior for θ under M_1.

If one considers the FBF, one can directly show that the FBF, with fraction b, is exactly equal to the Bayes factor with a $N(0,(b^{-1}-1)/n)$ prior.

Let us now consider the general case. Let B_{10} be the Bayes factor of M_1 to M_0 with noninformative priors $g_i(\boldsymbol{\theta}_i)$ for $\boldsymbol{\theta}_i$ under M_i, $i = 0,1$. We illustrate below with the AIBF. Treatment for the other IBFs and FBF will be similar. Recall that

$$AIBF_{10} = B_{10}\bar{B}_{01} \text{ where } \bar{B}_{01} = \frac{1}{L}\sum_{l=1}^{L}\frac{m_0(\boldsymbol{X}(l))}{m_1(\boldsymbol{X}(l))}.$$

Suppose for some priors π_i under M_i, $i = 0,1$, $AIBF_{10}$ is approximately equal to the Bayes factor based on π_0 and π_1, denoted $BF_{10}(\pi_0,\pi_1)$. Using Laplace approximation (Section 4.3) to both the numerator and denominator of B_{10} (see 6.36), $AIBF_{10}$ can be shown to be approximately equal to

$$\frac{f_1(\boldsymbol{X}|\hat{\boldsymbol{\theta}}_1)g_1(\hat{\boldsymbol{\theta}}_1)(2\pi/n)^{p_1/2}|\hat{I}_1|^{-1/2}}{f_0(\boldsymbol{X}|\hat{\boldsymbol{\theta}}_0)g_0(\hat{\boldsymbol{\theta}}_0)(2\pi/n)^{p_0/2}|\hat{I}_0|^{-1/2}} \times \bar{B}_{01} \qquad (6.44)$$

where n denotes the sample size, p_i is the dimension of $\boldsymbol{\theta}_i$, $\hat{\boldsymbol{\theta}}_i$ is the MLE of $\boldsymbol{\theta}_i$, and $\hat{I}_i$ is the observed Fisher information matrix under M_i, $i = 0,1$. The same approximation applied to $BF_{10}(\pi_0,\pi_1)$, yields the approximation

$$\frac{f_1(\boldsymbol{X}|\hat{\boldsymbol{\theta}}_1)\pi_1(\hat{\boldsymbol{\theta}}_1)(2\pi/n)^{p_1/2}|\hat{I}_1|^{-1/2}}{f_0(\boldsymbol{X}|\hat{\boldsymbol{\theta}}_0)\pi_0(\hat{\boldsymbol{\theta}}_0)(2\pi/n)^{p_0/2}|\hat{I}_0|^{-1/2}} \tag{6.45}$$

to $BF_{10}(\pi_0, \pi_1)$. We assume that conditions for the Laplace approximation hold for the given models.

To find the intrinsic priors, we equate (6.44) with (6.45) and this yields

$$\frac{\pi_1(\hat{\boldsymbol{\theta}}_1)g_0(\hat{\boldsymbol{\theta}}_0)}{\pi_0(\hat{\boldsymbol{\theta}}_0)g_1(\hat{\boldsymbol{\theta}}_1)} \cong \bar{B}_{01}. \tag{6.46}$$

Berger and Pericchi (1996a) obtain the intrinsic prior determining equations by taking limits on both sides of (6.46) under M_0 and M_1. Assume that, as $n \to \infty$,

under M_1, $\hat{\boldsymbol{\theta}}_1 \to \boldsymbol{\theta}_1$, $\hat{\boldsymbol{\theta}}_0 \to a(\boldsymbol{\theta}_1)$, and $\bar{B}_{01} \to B_1^*(\boldsymbol{\theta}_1)$;

under M_0, $\hat{\boldsymbol{\theta}}_0 \to \boldsymbol{\theta}_0$, $\hat{\boldsymbol{\theta}}_1 \to b(\boldsymbol{\theta}_0)$, and $\bar{B}_{01} \to B_0^*(\boldsymbol{\theta}_0)$.

The equations obtained by Berger and Pericchi (1996a) are

$$\frac{\pi_1(\boldsymbol{\theta}_1)g_0(a(\boldsymbol{\theta}_1))}{g_1(\theta_1)\pi_0(a(\boldsymbol{\theta}_1))} = B_1^*(\boldsymbol{\theta}_1) \text{ and } \frac{\pi_1(b(\boldsymbol{\theta}_0))g_0(\boldsymbol{\theta}_0)}{g_1(b(\boldsymbol{\theta}_0))\pi_0(\boldsymbol{\theta}_0)} = B_0^*(\boldsymbol{\theta}_0). \tag{6.47}$$

When M_0 is nested in M_1, Berger and Pericchi suggested the solution

$$\pi_0(\boldsymbol{\theta}_0) = g_0(\boldsymbol{\theta}_0), \quad \pi_1(\boldsymbol{\theta}_1) = g_1(\boldsymbol{\theta}_1)B_1^*(\boldsymbol{\theta}_1). \tag{6.48}$$

However, this may not be the unique solution to (6.47). See also Dmochowski (1994) in this context.

Example 6.30. (Example 6.27, continued.) A solution to the intrinsic prior determining equations suggested by Berger and Pericchi (see (6.48)) is

$$\pi_0(\sigma_0) = \frac{1}{\sigma_0}, \quad \pi_1(\mu, \sigma_1) = \frac{1}{\sigma_1}B_1^*(\mu, \sigma_1) \tag{6.49}$$

where

$$B_1^*(\mu, \sigma_1) = E_{\mu,\sigma_1}B_{01}(X_1, X_2) \text{ and } B_{01}(X_1, X_2) = \frac{|X_1 - X_2|}{\pi(X_1^2 + X_2^2)}.$$

Note that $B_{01}(X_1, X_2)$ can be expressed as

$$B_{01}(X_1, X_2) = \frac{Z_1^{1/2}}{(Z_1 + Z_2)} \frac{\sqrt{2}}{\pi\sigma_1}$$

where $Z_1 = (X_1 - X_2)^2/(2\sigma_1^2) \sim \chi_1^2$ and $Z_2 = (X_1 + X_2)^2/(2\sigma_1^2) \sim$ noncentral χ^2 with d.f. $= 1$ and noncentrality parameter $\lambda = 2\mu^2/\sigma_1^2$. Also, Z_1 and Z_2 are independent. Using the representation of a noncentral χ^2 density as an (infinite) weighted sum of central χ^2 densities, we have

$$E\left[\frac{Z_1^p}{Z_1 + Z_2}\right] = \sum_{j=0}^{\infty} e^{-\lambda/2}\frac{(\lambda/2)^j}{j!}E\left[\frac{Z_1^p}{Z_1 + W_j}\right] \qquad (6.50)$$

where $W_j \sim \chi_{1+2j}^2$ and is independent of Z_1. We then have

$$B_1^*(\mu,\sigma_1) = \frac{e^{-\lambda/2}}{\pi\sqrt{\pi}\sigma_1}\sum_{j=0}^{\infty}\frac{(\lambda/2)^j}{j!(j+\frac{1}{2})}, \quad \lambda = 2\mu^2/\sigma_1^2$$

and the intrinsic priors are given by

$$\pi_0(\sigma_0) = \frac{1}{\sigma_0}, \quad \pi_1(\mu,\sigma_1) = \frac{1}{\sigma_1}\pi_1(\mu|\sigma_1)$$

$$\text{with} \quad \pi_1(\mu|\sigma_1) = \frac{1}{\pi\sqrt{\pi}\sigma_1}\exp(-\mu^2/\sigma_1^2)\sum_{j=0}^{\infty}\frac{(\mu^2/\sigma_1^2)^j}{j!(j+\frac{1}{2})}.$$

It is to be noted that $\int_{-\infty}^{\infty}\pi_1(\mu|\sigma_1)d\mu = 1$.

Example 6.31. (Testing normal mean with variance unknown.) This is from Berger and Pericchi (1996a). Consider the setup of Example 6.25 with the same prior g_0 under M_0 but in place of the standard noninformative prior $g_1(\mu,\sigma_1) = 1/\sigma_1$ use the Jeffreys prior $g_1^*(\mu,\sigma_1) = 1/\sigma_1^2$. In this case, a minimal training sample consists of two distinct observations X_i, X_j for which

$$m_0(X_i, X_j) = \frac{1}{2\pi(X_i^2 + X_j^2)} \text{ and } m_1(X_i, X_j) = \frac{1}{\sqrt{\pi}(X_i - X_j)^2}.$$

Proceeding as in the previous example, noting that

$$\frac{m_0(X_1, X_2)}{m_1(X_1, X_2)} = \frac{Z_1}{\sqrt{\pi}(Z_1 + Z_2)},$$

where Z_1 and Z_2 are as above, and using (6.50), the intrinsic priors are obtained as

$$\pi_0(\sigma_0) = \frac{1}{\sigma_0}, \quad \pi_1(\mu,\sigma_1) = \frac{1}{\sigma_1}\pi_1(\mu|\sigma_1) = \frac{1}{\sigma_1}\frac{1 - \exp(-\mu^2/\sigma_1^2)}{2\sqrt{\pi}(\mu^2/\sigma_1)}.$$

Here $\pi_1(\mu|\sigma_1)$ is a proper prior, very close to the Cauchy $(0, \sigma_1)$ prior for μ, which was suggested by Jeffreys (1961) as a default proper prior for μ (given σ_1); see Subsection 2.7.2.

Example 6.32. Consider a negative binomial experiment; Bernoulli trials, each having probability θ of success, are independently performed until a total of n successes is accumulated. On the basis of the outcome of this experiment we want to test the null hypothesis $H_0 : \theta = \frac{1}{2}$ against the alternative $H_1 : \theta \neq \frac{1}{2}$. We consider this problem as choosing between the two models $M_0 : \theta = \frac{1}{2}$ and $M_1 = \theta \in (0, 1)$.

The data may be looked upon as n observations $X_1,\ldots,X_n$ where X_1 denotes number of failures before the first success, and for $i = 2,\cdots,n$, X_i denotes number of failures between $(i-1)$th success and ith success. The random variables $X_1,\ldots,X_n$ are i.i.d. with a common geometric distribution with probability mass function

$$P(X_i = x) = \theta^x(1 - \theta),\ x = 0,1,2,\ldots$$

The likelihood function is

$$f(X_1,\ldots,X_n|\theta) = \theta^{\sum_{i=1}^{n} X_i}(1 - \theta)^n.$$

We consider the Jeffreys prior

$$g(\theta) = \theta^{-1/2}(1 - \theta)^{-1},\ 0 < \theta < 1$$

which is improper. The Bayes factor with this prior is

$$B_{10} = 2^{\sum X_i + n} \int_0^1 \theta^{\sum X_i - 1/2}(1 - \theta)^{n-1}d\theta.$$

Minimal training samples are of size 1, and the AIBF is given by

$$AIBF_{10} = B_{10} \times \frac{1}{n}\sum_{i=1}^{n} \frac{2X_i + 1}{2^{X_i+2}}.$$

Let

$$B^*(\theta) = E_\theta\left[\frac{2X_1 + 1}{2^{X_1+2}}\right] = \sum_{x=0}^{\infty} \frac{(2x + 1)}{2^{x+2}}\theta^x(1 - \theta).$$

Then the intrinsic prior is

$$\pi(\theta) = \theta^{-1/2}(1 - \theta)^{-1}B^*(\theta) = \frac{1}{4}\sum_{x=0}^{\infty}(2x + 1)2^{-x}\theta^{x-1/2}.$$

Simplification yields

$$\pi(\theta) = (\theta^{-1/2} + \theta^{1/2}/2)(2 - \theta)^{-2}.$$

We now consider a simple example from Lindley and Phillips (1976), also presented in Carlin and Louis (1996, Chapter 1). In 12 independent tosses of a coin, one observes 9 heads and 3 tails, the last toss yielding a tail. It is shown that one gets different results according to a binomial or a negative binomial likelihood. Let us consider the problem of testing the null hypothesis $H_0 : \theta = 1/2$ against the alternative $H_1 : \theta \neq 1/2$ where θ denotes the probability of head in a trial. If a binomial model is assumed, the random observable X is the number of heads observed in a fixed number of 12 tosses. One rejects H_0 for large values of the statistic $|X - 6|$, and the corresponding

P-value is 0.150. On the other hand, if a negative binomial model is assumed, the random observable X is the number of heads before the third trial appears. Note that expected value of X under H_0 is 3. Suppose one rejects H_0 for large values of $|X - 3|$. Then the corresponding P-value is 0.0325. Thus with the usual 5% Type 1 error level, the two model assumptions lead to different decisions. Let us now use a Bayes test for this problem. For the binomial model, the Jeffreys prior is proportional to $\theta^{-1/2}(1 - \theta)^{-1/2}$, which can be normalized to get a proper prior. For the negative binomial model, the data can be treated as three i.i.d. geometrically distributed random variables, as described above. The Bayes factor under the binomial model (with Jeffreys prior) and the Bayes factor under the negative binomial model (with the intrinsic prior) are respectively 1.079 and 1.424. They are different as were the P-values of classical statistics, but unlike the P-values, the Bayes factors are quite close.

6.8 Exercises

1. Assume a sharp null and continuity of the null distribution of the test statistic.
   (a) Calculate $E_{H_0}(\text{P-value})$ and $E_{H_0}(\text{P-value}|\text{P-value} < \alpha)$, where $0 < \alpha < 1$ is the Type 1 error probability.
   (b) In view of your answer to (a), do you think 2(P-value) is a better measure of evidence against H_0 than P-value?
2. Suppose $X \sim N(\theta, 1)$ and consider the two hypothesis testing problems:

$$H_0 : \theta = -1 \quad \text{versus} \quad H_1 : \theta = 1;$$
$$H_0^* : \theta = 1 \quad \text{versus} \quad H_1^* : \theta = -1.$$

   Find the Bayes factor of H_0 relative to H_1 and that of H_0^* relative to H_1^* if (a) $x = 0$ is observed, and (b) $x = 1$ is observed. Compute the classical P-value in both cases.
3. Refer to Example 6.3. Take $\tau = 2\sigma$, but keep the other parameter values unchanged. Compute B_{01} for the same values of t and n as used in Table 6.1.
4. Suppose $X \sim N(\theta, 1)$ and consider testing

$$H_0 : \theta = 0 \quad \text{versus} \quad H_1 : \theta \neq 0.$$

   For three different values of x, $x = 0, 1, 2$, compute the upper and lower bounds on Bayes factors when the prior on θ under the alternative hypothesis lies in
   (a) $\Gamma_A = \{\text{all prior distributions on } \mathcal{R}\}$,
   (b) $\Gamma_N = \{N(0, \tau^2), \tau^2 > 0\}$,
   (c) $\Gamma_S = \{\text{all symmetric (about 0) prior distributions on } \mathcal{R}\}$,

(d) $\Gamma_{SU} = \{$all unimodal priors on $\mathcal{R}$, symmetric about $0\}$.
Compute the classical P-value for each x value. What is the implication
of $\Gamma_N \subset \Gamma_{SU} \subset \Gamma_S \subset \Gamma_A$?

5. Let $X \sim B(m, \theta)$, and let it be of interest to test

$$H_0 : \theta = \frac{1}{2} \quad \text{versus} \quad H_1 : \theta \neq \frac{1}{2}.$$

If $m = 10$ and observed data is $x = 8$, compute the upper and lower
bounds on Bayes factors when the prior on θ under the alternative hypothesis lies in

(a) $\Gamma_A = \{$all prior distributions on $(0, 1)\}$,
(b) $\Gamma_B = \{\text{Beta}(\alpha, \alpha), \alpha > 0\}$,
(c) $\Gamma_S = \{$all symmetric (about $\frac{1}{2}$) priors on $(0, 1)\}$,
(d) $\Gamma_{SU} = \{$all unimodal priors on $(0, 1)$, symmetric about $\frac{1}{2}\}$.
Compute the classical P-value also.

6. Refer to Example 6.7.
(a) Show that $\underline{B}(G_A, x) = \exp(-\frac{t^2}{2})$, $\underline{P}(H_0|G_A, x) = \left[1 + \frac{1-\pi_0}{\pi_0} \exp(\frac{t^2}{2})\right]^{-1}$.
(b) Show that, if $t \leq 1$, $\underline{B}(G_{US}, x) = 1$, and $\underline{P}(H_0|G_{US}, x) = \pi_0$.
(c) Show that, if $t \leq 1$, $\underline{B}(G_{Nor}, x) = 1$, and $\underline{P}(H_0|G_{Nor}, x) = \pi_0$. If $t > 1$,
$\underline{B}(G_{Nor}, x) = t \exp(-(t^2 - 1)/2)$.

7. Suppose $\mathbf{X}|\boldsymbol{\theta}$ has the $t_p(3, \boldsymbol{\theta}, I_p)$ distribution with density

$$f(\mathbf{x}|\boldsymbol{\theta}) \propto \left(1 + \frac{1}{3}(\mathbf{x} - \boldsymbol{\theta})'(\mathbf{x} - \boldsymbol{\theta})\right)^{-(3+p)/2},$$

and it is of interest to test $H_0 : \boldsymbol{\theta} = \mathbf{0}$ versus $H_1 : \boldsymbol{\theta} \neq \mathbf{0}$. Show that this
testing problem is invariant under the group of all orthogonal transforma-
tions.

8. Refer to Example 6.13. Show that the testing problem mentioned there is
invariant under the group of scale transformations.

9. In Example 6.16, find the maximal invariants in the sample space and the
parameter space.

10. In Example 6.17, find the maximal invariants in the sample space and the
parameter space.

11. Let $X|\theta \sim N(\theta, 1)$ and consider testing

$$H_0 : |\theta - \theta_0| \leq 0.1 \quad \text{versus} \quad H_1 : |\theta - \theta_0| > 0.1.$$

Suppose $x = \theta_0 + 1.97$ is observed.
(a) Compute the P-value.
(b) Compute B_{01} and $P(H_0|x)$ under the two priors, $N(\theta_0, \tau^2)$, with $\tau^2 = (0.148)^2$ and $U(\theta_0 - 1, \theta_0 + 1)$.

12. Let $X|p \sim \text{Binomial}(10, p)$. Consider the two models:

$$M_0 : p = \frac{1}{2} \quad \text{versus} \quad M_1 : p \neq \frac{1}{2}.$$

Under M_1, consider the following three priors for p: (i) $U(0,1)$, (ii) Beta(10, 10), and (iii) Beta(100, 10). If four observations, $x = 0, 3, 5, 7$, and 10 are available, compute k_d given in Equation (6.27) for each observation, and for each of the priors and check which of the observations may be considered outliers under M_0.

13. (Box (1980)) Let $X_1, X_2, \cdots, X_n$ be a random sample from $N\left(\theta, \sigma^2\right)$ with both θ and σ^2 unknown. It is of interest to detect discrepancy in the variance of the model with the target model being

$$M_0 : \sigma^2 = \sigma_0^2, \text{ and } \theta \sim N(\mu, \tau^2),$$

where μ and τ^2 are specified.

(a) Show that the predictive distribution of $(X_1, X_2, \cdots, X_n)$ under M_0 is multivariate normal with covariance matrix $\sigma_0^2 I_n + \tau^2 \mathbf{11}'$ and $E(X_i) = \mu$, for $i = 1, 2, \ldots, n$.

(b) Show that under this predictive distribution,

$$T(\mathbf{X}) = \frac{1}{\sigma_0^2} \sum_{i=1}^{n} (X_i - \bar{X})^2 + \frac{n(\bar{X} - \mu)^2}{n\tau^2 + \sigma_0^2} \sim \chi_n^2.$$

(c) Derive and justify the prior predictive P-value based on the model departure statistic $T(\mathbf{X})$. Apply this to data, $\mathbf{x} = (8, 5, 4, 7)$, and $\sigma_0^2 = 1$, $\mu = 0$, $\tau^2 = 2$.

(c) What is the classical P-value for testing $H_0 : \sigma^2 = \sigma_0^2$ in this problem?

14. (Box (1980)) Suppose that under the target model, for $i = 1, 2, \ldots, n$,

$$y_i | \beta_0, \boldsymbol{\theta}, \sigma^2 = \beta_0 + \mathbf{x}_i' \boldsymbol{\theta} + \epsilon_i, \epsilon_i \sim N(0, \sigma^2) \text{ i.i.d.,}$$
$$\beta_0 | \sigma^2 \sim N(\mu_0, c\sigma^2), \boldsymbol{\theta} | \sigma^2 \sim N_p(\boldsymbol{\theta}_0, \sigma^2 \Gamma),$$
$$\sigma^2 \sim \text{inverse Gamma}(\alpha, \gamma),$$

where $c, \mu_0, \boldsymbol{\theta}_0, \Gamma, \alpha$ and γ are specified. Assume the standard linear regression model notation of $\mathbf{y} = \beta_0 \mathbf{1} + X\boldsymbol{\theta} + \boldsymbol{\epsilon}$, and suppose that $X'\mathbf{1} = \mathbf{0}$. Further assume that, given σ^2, conditionally β_0, $\boldsymbol{\theta}$ and $\boldsymbol{\epsilon}$ are independent. Also, let $\hat{\beta}_0$ and $\hat{\boldsymbol{\theta}}$ be the least squares estimates of β_0 and $\boldsymbol{\theta}$, respectively, and $RSS = (\mathbf{y} - \hat{\beta}_0 \mathbf{1} - X\hat{\boldsymbol{\theta}})'(\mathbf{y} - \hat{\beta}_0 \mathbf{1} - X\hat{\boldsymbol{\theta}})$.

(a) Show that under the target model, conditionally on σ^2, the predictive density of $\mathbf{y}$ is proportional to

$$(\sigma^2)^{-n/2} \exp\left(-\frac{1}{2\sigma^2}\left(\frac{(\hat{\beta}_0 - \mu_0)^2}{c + 1/n} + RSS \right.\right.$$
$$\left.\left. + (\hat{\boldsymbol{\theta}} - \boldsymbol{\theta}_0)'((X'X)^{-1} + \Gamma^{-1})^{-1}(\hat{\boldsymbol{\theta}} - \boldsymbol{\theta}_0))\right)\right).$$

(b) Prove that the predictive distribution of $\mathbf{y}$ under the target model is a multivariate t.

(c) Show that the joint predictive density of $(RSS, \hat{\boldsymbol{\theta}})$ is proportional to

$$\left\{2\gamma + RSS + (\widehat{\boldsymbol{\theta}} - \boldsymbol{\theta}_0)'((X'X)^{-1} + \Gamma^{-1})^{-1}(\widehat{\boldsymbol{\theta}} - \boldsymbol{\theta}_0)\right\}^{-(n+\alpha-1)/2}.$$

(d) Derive the prior predictive distribution of

$$T(\mathbf{y}) = \frac{(\widehat{\boldsymbol{\theta}} - \boldsymbol{\theta}_0)'((X'X)^{-1} + \Gamma^{-1})^{-1}(\widehat{\boldsymbol{\theta}} - \boldsymbol{\theta}_0)}{2\gamma + RSS}.$$

(e) Using an appropriately scaled $T(\mathbf{y})$ as the model departure statistic derive the prior predictive P-value.

15. Consider the same linear regression set-up as in Exercise 14, but let the target model now be

$$M_0 : \boldsymbol{\theta} = \mathbf{0}, \beta_0|\sigma^2 \sim N(\mu_0, c\sigma^2), \sigma^2 \sim \text{inverse Gamma}(\alpha, \gamma).$$

Assuming γ to be close to 0, use

$$T(\mathbf{y}) = \frac{\widehat{\boldsymbol{\theta}}' X'X \widehat{\boldsymbol{\theta}}}{RSS}$$

as the model departure statistic to derive the prior predictive P-value. Compare it with the classical P-value for testing $H_0 : \boldsymbol{\theta} = \mathbf{0}$.

16. Consider the same problem as in Exercise 15, but let the target model be

$$M_0 : \boldsymbol{\theta} = \mathbf{0}, \beta_0|\sigma^2 \sim N(\mu_0, c\sigma^2), \pi(\sigma^2) \propto \frac{1}{\sigma^2}.$$

Using $T(\mathbf{y}) = \widehat{\boldsymbol{\theta}}' X'X \widehat{\boldsymbol{\theta}}$ as the model departure statistic and RSS as the conditioning statistic, derive the conditional predictive P-value. Compute the partial predictive P-value using the same model departure statistic. Compare these with the classical P-value for testing $H_0 : \boldsymbol{\theta} = \mathbf{0}$.

17. Let $X_1, X_2, \cdots, X_n$ be i.i.d. with density

$$f(x|\lambda, \theta) = \lambda \exp(-\lambda(x - \theta)), x > \theta,$$

where $\lambda > 0$ and $-\infty < \theta < \infty$ are both unknown. Let the target model be

$$M_0 : \theta = 0, \pi(\lambda) \propto \frac{1}{\lambda}.$$

Suppose the smallest order statistic, $T = X_{(1)}$ is considered a suitable model departure statistic for this problem.

(a) Show that $T|\lambda \sim \text{exponential}(n\lambda)$ under M_0.

(b) Show that $\lambda|\mathbf{x}_{obs} \sim \text{Gamma}(n, n\bar{x}_{obs})$ under M_0.

(c) Show that

$$m(t|\mathbf{x}_{obs}) = \frac{n\bar{x}_{obs}^n}{(t + \bar{x}_{obs})^{n+1}}.$$

(d) Compute the posterior predictive P-value.

(e) Show that as $t_{obs} \longrightarrow \infty$, the posterior predictive P-value does not necessarily approach 0. (Note that $t_{obs} \le \bar{x}_{obs} \longrightarrow \infty$ also.)

18. (Contingency table) Casella and Berger (1990) present the following two-way table, which is the outcome of a famous medical experiment conducted by Joseph Lister. Lister performed 75 amputations with and without using carbolic acid.

Patient	Carbolic Acid Used?	
Lived?	Yes	No
Yes	34	19
No	6	16

Test for association of patient mortality with the use of carbolic acid on the basis of the above data using (a) BIC and (b) the classical likelihood ratio test. Discuss the different probabilistic interpretations underlying the two tests.

19. On the basis of the data on food poisoning presented in Table 2.1, you have to test whether potato salad was the cause. (Do this separately for Crab-meat and No Crab-meat).
    (a) Formulate this as a problem of testing a sharp null against the alternative that the null is false.
    (b) Test the sharp null using BIC.
    (c) Test the same null using the classical likelihood ratio test.
    (d) Discuss whether the notions of classical Type 1 and Type 2 error probabilities make sense here.

20. Using the BIC analyze the data of Problem 19 to explore whether crab-meat also contributed to food poisoning.

21. (Goodness of fit test). Feller (1973) presents the following data on bombing of London during World War II. The entire area of South London is divided into 576 small regions of equal area and the number (n_k) of regions with exactly k bomb hits are recorded.

k	0	1	2	3	4	5 and above
n_k	229	211	93	35	7	1

Test the null hypothesis that bombing was at random rather than the general belief that special targets were being bombed.
(Hint: Under H_0 use the Poisson model, under the alternative use the full multinomial model with 5 parameters and use BIC.)

22. (Hald's regression data). We present below a small set of data on heat evolved during the hardening of Portland cement and four variables that may be related to it (Woods et al. (1932), pp. 635–649). The sample size (n) is 13. The regressor variables (in percent of the weight) are $x_1 =$ calcium aluminate ($3Cao.Al_2O_3$), $x_2 =$ tricalcium silicate ($3CaO.SiO_2$), $x_3 =$ tetracalcium alumino ferrite ($4CaO.Al_2O_3.Fe_2O_3$), and $x_4 =$ dicalcium

Table 6.8. Cement Hardening Data

x_1	x_2	x_3	x_4	y
7	26	6	60	78.6
1	29	15	52	74.3
11	56	8	20	104.3
11	31	8	47	87.6
7	52	6	33	95.9
11	55	9	22	109.2
3	71	17	6	102.7
1	31	22	44	72.5
2	54	18	22	93.1
21	47	4	26	115.9
1	40	23	34	83.8
11	66	9	12	113.3
10	68	8	12	109.4

silicate (2CaO.SiO_2). The response variable is $y =$ total calories given off during hardening per gram of cement after 180 days.

Usually such a data set is analyzed using normal linear regression model of the form

$$y_i = \beta_0 + \beta_1 x_{1i} + \beta_2 x_{2i} \cdots + \beta_p x_{pi} + \epsilon_i, \ i = 1, \ldots, n,$$

where p is the number of regressor variables in the model, $\beta_0, \beta_1, \ldots \beta_p$ are unknown parameters, and ϵ_i's are independent errors having a $N(0, \sigma^2)$ distribution. There are a number of possible models depending on which regressor variables are kept in the model. Analyze the data and choose one from this set of possible models using (a) BIC, (b) AIBF of the full model relative to all possible models.

7

Bayesian Computations

Bayesian analysis requires computation of expectations and quantiles of probability distributions that arise as posterior distributions. Modes of the densities of such distributions are also sometimes used. The standard Bayes estimate is the posterior mean, which is also the Bayes rule under the squared error loss. Its accuracy is assessed using the posterior variance, which is again an expected value. Posterior median is sometimes utilized, and to provide Bayesian credible regions, quantiles of posterior distributions are needed. If conjugate priors are not used, as is mostly the case these days, posterior distributions will not be standard distributions and hence the required Bayesian quantities (i.e., posterior quantities of inferential interest) cannot be computed in closed form. Thus special techniques are needed for Bayesian computations.

Example 7.1. Suppose X is $N(\theta, \sigma^2)$ with known σ^2 and a Cauchy(μ, τ) prior on θ is considered appropriate from robustness considerations (see Chapter 3, Example 3.20). Then

$$\pi(\theta|x) \propto \exp\left(-(\theta - x)^2/(2\sigma^2)\right) \left(\tau^2 + (\theta - \mu)^2\right)^{-1},$$

and hence the posterior mean and variance are

$$E^\pi(\theta|x) = \frac{\int_{-\infty}^{\infty} \theta \exp\left(-\frac{(\theta-x)^2}{2\sigma^2}\right) \left(\tau^2 + (\theta - \mu)^2\right)^{-1} d\theta}{\int_{-\infty}^{\infty} \exp\left(-\frac{(\theta-x)^2}{2\sigma^2}\right) \left(\tau^2 + (\theta - \mu)^2\right)^{-1} d\theta}, \text{ and}$$

$$V^\pi(\theta|x) = \frac{\int_{-\infty}^{\infty} \theta^2 \exp\left(-\frac{(\theta-x)^2}{2\sigma^2}\right) \left(\tau^2 + (\theta - \mu)^2\right)^{-1} d\theta}{\int_{-\infty}^{\infty} \exp\left(-\frac{(\theta-x)^2}{2\sigma^2}\right) \left(\tau^2 + (\theta - \mu)^2\right)^{-1} d\theta} - \left(E^\pi(\theta|x)\right)^2.$$

Note that the above integrals cannot be computed in closed form, but various numerical integration techniques such as IMSL routines or Gaussian quadrature can be efficiently used to obtain very good approximations of these. On the other hand, the following example provides a more difficult problem.

Example 7.2. Suppose $X_1, X_2, \ldots, X_k$ are independent Poisson counts with $X_i \sim \text{Poisson}(\theta_i)$. θ_i are *a priori* considered related, and a joint multivariate normal prior distribution on their logarithm is assumed. Specifically, let $\nu_i = \log(\theta_i)$ be the ith element of $\boldsymbol{\nu}$ and suppose

$$\boldsymbol{\nu} \sim N_k \left(\mu\mathbf{1}, \tau^2 \left\{ (1-\rho)I_k + \rho\mathbf{11}' \right\} \right),$$

where $\mathbf{1}$ is the k-vector with all elements being 1, and μ, τ^2 and ρ are known constants. Then, because

$$f(\mathbf{x}|\boldsymbol{\nu}) = \exp\left(-\sum_{i=1}^{k}\{e^{\nu_i} - \nu_i x_i\} \right) \Big/ \prod_{i=1}^{k} x_i!,$$

and

$$\pi(\boldsymbol{\nu}) \propto \exp\left(-\frac{1}{2\tau^2}(\boldsymbol{\nu} - \mu\mathbf{1})' \left((1-\rho)I_k + \rho\mathbf{11}'\right)^{-1} (\boldsymbol{\nu} - \mu\mathbf{1}) \right)$$

we have that

$$\pi(\boldsymbol{\nu}|\mathbf{x}) \propto$$
$$\exp\left\{ -\sum_{i=1}^{k}\{e^{\nu_i} - \nu_i x_i\} - \frac{1}{2\tau^2}(\boldsymbol{\nu} - \mu\mathbf{1})' \left((1-\rho)I_k + \rho\mathbf{11}'\right)^{-1} (\boldsymbol{\nu} - \mu\mathbf{1}) \right\}.$$

Therefore, if the posterior mean of θ_j is of interest, we need to compute

$$E^\pi(\theta_j|x) = E^\pi(\exp(\nu_j)|x) = \frac{\int_{\mathcal{R}^k} \exp(\nu_j) g(\boldsymbol{\nu}|\mathbf{x})\, d\boldsymbol{\nu}}{\int_{\mathcal{R}^k} g(\boldsymbol{\nu}|\mathbf{x})\, d\boldsymbol{\nu}}$$

where $g(\boldsymbol{\nu}|\mathbf{x}) =$
$$\exp\left\{ -\sum_{i=1}^{k}\{e^{\nu_i} - \nu_i x_i\} - \frac{1}{2\tau^2}(\boldsymbol{\nu} - \mu\mathbf{1})' \left((1-\rho)I_k + \rho\mathbf{11}'\right)^{-1} (\boldsymbol{\nu} - \mu\mathbf{1}) \right\}.$$

This is a ratio of two k-dimensional integrals, and as k grows, the integrals become less and less easy to work with. Numerical integration techniques fail to be an efficient technique in this case. This problem, known as the *curse of dimensionality*, is due to the fact that the size of the part of the space that is not relevant for the computation of the integral grows very fast with the dimension. Consequently, the error in approximation associated with this numerical method increases as the power of the dimension k, making the technique inefficient. In fact, numerical integration techniques are presently not preferred except for single and two-dimensional integrals.

The recent popularity of Bayesian approach to statistical applications is mainly due to advances in statistical computing. These include the E-M algorithm discussed in Section 7.2 and the Markov chain Monte Carlo (MCMC) sampling techniques that are discussed in Section 7.4. As we see later, Bayesian analysis of real-life problems invariably involves difficult computations while MCMC techniques such as Gibbs sampling (Section 7.4.4) and Metropolis-Hastings algorithm (M-H) (Section 7.4.3) have rendered some of these very difficult computational tasks quite feasible.

7.1 Analytic Approximation

This is exactly what we saw in Section 4.3.2 where we derived analytic large sample approximations for certain integrals using the Laplace approximation. Specifically, suppose

$$E^{\pi}(g(\boldsymbol{\theta})|\mathbf{x}) = \frac{\int_{\mathcal{R}^k} g(\boldsymbol{\theta}) f(\mathbf{x}|\boldsymbol{\theta}) \pi(\boldsymbol{\theta})\, d\boldsymbol{\theta}}{\int_{\mathcal{R}^k} f(\mathbf{x}|\boldsymbol{\theta}) \pi(\boldsymbol{\theta})\, d\boldsymbol{\theta}} \tag{7.1}$$

is the Bayesian quantity of interest where g, f, and π are smooth functions of θ.

First, consider any integral of the form

$$I = \int_{\mathcal{R}^k} q(\boldsymbol{\theta}) \exp\left(-nh(\boldsymbol{\theta})\right)\, d\boldsymbol{\theta},$$

where h is a smooth function with $-h$ having its unique maximum at $\widehat{\theta}$. Then, as indicated in Section 4.3.1 for the univariate case, the Laplace method involves expanding q and h about $\widehat{\theta}$ in a Taylor series. Let $\mathbf{h}'$ and $\mathbf{q}'$ denote the vectors of partial derivatives of h and q, respectively, and Δ_h and Δ_q denote the Hessians of h and q. Then writing

$$h(\boldsymbol{\theta}) = h(\widehat{\boldsymbol{\theta}}) + (\boldsymbol{\theta} - \widehat{\boldsymbol{\theta}})'\mathbf{h}'(\widehat{\boldsymbol{\theta}}) + \frac{1}{2}(\boldsymbol{\theta} - \widehat{\boldsymbol{\theta}})'\Delta_h(\widehat{\boldsymbol{\theta}})(\boldsymbol{\theta} - \widehat{\boldsymbol{\theta}}) + \cdots$$

$$= h(\widehat{\boldsymbol{\theta}}) + \frac{1}{2}(\boldsymbol{\theta} - \widehat{\boldsymbol{\theta}})'\Delta_h(\widehat{\boldsymbol{\theta}})(\boldsymbol{\theta} - \widehat{\boldsymbol{\theta}}) + \cdots \text{ and}$$

$$q(\boldsymbol{\theta}) = q(\widehat{\boldsymbol{\theta}}) + (\boldsymbol{\theta} - \widehat{\boldsymbol{\theta}})'\mathbf{q}'(\widehat{\boldsymbol{\theta}}) + \frac{1}{2}(\boldsymbol{\theta} - \widehat{\boldsymbol{\theta}})'\Delta_q(\widehat{\boldsymbol{\theta}})(\boldsymbol{\theta} - \widehat{\boldsymbol{\theta}}) + \cdots,$$

we obtain

$$I = \int_{\mathcal{R}^k} \left\{ q(\widehat{\boldsymbol{\theta}}) + (\boldsymbol{\theta} - \widehat{\boldsymbol{\theta}})'\mathbf{q}'(\widehat{\boldsymbol{\theta}}) + \frac{1}{2}(\boldsymbol{\theta} - \widehat{\boldsymbol{\theta}})'\Delta_q(\widehat{\boldsymbol{\theta}})(\boldsymbol{\theta} - \widehat{\boldsymbol{\theta}}) + \cdots \right\}$$

$$\times e^{-nh(\widehat{\boldsymbol{\theta}})} \exp\left(-\frac{n}{2}(\boldsymbol{\theta} - \widehat{\boldsymbol{\theta}})'\Delta_h(\widehat{\boldsymbol{\theta}})(\boldsymbol{\theta} - \widehat{\boldsymbol{\theta}}) + \cdots\right)\, d\boldsymbol{\theta}$$

$$= e^{-nh(\widehat{\boldsymbol{\theta}})}(2\pi)^{k/2} n^{-k/2} |\Delta_h(\widehat{\boldsymbol{\theta}})|^{-1/2} \left\{ q(\widehat{\boldsymbol{\theta}}) + O(n^{-1}) \right\}, \tag{7.2}$$

which is exactly (4.16). Upon applying this to both the numerator and denominator of (7.1) separately (with q equal to g and 1), a first-order approximation

$$E^{\pi}(g(\boldsymbol{\theta})|\mathbf{x}) = g(\widehat{\boldsymbol{\theta}}) \left\{ 1 + O(n^{-1}) \right\}$$

easily emerges. It also indicates that a second-order approximation may be available if further terms in the Taylor series expansion are retained in the approximation.

Suppose that g in (7.1) is positive, and let $-nh(\boldsymbol{\theta}) = \log f(\mathbf{x}|\boldsymbol{\theta}) + \log \pi(\boldsymbol{\theta})$, $-nh^*(\boldsymbol{\theta}) = -nh(\boldsymbol{\theta}) + \log g(\boldsymbol{\theta})$. Now apply (7.2) to both the numerator and denominator of (7.1) with q equal to 1. Then, letting $\boldsymbol{\theta}^*$ denote the maximum of $-h^*$, $\Sigma = \Delta_h^{-1}(\widehat{\boldsymbol{\theta}})$, $\Sigma^* = \Delta_{h^*}^{-1}(\widehat{\boldsymbol{\theta}^*})$, as mentioned in Section 4.3.2, Tierney and Kadane (1986) obtain the fantastic approximation

$$E^{\pi}(g(\boldsymbol{\theta})|\mathbf{x}) = \frac{|\Sigma^*|^{1/2}\exp\left(-nh^*(\widehat{\boldsymbol{\theta}^*})\right)}{|\Sigma|^{1/2}\exp\left(-nh(\widehat{\boldsymbol{\theta}})\right)}\left\{1 + O(n^{-2})\right\}, \qquad (7.3)$$

which they call *fully exponential*. This technique can be used in Example 7.2. Note that to derive the approximation in (7.3), it is enough to have the probability distribution of $g(\boldsymbol{\theta})$ concentrate away from the origin on the positive side. Therefore, often when g is non-positive, (7.3) can be applied after adding a large positive constant to g, and this constant is to be subtracted after obtaining the approximation. Some other analytic approximations are also available. Angers and Delampady (1997) use an *exponential* approximation for a probability distribution that concentrates near the origin. We will not be emphasizing any of these techniques here, including the many numerical integration methods mentioned previously, because the availability of powerful and all-purpose simulation methods have rendered them less powerful.

7.2 The E-M Algorithm

We shall use a slightly different notation here. Suppose $\mathbf{Y}|\boldsymbol{\theta}$ has density $f(\mathbf{y}|\boldsymbol{\theta})$, and suppose the prior on $\boldsymbol{\theta}$ is $\pi(\boldsymbol{\theta})$, resulting in the posterior density $\pi(\boldsymbol{\theta}|\mathbf{y})$. When $\pi(\boldsymbol{\theta}|\mathbf{y})$ is computationally difficult to handle, as is usually the case, there are some 'data augmentation' methods that can help. The idea is to augment the observed data $\mathbf{y}$ with missing or latent data $\mathbf{z}$ to obtain the 'complete' data $\mathbf{x} = (\mathbf{y}, \mathbf{z})$ so that the augmented posterior density $\pi(\boldsymbol{\theta}|\mathbf{x}) = \pi(\boldsymbol{\theta}|\mathbf{y}, \mathbf{z})$ is computationally easy to handle. The E-M algorithm (see Dempster et al. (1977), Tanner (1991), or McLachlan and Krishnan (1997)) is the simplest among such data augmentation methods. In our context, the E-M algorithm is meant for computing the posterior mode. However, if data augmentation yields a computationally simple posterior distribution, there are more powerful computational tools available that can provide a lot more information on the posterior distribution as will be seen later in this chapter.

The basic steps in the iterations of the E-M algorithm are the following. Let $p(\mathbf{z}|\mathbf{y}, \hat{\boldsymbol{\theta}})$ be the predictive density of $\mathbf{Z}$ given $\mathbf{y}$ and an estimate $\hat{\boldsymbol{\theta}}$ of $\boldsymbol{\theta}$. Find $\mathbf{z}^{(i)} = E(\mathbf{Z}|\mathbf{y}, \hat{\boldsymbol{\theta}}^{(i)})$, where $\hat{\boldsymbol{\theta}}^{(i)}$ is the estimate of $\boldsymbol{\theta}$ used at the ith step of the iteration. Note the similarity with estimating missing values. Use $\mathbf{z}^{(i)}$ to augment $\mathbf{y}$ and maximize $\pi(\boldsymbol{\theta}|\mathbf{y}, \mathbf{z}^{(i)})$ to obtain $\hat{\boldsymbol{\theta}}^{(i+1)}$. Then find $\mathbf{z}^{(i+1)}$ using $\hat{\boldsymbol{\theta}}^{(i+1)}$ and continue this iteration. This combination of expectation followed by maximization in each iteration gives its name to the E-M algorithm.

Implementation of the E-M Algorithm

Note that because $\pi(\boldsymbol{\theta}|\mathbf{y}) = \pi(\boldsymbol{\theta}, \mathbf{z}|\mathbf{y})/p(\mathbf{z}|\mathbf{y}, \boldsymbol{\theta})$, we have that

$$\log \pi(\boldsymbol{\theta}|\mathbf{y}) = \log \pi(\boldsymbol{\theta}, \mathbf{z}|\mathbf{y}) - \log p(\mathbf{z}|\mathbf{y}, \boldsymbol{\theta}).$$

Taking expectation with respect to $\mathbf{Z}|\hat{\boldsymbol{\theta}}^{(i)}, \mathbf{y}$ on both sides, we get

$$\log \pi(\boldsymbol{\theta}|\mathbf{y}) = \int \log \pi(\boldsymbol{\theta}, \mathbf{z}|\mathbf{y}) p(\mathbf{z}|\mathbf{y}, \hat{\boldsymbol{\theta}}^{(i)}) \, d\mathbf{z} - \int \log p(\mathbf{z}|\mathbf{y}, \boldsymbol{\theta}) p(\mathbf{z}|\mathbf{y}, \hat{\boldsymbol{\theta}}^{(i)}) \, d\mathbf{z}$$

$$= Q(\boldsymbol{\theta}, \hat{\boldsymbol{\theta}}^{(i)}) - H(\boldsymbol{\theta}, \hat{\boldsymbol{\theta}}^{(i)}) \tag{7.4}$$

(where Q and H are according to the notation of Dempster et al. (1977)). Then, the general E-M algorithm involves the following two steps in the ith iteration:

E-Step: Calculate $Q(\boldsymbol{\theta}, \hat{\boldsymbol{\theta}}^{(i)})$;

M-Step: Maximize $Q(\boldsymbol{\theta}, \hat{\boldsymbol{\theta}}^{(i)})$ with respect to $\boldsymbol{\theta}$ and obtain $\hat{\boldsymbol{\theta}}^{(i)}$ such that

$$\max_{\boldsymbol{\theta}} Q(\boldsymbol{\theta}, \hat{\boldsymbol{\theta}}^{(i)}) = Q(\boldsymbol{\theta}^{(i+1)}, \hat{\boldsymbol{\theta}}^{(i)}).$$

Note that

$$\log \pi(\hat{\boldsymbol{\theta}}^{(i+1)}|\mathbf{y}) - \log \pi(\hat{\boldsymbol{\theta}}^{(i)}|\mathbf{y}) = \left\{ Q(\boldsymbol{\theta}^{(i+1)}, \hat{\boldsymbol{\theta}}^{(i)}) - Q(\boldsymbol{\theta}^{(i)}, \hat{\boldsymbol{\theta}}^{(i)}) \right\}$$

$$- \left\{ H(\boldsymbol{\theta}^{(i+1)}, \hat{\boldsymbol{\theta}}^{(i)}) - H(\boldsymbol{\theta}^{(i)}, \hat{\boldsymbol{\theta}}^{(i)}) \right\}.$$

From the E-M algorithm, we have that $Q(\boldsymbol{\theta}^{(i+1)}, \hat{\boldsymbol{\theta}}^{(i)}) \geq Q(\boldsymbol{\theta}^{(i)}, \hat{\boldsymbol{\theta}}^{(i)})$. Further, for any $\boldsymbol{\theta}$,

$$H(\boldsymbol{\theta}, \hat{\boldsymbol{\theta}}^{(i)}) - H(\boldsymbol{\theta}^{(i)}, \hat{\boldsymbol{\theta}}^{(i)})$$

$$= \int \log p(\mathbf{z}|\mathbf{y}, \boldsymbol{\theta}) p(\mathbf{z}|\mathbf{y}, \hat{\boldsymbol{\theta}}^{(i)}) \, d\mathbf{z} - \int \log p(\mathbf{z}|\mathbf{y}, \boldsymbol{\theta}^{(i)}) p(\mathbf{z}|\mathbf{y}, \hat{\boldsymbol{\theta}}^{(i)}) \, d\mathbf{z}$$

$$= \int \log \left[\frac{p(\mathbf{z}|\mathbf{y}, \boldsymbol{\theta})}{p(\mathbf{z}|\mathbf{y}, \boldsymbol{\theta}^{(i)})} \right] p(\mathbf{z}|\mathbf{y}, \boldsymbol{\theta}^{(i)}) \, d\mathbf{z}$$

$$= - \int \log \left[\frac{p(\mathbf{z}|\mathbf{y}, \boldsymbol{\theta}^{(i)})}{p(\mathbf{z}|\mathbf{y}, \boldsymbol{\theta})} \right] p(\mathbf{z}|\mathbf{y}, \boldsymbol{\theta}^{(i)}) \, d\mathbf{z}$$

$$\leq 0,$$

because, for any two densities p_1 and p_2, $\int \log(p_1(x)/p_2(x)) p_1(x) \, dx$ is the Kullback-Leibler distance between p_1 and p_2, which is at least 0. Therefore,

$$H(\boldsymbol{\theta}^{(i+1)}, \hat{\boldsymbol{\theta}}^{(i)}) - H(\boldsymbol{\theta}^{(i)}, \hat{\boldsymbol{\theta}}^{(i)}) \leq 0,$$

and hence

$$\pi(\hat{\boldsymbol{\theta}}^{(i+1)}|\mathbf{y}) \geq \pi(\hat{\boldsymbol{\theta}}^{(i)}|\mathbf{y})$$

for any iteration i. Therefore, starting from any point, the E-M algorithm can usually be expected to converge to a local maximum.

Table 7.1. *Genetic Linkage* Data

Cell Count		Probability
$y_1 =$	125	$\frac{1}{2} + \frac{\theta}{4}$
$y_2 =$	18	$\frac{1}{4}(1 - \theta)$
$y_3 =$	20	$\frac{1}{4}(1 - \theta)$
$y_4 =$	34	$\frac{\theta}{4}$

Example 7.3. (genetic linkage model.) Consider the data from Rao (1973) on a certain recombination rate in genetics (see Sorensen and Gianola (2002) for details). Here 197 counts are classified into 4 categories as shown in Table 7.1, along with the corresponding theoretical cell probabilities.

The multinomial mass function in this example is given by $f(\mathbf{y}|\theta) \propto (2 + \theta)^{y_1}(1 - \theta)^{y_2 + y_3}\theta^{y_4}$, so that under the uniform$(0, 1)$ prior on θ, the observed posterior density is given by

$$\pi(\theta|\mathbf{y}) \propto (2 + \theta)^{y_1}(1 - \theta)^{y_2 + y_3}\theta^{y_4}.$$

This is not a standard density due to the presence of $2 + \theta$. If we split the first cell into two with probabilities $1/2$ and $\theta/4$, respectively, the complete data will be given by $\mathbf{x} = (x_1, x_2, x_3, x_4, x_5)$, where $x_1 + x_2 = y_1$, $x_3 = y_2$, $x_4 = y_3$ and $x_5 = y_4$. The augmented posterior density will then be given by

$$\pi(\theta|\mathbf{x}) \propto \theta^{x_2 + x_5}(1 - \theta)^{x_3 + x_4},$$

which corresponds with the Beta density.

The E-step of E-M consists of obtaining

$$Q(\theta, \hat{\theta}^{(i)}) = E\left[(X_2 + X_5)\log\theta + (X_3 + X_4)\log(1 - \theta)|\mathbf{y}, \hat{\theta}^{(i)}\right]$$

$$= \left\{E\left[X_2|\mathbf{y}, \hat{\theta}^{(i)}\right] + y_4\right\}\log\theta + (y_2 + y_3)\log(1 - \theta). \quad (7.5)$$

The M-step involves finding $\hat{\theta}^{(i+1)}$ to maximize (7.5). We can do this by solving $\frac{\partial}{\partial\theta}Q(\theta, \hat{\theta}^{(i)}) = 0$, so that

$$\hat{\theta}^{(i+1)} = \frac{E\left[X_2|\mathbf{y}, \hat{\theta}^{(i)}\right] + y_4}{E\left[X_2|\mathbf{y}, \hat{\theta}^{(i)}\right] + y_4 + y_2 + y_3}. \quad (7.6)$$

Now note that $E\left[X_2|\mathbf{y}, \hat{\theta}^{(i)}\right] = E\left[X_2|X_1 + X_2, \hat{\theta}^{(i)}\right]$, and that

$X_2|X_1 + X_2, \hat{\theta}^{(i)} \sim \text{binomial}(X_1 + X_2, \frac{\hat{\theta}^{(i)}/4}{1/2 + \hat{\theta}^{(i)}/4})$. Therefore,

$$E\left[X_2|X_1 + X_2 = y_1, \hat{\theta}^{(i)}\right] = y_1\frac{\hat{\theta}^{(i)}}{2 + \hat{\theta}^{(i)}},$$

and hence

Table 7.2. E-M Iterations for *Genetic Linkage* Data Example

Iteration i	$\hat{\theta}^{(i)}$
1	.60825
2	.62432
3	.62648
4	.62678
5	.62682
6	.62682

$$\hat{\theta}^{(i+1)} = \frac{y_1 \frac{\hat{\theta}^{(i)}}{2+\hat{\theta}^{(i)}} + y_4}{y_1 \frac{\hat{\theta}^{(i)}}{2+\hat{\theta}^{(i)}} + y_2 + y_3 + y_4}. \tag{7.7}$$

In our example, (7.7) converges to $\hat{\theta} = .62682$ in 5 iterations starting from $\hat{\theta}^{(0)} = .5$ as shown in Table 7.2.

7.3 Monte Carlo Sampling

Consider an expectation that is not available in closed form. An alternative to numerical integration or analytic approximation to compute this is statistical sampling. This probabilistic technique is a familiar tool in statistical inference. To estimate a population mean or a population proportion, a natural approach is to gather a large sample from this population and to consider the corresponding sample mean or the sample proportion. The law of large numbers guarantees that the estimates so obtained will be *good* provided the sample is large enough. Specifically, let f be a probability density function (or a mass function) and suppose the quantity of interest is a finite expectation of the form

$$E_f h(\mathbf{X}) = \int_{\mathcal{X}} h(\mathbf{x}) f(\mathbf{x}) \, d\mathbf{x} \tag{7.8}$$

(or the corresponding sum in the discrete case). If i.i.d. observations $\mathbf{X}_1, \mathbf{X}_2, \ldots$ can be generated from the density f, then

$$\bar{h}_m = \frac{1}{m} \sum_{i=1}^{m} h(\mathbf{X}_i) \tag{7.9}$$

converges in probability (or even almost surely) to $E_f h(\mathbf{X})$. This justifies using $\bar{h}_m$ as an approximation for $E_f h(\mathbf{X})$ for large m. To provide a measure of accuracy or the extent of error in the approximation, we can again use a statistical technique and compute the standard error. If $\mathrm{Var}_f h(\mathbf{X})$ is finite, then $\mathrm{Var}_f(\bar{h}_m) = \mathrm{Var}_f h(\mathbf{X})/m$. Further, $\mathrm{Var}_f h(\mathbf{X}) = E_f h^2(\mathbf{X}) - \left(E_f h(\mathbf{X})\right)^2$ can be estimated by

$$s_m^2 = \frac{1}{m} \sum_{i=1}^{m} \left(h(\mathbf{X}_i) - \bar{h}_m \right)^2,$$

and hence the standard error of $\bar{h}_m$ can be estimated by

$$\frac{1}{\sqrt{m}} s_m = \frac{1}{m} \left(\sum_{i=1}^{m} \left(h(\mathbf{X}_i) - \bar{h}_m \right)^2 \right)^{1/2}.$$

If one wishes, confidence intervals for $E_f h(\mathbf{X})$ can also be provided using the central limit theorem. Because

$$\frac{\sqrt{m} \left(\bar{h}_m - E_f h(\mathbf{X}) \right)}{s_m} \xrightarrow[m \to \infty]{} N(0,1)$$

in distribution, $\left(\bar{h}_m - z_{\alpha/2} s_m / \sqrt{m}, \bar{h}_m + z_{\alpha/2} s_m / \sqrt{m} \right)$ can be used as an approximate $100(1 - \alpha)\%$ confidence interval for $E_f h(\mathbf{X})$, with $z_{\alpha/2}$ denoting the $100(1 - \alpha/2)\%$ quantile of standard normal.

The above discussion suggests that if we want to approximate the posterior mean, we could try to generate i.i.d. observations from the posterior distribution and consider the mean of this sample. This is rarely useful because most often the posterior distribution will be a non-standard distribution which may not easily allow sampling from it. Note that there are other possibilities as seen below.

Example 7.4. (Example 7.1 continued.) Recall that

$$E^{\pi}(\theta|x) = \frac{\int_{-\infty}^{\infty} \theta \exp\left(-\frac{(\theta-x)^2}{2\sigma^2} \right) \left(\tau^2 + (\theta - \mu)^2 \right)^{-1} d\theta}{\int_{-\infty}^{\infty} \exp\left(-\frac{(\theta-x)^2}{2\sigma^2} \right) \left(\tau^2 + (\theta - \mu)^2 \right)^{-1} d\theta}$$

$$= \frac{\int_{-\infty}^{\infty} \theta \left\{ \frac{1}{\sigma} \phi\left(\frac{\theta-x}{\sigma} \right) \right\} \left(\tau^2 + (\theta - \mu)^2 \right)^{-1} d\theta}{\int_{-\infty}^{\infty} \left\{ \frac{1}{\sigma} \phi\left(\frac{\theta-x}{\sigma} \right) \right\} \left(\tau^2 + (\theta - \mu)^2 \right)^{-1} d\theta},$$

where ϕ denotes the density of standard normal. Thus $E^{\pi}(\theta|x)$ is the ratio of expectation of $h(\theta) = \theta/(\tau^2 + (\theta - \mu)^2)$ to that of $h(\theta) = 1/(\tau^2 + (\theta - \mu)^2)$, both expectations being with respect to the $N(x, \sigma^2)$ distribution. Therefore, we simply sample $\theta_1, \theta_2, \ldots$ from $N(x, \sigma^2)$ and use

$$\widehat{E^{\pi}(\theta|x)} = \frac{\sum_{i=1}^{m} \theta_i \left(\tau^2 + (\theta_i - \mu)^2 \right)^{-1}}{\sum_{i=1}^{m} \left(\tau^2 + (\theta_i - \mu)^2 \right)^{-1}}$$

as our Monte Carlo estimate of $E^{\pi}(\theta|x)$. Note that (7.8) and (7.9) are applied separately to both the numerator and denominator, but using the same sample of θ's.

It is unwise to assume that the problem has been completely solved. The sample of θ's generated from $N(x, \sigma^2)$ will tend to concentrate around x, whereas to satisfactorily account for the contribution of the Cauchy prior to the posterior mean, a significant portion of the θ's should come from the tails of the posterior distribution. It may therefore appear that it is perhaps better to express the posterior mean in the form

$$E^\pi(\theta|x) = \frac{\int_{-\infty}^{\infty} \theta \exp\left(-\frac{(\theta-x)^2}{2\sigma^2}\right) \pi(\theta)\, d\theta}{\int_{-\infty}^{\infty} \exp\left(-\frac{(\theta-x)^2}{2\sigma^2}\right) \pi(\theta)\, d\theta},$$

then sample θ's from Cauchy(μ, τ) and use the approximation

$$\widehat{E^\pi(\theta|x)} = \frac{\sum_{i=1}^{m} \theta_i \exp\left(-\frac{(\theta_i-x)^2}{2\sigma^2}\right)}{\sum_{i=1}^{m} \exp\left(-\frac{(\theta_i-x)^2}{2\sigma^2}\right)}.$$

However, this is also not totally satisfactory because the tails of the posterior distribution are not as heavy as those of the Cauchy prior, and hence there will be excess sampling from the tails relative to the center. The implication is that the convergence of the approximation is slower and hence a larger error in approximation (for a fixed m). Ideally, therefore, sampling should be from the posterior distribution itself for a satisfactory approximation. With this view in mind, a variation of the above theme has been developed. This is called the Monte Carlo importance sampling.

Consider (7.8) again. Suppose that it is difficult or expensive to sample directly from f, but there exists a probability density u that is very *close* to f from which it is easy to sample. Then we can rewrite (7.8) as

$$\begin{aligned}
E_f h(\mathbf{X}) &= \int_{\mathcal{X}} h(\mathbf{x}) f(\mathbf{x})\, d\mathbf{x} \\
&= \int_{\mathcal{X}} h(\mathbf{x}) \frac{f(\mathbf{x})}{u(\mathbf{x})} u(\mathbf{x})\, d\mathbf{x} \\
&= \int_{\mathcal{X}} \{h(\mathbf{x}) w(\mathbf{x})\}\, u(\mathbf{x})\, d\mathbf{x} \\
&= E_u \{h(\mathbf{X}) w(\mathbf{X})\},
\end{aligned}$$

where $w(\mathbf{x}) = f(\mathbf{x})/u(\mathbf{x})$. Now apply (7.9) with f replaced by u and h replaced by hw. In other words, generate i.i.d. observations $\mathbf{X}_1, \mathbf{X}_2, \ldots$ from the density u and compute

$$\overline{hw}_m = \frac{1}{m} \sum_{i=1}^{m} h(\mathbf{X}_i) w(\mathbf{X}_i).$$

The sampling density u is called the *importance* function. We illustrate importance sampling with the following example.

Example 7.5. Suppose $X_1, X_2, \ldots, X_n$ are i.i.d. $N(\theta, \sigma^2)$, where both θ and σ^2 are unknown. Independent priors are assumed for θ and σ^2, where θ has a double exponential distribution with density $\exp(-|\theta|)/2$ and σ^2 has the prior density of $(1 + \sigma^2)^{-2}$. Neither of these is a standard prior, but robust choice of proper prior all the same. If the posterior mean of θ is of interest, then it is necessary to compute

$$E^\pi(\theta|\mathbf{x}) = \int_{-\infty}^{\infty} \int_0^{\infty} \theta \pi(\theta, \sigma^2|\mathbf{x}) \, d\theta \, d\sigma^2.$$

Because $\pi(\theta, \sigma^2|\mathbf{x})$ is not a standard density, let us look for a standard density close to it. Letting $\bar{x}$ denote the mean of the sample $x_1, x_2, \ldots, x_n$ and $s_n^2 = \sum_{i=1}^{n}(x_i - \bar{x})^2/n$, note that

$$\pi(\theta, \sigma^2|\mathbf{x}) \propto (\sigma^2)^{-n/2} \exp\left(-\frac{n}{2\sigma^2}\left\{(\theta - \bar{x})^2 + s_n^2\right\}\right) \exp(-|\theta|)(1 + \sigma^2)^{-2}$$

$$= \left[s_n^2 + (\theta - \bar{x})^2\right]^{n/2+1} (\sigma^2)^{-(n/2+2)} \exp\left(-\frac{n}{2\sigma^2}\left\{(\theta - \bar{x})^2 + s_n^2\right\}\right)$$

$$\times \left\{\left[s_n^2 + (\theta - \bar{x})^2\right]^{-(n/2+1)}\right\} \exp(-|\theta|)\left(\frac{\sigma^2}{1 + \sigma^2}\right)^2$$

$$\propto u_1(\sigma^2|\theta)u_2(\theta) \exp(-|\theta|)\left(\frac{\sigma^2}{1 + \sigma^2}\right)^2,$$

where $u_1(\sigma^2|\theta)$ is the density of inverse Gamma with shape parameter $n/2+1$ and scale parameter $\frac{n}{2}\{(\theta - \bar{x})^2 + s_n^2\}$, and u_2 is the Student's t density with d.f. $n + 1$, location $\bar{x}$ and scale a multiple of s_n. It may be noted that the tails of $\exp(-|\theta|)(\frac{\sigma^2}{1+\sigma^2})^2$ do not have much of an influence in the presence of $u_1(\sigma^2|\theta)u_2(\theta)$. Therefore, $u(\theta, \sigma^2) = u_1(\sigma^2|\theta)u_2(\theta)$ may be chosen as a suitable importance function. This involves sampling θ first from the density $u_2(\theta)$, and given this θ, sampling σ^2 from $u_1(\sigma^2|\theta)$. This is repeated to generate further values of (θ, σ^2). Finally, after generating m of these pairs (θ, σ^2), the required posterior mean of θ is approximated by

$$\widehat{E^\pi(\theta|\mathbf{x})} = \frac{\sum_{i=1}^{m} \theta_i w(\theta_i, \sigma_i^2)}{\sum_{i=1}^{m} w(\theta_i, \sigma_i^2)},$$

where $w(\theta, \sigma^2) = f(\mathbf{x}|\theta, \sigma^2)\pi(\theta, \sigma^2)/u(\theta, \sigma^2)$.

In some high-dimensional problems, a combination of numerical integration, Laplace approximation and Monte Carlo sampling seems to give appealing results. Delampady et al. (1993) use a Laplace-type approximation to obtain a suitable importance function in a high-dimensional problem.

One area that we have not touched upon is how to generate random deviates from a given probability distribution. Clearly, this is a very important subject being the basis of any Monte Carlo sampling technique. Instead of providing a sketchy discussion from this vast area, we refer the reader to the excellent book by Robert and Casella (1999). We would, however, like to mention one recent and very important development in this area. This is the discovery of a very efficient algorithm to generate a sequence of uniform random deviates with a very big period of $2^{19937} - 1$. This algorithm, known as the Mersenne twister (MT), has many other desirable features as well, details on which may be found in Matsumoto and Nishimura (1998). The property of having a very large period is especially important because Monte Carlo simulation methods, especially MCMC, require very long sequences of random deviates for proper implementation.

7.4 Markov Chain Monte Carlo Methods

7.4.1 Introduction

A severe drawback of the standard Monte Carlo sampling or Monte Carlo importance sampling is that complete determination of the functional form of the posterior density is needed for their implementation. Situations where posterior distributions are incompletely specified or are specified indirectly cannot be handled. One such instance is where the joint posterior distribution of the vector of parameters is specified in terms of several conditional and marginal distributions, but not directly. This actually covers a very large range of Bayesian analysis because a lot of Bayesian modeling is hierarchical so that the joint posterior is difficult to calculate but the conditional posteriors given parameters at different levels of hierarchy are easier to write down (and hence sample from). For instance, consider the normal-Cauchy problem of Example 7.1. As shown later in Section 7.4.6, this problem can be given a hierarchical structure wherein we have the normal model, the conjugate normal prior in the first stage with a hyperparameter for its variance and this hyperparameter again has the conjugate prior. Similarly, consider Example 7.2 where we have independent observations $X_i \sim \text{Poisson}(\theta_i)$. Now suppose the prior on the θ_i's is a conjugate mixture. We again see (Problem 14) that a hierarchical prior structure can lead to analytically tractable conditional posteriors. It turns out that it is indeed possible in such cases to adopt an iterative Monte Carlo sampling scheme, which at the point of convergence will guarantee a random draw from the target joint posterior distribution. These iterative Monte Carlo procedures typically generate a random sequence with the Markov property such that this Markov chain is ergodic with the limiting distribution being the target posterior distribution. There is actually a whole class of such iterative procedures collectively called Markov chain Monte Carlo (MCMC) procedures. Different procedures from this class are suitable for different situations.

As mentioned above, convergence of a random sequence with the Markov property is being utilized in this procedure, and hence some basic understanding of Markov chains is required. This material is presented below. This discussion as well as the following sections are mainly based on Athreya et al. (2003).

7.4.2 Markov Chains in MCMC

A sequence of random variables $\{X_n\}_{n \geq 0}$ is a *Markov chain* if for any n, given the current value, X_n, the *past* $\{X_j, j \leq n-1\}$ and the *future* $\{X_j : j \geq n+1\}$ are *independent*. In other words,

$$P(A \cap B | X_n) = P(A | X_n) P(B | X_n), \tag{7.10}$$

where A and B are events defined respectively in terms of the past and the future. Among Markov chains there is a subclass that has wide applicability. They are Markov chains with time homogeneous or *stationary transition probabilities*, meaning that the probability distribution of X_{n+1} given $X_n = x$, and the past, $X_j : j \leq n-1$ depends only on x and does not depend on the values of $X_j : j \leq n-1$ or n. If the set S of values $\{X_n\}$ can take, known as the *state space*, is countable, this reduces to specifying the transition probability matrix $P \equiv ((p_{ij}))$ where for any two values i, j in S, p_{ij} is the probability that $X_{n+1} = j$ given $X_n = i$, i.e., of moving from state i to state j in one time unit. For state space S that is not countable, one has to specify a *transition kernel* or *transition function* $P(x, \cdot)$ where $P(x, A)$ is the probability of moving from x into A in one step, i.e., $P(X_{n+1} \in A | X_n = x)$. Given the transition probability and the probability distribution of the initial value X_0, one can construct the joint probability distribution of $\{X_j : 0 \leq j \leq n\}$ for any finite n. For example, in the countable state space case

$$\begin{aligned}
P(X_0 &= i_0, X_1 = i_1, \ldots, X_{n-1} = i_{n-1}, X_n = i_n) \\
&= P(X_n = i_n | X_0 = i_0, \ldots, X_{n-1} = i_{n-1}) \\
&\quad \times P(X_0 = i_0, X_1 = i_1, \ldots X_{n-1} = i_{n-1}) \\
&= p_{i_{n-1} i_n} P(X_0 = i_0, \ldots, X_{n-1} = i_{n-1}) \\
&= P(X_0 = i_0) p_{i_0 i_1} p_{i_1 i_2} \cdots p_{i_{n-1} i_n}.
\end{aligned}$$

A probability distribution π is called *stationary* or *invariant* for a transition probability P or the associated Markov chain $\{X_n\}$ if it is the case that when the probability distribution of X_0 is π then the same is true for X_n for all $n \geq 1$. Thus in the countable state space case a probability distribution $\pi = \{\pi_i : i \in S\}$ is stationary for a transition probability matrix P if for each j in S,

$$\begin{aligned}
P(X_1 = j) &= \sum_i P(X_1 = j | X_0 = i) P(X_0 = i) \\
&= \sum_i \pi_i p_{ij} = P(X_0 = j) = \pi_j. \tag{7.11}
\end{aligned}$$

In vector notation it says $\boldsymbol{\pi} = (\pi_1, \pi_2, \ldots)$ is a left eigenvector of the matrix P with eigenvalue 1 and

$$\boldsymbol{\pi} = \boldsymbol{\pi} P. \tag{7.12}$$

Similarly, if S is a continuum, a probability distribution π with density $p(x)$ is *stationary* for the transition kernel $P(\cdot, \cdot)$ if

$$\pi(A) = \int_A p(x)\, dx = \int_S P(x, A) p(x)\, dx$$

for all $A \subset S$.

A Markov chain $\{X_n\}$ with a countable state space S and transition probability matrix $P \equiv ((p_{ij}))$ is said to be *irreducible* if for any two states i and j the probability of the Markov chain visiting j starting from i is positive, i.e., for some $n \geq 1, p_{ij}^{(n)} \equiv P(X_n = j | X_0 = i) > 0$. A similar notion of *irreducibility*, known as Harris or Doeblin irreducibility exists for the general state space case also. For details on this somewhat advanced notion as well as other results that we state here without proof, see Robert and Casella (1999) or Meyn and Tweedie (1993). In addition, Tierney(1994) and Athreya et al. (1996) may be used as more advanced references on irreducibility and MCMC. In particular, the last reference uses the fact that there is a stationary distribution of the Markov chain, namely, the joint posterior, and thus provides better and very explicit conditions for the MCMC to converge.

Theorem 7.6. *(law of large numbers for Markov chains) Let $\{X_n\}_{n \geq 0}$ be a Markov chain with a countable state space S and a transition probability matrix P. Further, suppose it is* irreducible *and has a stationary probability distribution $\pi \equiv (\pi_i : i \in S)$ as defined in (7.11). Then, for any bounded function $h : S \to R$ and for any initial distribution of X_0*

$$\frac{1}{n} \sum_{i=0}^{n-1} h(X_i) \to \sum_j h(j)\pi_j \tag{7.13}$$

in probability as $n \to \infty$.

A similar law of large numbers (LLN) holds when the state space S is not countable. The limit value in (7.13) will be the integral of h with respect to the stationary distribution π. A sufficient condition for the validity of this LLN is that the Markov chain $\{X_n\}$ be Harris irreducible and have a stationary distribution π.

To see how this is useful to us, consider the following. Given a probability distribution π on a set S, and a function h on S, suppose it is desired to compute the "integral of h with respect to π", which reduces to $\sum_j h(j)\pi_j$ in the countable case. *Look for an irreducible Markov chain $\{X_n\}$ with state space S and stationary distribution π.* Then, starting from some initial value

X_0, run the Markov chain $\{X_j\}$ for a period of time, say $0, 1, 2, \ldots n-1$ and consider as an estimate

$$\mu_n = \frac{1}{n} \sum_0^{n-1} h(X_j). \tag{7.14}$$

By the LLN (7.13), this estimate μ_n will be close to $\sum_j h(j)\pi_j$ for large n.

This technique is called *Markov chain Monte Carlo* (MCMC). For example, if one is interested in $\pi(A) \equiv \sum_{j \in A} \pi_j$ for some $A \subset S$ then by LLN (7.13) this reduces to

$$\pi_n(A) \equiv \frac{1}{n} \sum_0^{n-1} I_A(X_j) \to \pi(A)$$

in probability as $n \to \infty$, where $I_A(X_j) = 1$ if $X_j \in A$ and 0 otherwise.

An irreducible Markov chain $\{X_n\}$ with a countable state space S is called *aperiodic* if for some $i \in S$ the greatest common divisor, g.c.d. $\{n : p_{ii}^{(n)} > 0\} = 1$. Then, in addition to the LLN (7.13), the following result on the convergence of $P(X_n = j)$ holds.

$$\sum_j |P(X_n = j) - \pi_j| \to 0 \tag{7.15}$$

as $n \to \infty$, for any initial distribution of X_0. In other words, for large n the probability distribution of X_n will be close to π. There exists a result similar to (7.15) for the general state space case that asserts that under suitable conditions, the probability distribution of X_n will be close to π as $n \to \infty$.

This suggests that instead of doing one run of length n, one could do N independent runs each of length m so that $n = Nm$ and then from the ith run use only the mth observation, say, $X_{m,i}$ and consider the estimate

$$\tilde{\mu}_{N,m} \equiv \frac{1}{N} \sum_{i=1}^{N} h(X_{m,i}). \tag{7.16}$$

Other variations exist as well. Some of the special Markov chains used in MCMC are discussed in the next two sections.

7.4.3 Metropolis-Hastings Algorithm

In this section, we discuss a very general MCMC method with wide applications. It will soon become clear why this important discovery has led to very considerable progress in simulation-based inference, particularly in Bayesian analysis. The idea here is not to directly simulate from the given target density (which may be computationally very difficult) at all, but to simulate an easy Markov chain that has this target density as the density of its stationary distribution. We begin with a somewhat abstract setting but very soon will get to practical implementation.

Let S be a finite or countable set. Let π be a probability distribution on S. We shall call π the target distribution. (There is room for slight confusion here because in our applications the target distribution will always be the posterior distribution, so let us note that π here does not denote the prior distribution, but just a standard notation for a generic target.) Let $Q \equiv ((q_{ij}))$ be a transition probability matrix such that for each i, it is computationally easy to generate a sample from the distribution $\{q_{ij} : j \in S\}$. Let us generate a Markov chain $\{X_n\}$ as follows. If $X_n = i$, first sample from the distribution $\{q_{ij} : j \in S\}$ and denote that observation Y_n. Then, choose X_{n+1} from the two values X_n and Y_n according to

$$P(X_{n+1} = Y_n | X_n, Y_n) = \rho(X_n, Y_n)$$
$$P(X_{n+1} = X_n | X_n, Y_n) = 1 - \rho(X_n, Y_n), \tag{7.17}$$

where the "acceptance probability" $\rho(\cdot, \cdot)$ is given by

$$\rho(i, j) = \min\left\{ \frac{\pi_j}{\pi_i} \frac{q_{ji}}{q_{ij}}, 1 \right\} \tag{7.18}$$

for all (i, j) such that $\pi_i q_{ij} > 0$. Note that $\{X_n\}$ is a Markov chain with transition probability matrix $P = ((p_{ij}))$ given by

$$p_{ij} = \begin{cases} q_{ij}\rho_{ij} & j \neq i, \\ 1 - \sum_{k \neq i} p_{ik}, & j = i. \end{cases} \tag{7.19}$$

Q is called the "proposal transition probability" and ρ the "acceptance probability". A significant feature of this transition mechanism P is that P and π satisfy

$$\pi_i p_{ij} = \pi_j p_{ji} \quad \text{for all } i, j. \tag{7.20}$$

This implies that for any j

$$\sum_i \pi_i p_{ij} = \pi_j \sum_i p_{ji} = \pi_j, \tag{7.21}$$

or, π is a stationary probability distribution for P.

Now assume that S is irreducible with respect to Q and $\pi_i > 0$ for all i in S. It can then be shown that P is irreducible, and because it has a stationary distribution π, LLN (7.13) is available. This algorithm is thus a very flexible and useful one. The choice of Q is subject only to the condition that S is irreducible with respect to Q. Clearly, it is no loss of generality to assume that $\pi_i > 0$ for all i in S. A sufficient condition for the aperiodicity of P is that $p_{ii} > 0$ for some i or equivalently

$$\sum_{j \neq 1} q_{ij}\rho_{ij} < 1.$$

A sufficient condition for this is that there exists a pair (i, j) such that $\pi_i q_{ij} > 0$ and $\pi_j q_{ji} < \pi_i q_{ij}$.

Recall that if P is aperiodic, then both the LLN (7.13) and (7.15) hold. If S is not finite or countable but is a continuum and the target distribution $\pi(\cdot)$ has a density $p(\cdot)$, then one proceeds as follows: Let Q be a transition function such that for each x, $Q(x, \cdot)$ has a density $q(x, y)$. Then proceed as in the discrete case but set the "acceptance probability" $\rho(x, y)$ to be

$$\rho(x, y) = \min \left\{ \frac{p(y)q(y, x)}{p(x)q(x, y)}, 1 \right\}$$

for all (x, y) such that $p(x)q(x, y) > 0$. A particularly useful feature of the above algorithm is that it is enough to know $p(\cdot)$ upto a multiplicative constant as in the definition of the "acceptance probability" $\rho(\cdot, \cdot)$, only the ratios $p(y)/p(x)$ need to be calculated. (In the discrete case, it is enough to know $\{\pi_i\}$ upto a multiplicative constant because the "acceptance probability" $\rho(\cdot, \cdot)$ needs only the ratios π_i/π_j.) This assures us that in Bayesian applications it is not necessary to have the normalizing constant of the posterior density available for computation of the posterior quantities of interest.

7.4.4 Gibbs Sampling

As was pointed out in Chapter 2, most of the new problems that Bayesians are asked to solve are high-dimensional. Applications to areas such as micro-arrays and image processing are some examples. Bayesian analysis of such problems invariably involve target (posterior) distributions that are high-dimensional multivariate distributions. In image processing, for example, typically one has $N \times N$ square grid of pixels with $N = 256$ and each pixel has $k \geq 2$ possible values. Thus each configuration has $(256)^2$ components and the state space S has $k^{(256)^2}$ configurations. To simulate a random configuration from a target distribution over such a large S is not an easy task. The *Gibbs sampler* is a technique especially suitable for generating an irreducible aperiodic Markov chain that has as its stationary distribution a target distribution in a high-dimensional space but having some special structure. The most interesting aspect of this technique is that to run this Markov chain, it suffices to generate observations from *univariate distributions*.

The *Gibbs sampler* in the context of a bivariate probability distribution can be described as follows. Let π be a target probability distribution of a bivariate random vector (X, Y). For each x, let $P(x, \cdot)$ be the conditional probability distribution of Y given $X = x$. Similarly, let $Q(y, \cdot)$ be the conditional probability distribution of X given $Y = y$. Note that for each x, $P(x, \cdot)$ is a univariate distribution, and for each y, $Q(y, \cdot)$ is also a univariate distribution. Now generate a bivariate Markov chain $Z_n = (X_n, Y_n)$ as follows:

Start with some $X_0 = x_0$. Generate an observation Y_0 from the distribution $P(x_0, \cdot)$. Then generate an observation X_1 from $Q(Y_0, \cdot)$. Next generate an observation Y_1 from $P(X_1, \cdot)$ and so on. At stage n if $Z_n = (X_n, Y_n)$ is known, then generate X_{n+1} from $Q(Y_n, \cdot)$ and Y_{n+1} from $P(X_{n+1}, \cdot)$.

If π is a discrete distribution concentrated on $\{(x_i, y_j) : 1 \leq i \leq K, 1 \leq j \leq L\}$ and if $\pi_{ij} = \pi(x_i, y_j)$ then $P(x_i, y_j) = \pi_{ij}/\pi_{i\cdot}$ and

$$Q(y_j, x_i) = \frac{\pi_{ij}}{\pi_{\cdot j}},$$

where $\pi_{i\cdot} = \sum_j \pi_{ij}$, $\pi_{\cdot j} = \sum_i \pi_{ij}$. Thus the transition probability matrix $R = ((r_{(ij),(k\ell)}))$ for the $\{Z_n\}$ chain is given by

$$\begin{aligned}
r_{(ij),(k\ell)} &= Q(y_j, x_k)P(x_k, y_\ell) \\
&= \frac{\pi_{kj}}{\pi_{\cdot j}} \frac{\pi_{k\ell}}{\pi_{k\cdot}}.
\end{aligned}$$

It can be verified that this chain is irreducible, aperiodic, and has π as its stationary distribution. Thus LLN (7.13) and (7.15) hold in this case. Thus for large n, Z_n can be viewed as a sample from a distribution that is close to π and one can approximate $\sum_{i,j} h(i,j)\pi_{ij}$ by $\sum_{1=1}^{n} h(X_i, Y_i)/n$.

As an illustration, consider sampling from $\begin{pmatrix} X \\ Y \end{pmatrix} \sim N_2\left(\begin{pmatrix} 0 \\ 0 \end{pmatrix}, \begin{bmatrix} 1 & \rho \\ \rho & 1 \end{bmatrix}\right)$. Note that the conditional distribution of X given $Y = y$ and that of Y given $X = x$ are

$$X|Y = y \sim N(\rho y, 1 - \rho^2) \text{ and } Y|X = x \sim N(\rho x, 1 - \rho^2). \tag{7.22}$$

Using this property, Gibbs sampling proceeds as described below to generate (X_n, Y_n), $n = 0, 1, 2, \ldots$, by starting from an arbitrary value x_0 for X_0, and repeating the following steps for $i = 0, 1, \ldots, n$.

1. Given x_i for X, draw a random deviate from $N(\rho x_i, 1 - \rho^2)$ and denote it by Y_i.
2. Given y_i for Y, draw a random deviate from $N(\rho y_i, 1 - \rho^2)$ and denote it by X_{i+1}.

The theory of Gibbs sampling tells us that if n is large, then (x_n, y_n) is a random draw from a distribution that is close to $N_2\left(\begin{pmatrix} 0 \\ 0 \end{pmatrix}, \begin{bmatrix} 1 & \rho \\ \rho & 1 \end{bmatrix}\right)$. To see why Gibbs sampler works here, recall that a sufficient condition for the LLN (7.13) and the limit result (7.15) is that an appropriate irreducibility condition holds and a stationary distribution exists. From steps 1 and 2 above and using (7.22), one has

$$Y_i = \rho X_i + \sqrt{1 - \rho^2}\, \eta_i$$

and

$$X_{i+1} = \rho Y_i + \sqrt{1 - \rho^2}\, \xi_i,$$

where η_i and ξ_i are independent standard normal random variables independent of X_i. Thus the sequence $\{X_i\}$ satisfies the stochastic difference equation

$$X_{i+1} = \rho^2 X_i + U_{i+1},$$

where

$$U_{i+1} = \rho\sqrt{1 - \rho^2}\ \eta_i + \sqrt{1 - \rho^2}\ \xi_i.$$

Because η_i, ξ_i are independent $N(0,1)$ random variables, U_{i+1} is also a normally distributed random variable with mean 0 and variance $\rho^2(1 - \rho^2) + (1 - \rho^2) = 1 - \rho^4$. Also $\{U_i\}_{i \geq 1}$ being i.i.d., makes $\{X_i\}_{i \geq 0}$ a Markov chain. It turns out that the irreducibility condition holds here. Turning to stationarity, note that if X_0 is a $N(0,1)$ random variable, then $X_1 = \rho^2 X_0 + U_1$ is also a $N(0,1)$ random variable, because the variance of $X_1 = \rho^4 + 1 - \rho^4 = 1$ and the mean of X_1 is 0. This makes the standard $N(0,1)$ distribution a stationary distribution for $\{X_n\}$.

The multivariate extension of the above-mentioned bivariate case is very straightforward. Suppose π is a probability distribution of a k-dimensional random vector $(X_1, X_2, \ldots, X_k)$. If $\mathbf{u} = (u_1, u_2, \ldots, u_k)$ is any k-vector, let $\mathbf{u}_{-i} = (u_1, u_2, \ldots, u_{i-1}, u_{i+1}, \ldots, u_k)$ be the $k-1$ dimensional vector resulting by dropping the ith component u_i. Let $\pi_i(\cdot | \mathbf{x}_{-i})$ denote the univariate conditional distribution of X_i given that $\mathbf{X}_{-i} \equiv (X_1, X_2, X_{i-1}, X_{i+1}, \ldots, X_k) = \mathbf{x}_{-i}$. Now starting with some initial value for $\mathbf{X}_0 = (x_{01}, x_{02}, \ldots, x_{0k})$ generate $\mathbf{X}_1 = (X_{11}, X_{12}, \ldots, X_{1k})$ sequentially by generating X_{11} according to the univariate distribution $\pi_1(\cdot | \mathbf{x}_{0_{-1}})$ and then generating X_{12} according to $\pi_2(\cdot | (X_{11}, x_{03}, x_{04}, \ldots, x_{0k})$ and so on. The most important feature to recognize here is that all the univariate conditional distributions, $X_i | \mathbf{X}_{-i} = \mathbf{x}_{-i}$, known as *full conditionals* should easily allow sampling from them. This turns out to be the case in most hierarchical Bayes problems. Thus, the Gibbs sampler is particularly well adapted for Bayesian computations with hierarchical priors. This was the motivation for some vigorous initial development of Gibbs sampling as can be seen in Gelfand and Smith (1990).

The Gibbs sampler can be justified without showing that it is a special case of the Metropolis-Hastings algorithm. Even if it is considered a special case, it still has special features that need recognition. One such feature is that full conditionals have sufficient information to uniquely determine a multivariate joint distribution. This is the famous *Hammersley-Clifford* theorem. The following condition introduced by Besag (1974) is needed to state this result.

Definition 7.7. *Let $p(y_1, \ldots, y_k)$ be the joint density of a random vector $\mathbf{Y} = (Y_1, \ldots, Y_k)$ and let $p^{(i)}(y_i)$ denote the marginal density of Y_i, $i = 1, \ldots, k$. If $p^{(i)}(y_i) > 0$ for every $i = 1, \ldots, k$ implies that $p(y_1, \ldots, y_k) > 0$, then the joint density p is said to satisfy the positivity condition.*

Let us use the notation $p_i(y_i | y_1, \ldots, y_{i-1}, y_{i+1}, \ldots, y_k)$ for the conditional density of $Y_i | \mathbf{Y}_{-i} = \mathbf{y}_{-i}$.

Theorem 7.8. (Hammersley-Clifford) *Under the positivity condition, the joint density p satisfies*

$$p(y_1, \ldots, y_k) \propto \prod_{j=1}^{k} \frac{p_j(y_j | y_1, \ldots, y_{j-1}, y'_{j+1}, \ldots, y'_k)}{p_j(y'_j | y_1, \ldots, y_{j-1}, y'_{j+1}, \ldots, y'_k)},$$

for every $\mathbf{y}$ *and* $\mathbf{y}'$ *in the support of* p.

Proof. For $\mathbf{y}$ and $\mathbf{y}'$ in the support of p,

$$\begin{aligned}
p(y_1, \ldots, y_k) &= p_k(y_k | y_1, \ldots, y_{k-1}) p(y_1, \ldots, y_{k-1}) \\
&= \frac{p_k(y_k | y_1, \ldots, y_{k-1})}{p_k(y'_k | y_1, \ldots, y_{k-1})} p(y_1, \ldots, y_{k-1}, y'_k) \\
&= \frac{p_k(y_k | y_1, \ldots, y_{k-1})}{p_k(y'_k | y_1, \ldots, y_{k-1})} \frac{p_{k-1}(y_{k-1} | y_1, \ldots, y_{k-2}, y'_k)}{p_{k-1}(y'_{k-1} | y_1, \ldots, y_{k-2}, y'_k)} \\
&\quad \times p(y_1, \ldots, y'_{k-1}, y'_k) \\
&= \cdots \\
&= \prod_{j=1}^{k} \frac{p_j(y_j | y_1, \ldots, y_{j-1}, y'_{j+1}, \ldots, y'_k)}{p_j(y'_j | y_1, \ldots, y_{j-1}, y'_{j+1}, \ldots, y'_k)} p(y'_1, \ldots, y'_k). \qquad \square
\end{aligned}$$

It can be shown also that under the positivity condition, the Gibbs sampler generates an irreducible Markov chain, thus providing the necessary convergence properties without recourse to the M-H algorithm. Additional conditions are, however, required to extend the above theorem to the non-positive case, details of which may be found in Robert and Casella (1999).

7.4.5 Rao-Blackwellization

The variance reduction idea of the famous *Rao-Blackwell theorem* in the presence of auxiliary information can be used to provide improved estimators when MCMC procedures are adopted. Let us first recall this theorem.

Theorem 7.9. *(***Rao-Blackwell theorem***) Let* $\delta(X_1, X_2, \ldots, X_n)$ *be an estimator of* θ *with finite variance. Suppose that* T *is sufficient for* θ, *and let* $\delta^*(T)$, *defined by* $\delta^*(t) = E(\delta(X_1, X_2, \ldots, X_n) | T = t)$, *be the conditional expectation of* $\delta(X_1, X_2, \ldots, X_n)$ *given* $T = t$. *Then*

$$E(\delta^*(T) - \theta)^2 \leq E(\delta(X_1, X_2, \ldots, X_n) - \theta)^2.$$

The inequality is strict unless $\delta = \delta^*$, *or equivalently,* δ *is already a function of* T.

Proof. By the property of iterated conditional expectation,

$$E(\delta^*(T)) = E\left[E(\delta(X_1, X_2, \ldots, X_n) | T)\right] = E(\delta(X_1, X_2, \ldots, X_n)).$$

Therefore, to compare the mean squared errors (MSE) of the two estimators, we need to compare their variances only. Now,

$$\mathrm{Var}(\delta(X_1, X_2, \ldots, X_n)) = \mathrm{Var}\left[E(\delta|T)\right] + E\left[\mathrm{Var}(\delta|T)\right]$$
$$= \mathrm{Var}(\delta^*) + E\left[\mathrm{Var}(\delta|T)\right]$$
$$> \mathrm{Var}(\delta^*),$$

unless $\mathrm{Var}(\delta|T) = 0$, which is the case only if δ itself is a function of T. $\square$

The Rao–Blackwell theorem involves two key steps: variance reduction by conditioning and conditioning by a sufficient statistic. The first step is based on the *analysis of variance* formula: For any two random variables S and T, because

$$\mathrm{Var}(S) = \mathrm{Var}(E(S|T)) + E(\mathrm{Var}(S|T)),$$

one can reduce the variance of a random variable S by taking conditional expectation given some auxiliary information T. This can be exploited in MCMC.

Let $(X_j, Y_j), j = 1, 2, \ldots, N$ be the data generated by a single run of the Gibbs sampler algorithm with a target distribution of a bivariate random vector (X, Y). Let $h(X)$ be a function of the X component of (X, Y) and let its mean value be μ. Suppose the goal is to estimate μ. A first estimate is the sample mean of the $h(X_j), j = 1, 2, \ldots, N$. From the MCMC theory, it can be shown that as $N \to \infty$, this estimate will converge to μ in probability. The computation of variance of this estimator is not easy due to the (Markovian) dependence of the sequence $\{X_j, j = 1, 2, \ldots, N\}$. Now suppose we make n independent runs of Gibbs sampler and generate $(X_{ij}, Y_{ij}), j = 1, 2, \ldots, N; i = 1, 2, \ldots, n$. Now suppose that N is sufficiently large so that (X_{iN}, Y_{iN}) can be regarded as a sample from the limiting target distribution of the Gibbs sampling scheme. Thus $(X_{iN}, Y_{iN}), i = 1, 2, \ldots, n$ are i.i.d. and hence form a random sample from the target distribution. Then one can offer a second estimate of μ—the sample mean of $h(X_{iN}), i = 1, 2, \ldots, n$. This estimator ignores a good part of the MCMC data but has the advantage that the variables $h(X_{iN})$, $i = 1, 2, \ldots, n$ are independent and hence the variance of their mean is of order n^{-1}. Now applying the variance reduction idea of the Rao-Blackwell theorem by using the auxiliary information Y_{iN}, $i = 1, 2, \ldots, n$, one can improve this estimator as follows:

Let $k(y) = E(h(X)|Y = y)$. Then for each i, $k(Y_{iN})$ has a smaller variance than $h(X_{iN})$ and hence the following third estimator,

$$\frac{1}{n} \sum_{i=1}^{n} k(Y_{iN}),$$

has a smaller variance than the second one. A crucial fact to keep in mind here is that the exact functional form of $k(y)$ be available for implementing this improvement.

7.4.6 Examples

Example 7.10. (**Example 7.1 continued.**) Recall that $X|\theta \sim N(\theta, \sigma^2)$ with known σ^2 and $\theta \sim \text{Cauchy}(\mu, \tau)$. The task is to simulate θ from the posterior distribution, but we have already noted that sampling directly from the posterior distribution is difficult. What facilitates Gibbs sampling here is the result that the Student's t density, of which Cauchy is a special case, is a scale mixture of normal densities, with the scale parameter having a Gamma distribution (see Section 2.7.2, Jeffreys test). Specifically,

$$\pi(\theta) \propto \left(\tau^2 + (\theta - \mu)^2\right)^{-1}$$

$$\propto \int_0^\infty (\frac{\lambda}{2\pi\tau^2})^{1/2} \exp\left(-\frac{\lambda}{2\tau^2}(\theta - \mu)^2\right) \lambda^{1/2 - 1} \exp(-\frac{\lambda}{2})\, d\lambda,$$

so that $\pi(\theta)$ may be considered the marginal prior density from the joint prior density of (θ, λ) where

$$\theta|\lambda \sim N(\mu, \tau^2/\lambda) \text{ and } \lambda \sim \text{Gamma}(1/2, 1/2).$$

It can be noted that this leads to an implicit hierarchical prior structure with λ being the hyperparameter. Consequently, $\pi(\theta|x)$ may be treated as the marginal density from $\pi(\theta, \lambda|x)$. Now note that the full conditionals of $\pi(\theta, \lambda|x)$ are standard distributions from which sampling is easy. In particular,

$$\theta|\lambda, x \sim N\left(\frac{\tau^2}{\tau^2 + \lambda\sigma^2}x + \frac{\lambda\sigma^2}{\tau^2 + \lambda\sigma^2}\mu, \frac{\tau^2\sigma^2}{\tau^2 + \lambda\sigma^2}\right), \quad (7.23)$$

$$\lambda|\theta, x \sim \lambda|\theta \sim \text{Exponential}\left(\frac{\tau^2 + (\theta - \mu)^2}{2\tau^2}\right). \quad (7.24)$$

Thus, the Gibbs sampler will use (7.23) and (7.24) to generate (θ, λ) from $\pi(\theta, \lambda|x)$.

Example 7.11. Consider the following example due to Casella and George given in Arnold (1993). Suppose we are studying the distribution of the number of defectives X in the daily production of a product. Consider the model $(X \mid Y, \theta) \sim \text{binomial}(Y, \theta)$, where Y, a day's production, is a random variable with a Poisson distribution with known mean λ, and θ is the probability that any product is defective. The difficulty, however, is that Y is not observable, and inference has to be made on the basis of X only. The prior distribution is such that $(\theta \mid Y = y) \sim \text{Beta}(\alpha, \gamma)$, with known α and γ independent of Y. Bayesian analysis here is not a particularly difficult problem because the posterior distribution of $\theta|X = x$ can be obtained as follows. First, note that $X|\theta \sim \text{Poisson}(\lambda\theta)$. Next, $\theta \sim \text{Beta}(\alpha, \gamma)$. Therefore,

$$\pi(\theta|X = x) \propto \exp(-\lambda\theta)\theta^{x + \alpha - 1}(1 - \theta)^{\gamma - 1}, 0 < \theta < 1. \quad (7.25)$$

The only difficulty is that this is not a standard distribution, and hence posterior quantities cannot be obtained in closed form. Numerical integration is quite simple to perform with this density. However, Gibbs sampling provides an excellent alternative. Instead of focusing on $\theta|X$ directly, view it as a marginal component of $(Y, \theta \mid X)$. It can be immediately checked that the full conditionals of this are given by

$Y|X = x, \theta \sim x + \text{Poisson}(\lambda(1 - \theta))$, and
$\theta|X = x, Y = y \sim \text{Beta}(\alpha + x, \gamma + y - x)$

both of which are standard distributions.

Example 7.12. (Example 7.11 continued.) It is actually possible here to sample from the posterior distribution using what is known as the *accept-reject* Monte Carlo method. This widely applicable method operates as follows. Let $g(\mathbf{x})/K$ be the target density, where K is the possibly unknown normalizing constant of the unnormalized density g. Suppose $h(\mathbf{x})$ is a density that can be simulated by a known method and is close to g, and suppose there exists a known constant $c > 0$ such that $g(\mathbf{x}) < ch(\mathbf{x})$ for all $\mathbf{x}$. Then, to simulate from the target density, the following two steps suffice. (See Robert and Casella (1999) for details.)

Step 1. Generate $\mathbf{Y} \sim h$ and $U \sim U(0, 1)$;
Step 2. Accept $\mathbf{X} = \mathbf{Y}$ if $U \leq g(\mathbf{Y})/\{ch(\mathbf{Y})\}$; return to Step 1 otherwise.
The optimal choice for c is $\sup\{g(\mathbf{x})/h(\mathbf{x})$, but even this choice may result in undesirably large number of rejections.

In our example, from (7.25),

$$g(\theta) = \exp(-\lambda\theta)\theta^{x+\alpha-1}(1 - \theta)^{\gamma-1}I\{0 \leq \theta \leq 1\},$$

so that $h(\theta)$ may be chosen to be the density of $\text{Beta}(x + \alpha, \gamma)$. Then, with the above-mentioned choice for c, if $\theta \sim \text{Beta}(x + \alpha, \gamma)$ is generated in Step 1, its 'acceptance probability' in Step 2 is simply $\exp(-\lambda\theta)$. Even though this method can be employed here, we, however, would like to use this technique to illustrate the Metropolis-Hastings algorithm. The required Markov chain is generated by taking the transition density $q(z, y) = q(y|z) = h(y)$, independently of z. Then the acceptance probability is

$$\rho(z, y) = \min\left\{\frac{g(y)h(z)}{g(z)h(y)}, 1\right\}$$
$$= \min\left\{\exp\left(-\lambda(y - z)\right), 1\right\}.$$

Thus the steps involved in this "independent" M-H algorithm are as follows. Start at $t = 0$ with a value x_0 in the support of the target distribution; in this case, $0 < x_0 < 1$. Given x_t, generate the next value in the chain as given below.

   (a) Draw Y_t from $\text{Beta}(x + \alpha, \gamma)$.
   (b) Let

$$x_{(t+1)} = \begin{cases} Y_t \text{ with probability } \rho_t \\ x_t \text{ otherwise,} \end{cases}$$

where $\rho_t = \min\{\exp\left(-\lambda(Y_t - x_t)\right), 1\}$.

(c) Set $t = t + 1$ and go to step (a).

Run this chain until $t = n$, a suitably chosen large integer. Details on its convergence as well as why independent M-H is more efficient than accept-reject Monte Carlo can be found in Robert and Casella (1999). In our example, for $x = 1$, $\alpha = 1$, $\gamma = 49$ and $\lambda = 100$, we simulated such a Markov chain. The resulting frequency histogram is shown in Figure 7.1, with the true posterior density super-imposed on it.

Example 7.13. In this example, we discuss the hierarchical Bayesian analysis of the usual one-way ANOVA. Consider the model

$$y_{ij} = \theta_i + \epsilon_{ij}, j = 1, \ldots, n_i; i = 1, \ldots, k;$$
$$\epsilon_{ij} \sim N(0, \sigma_i^2), j = 1, \ldots, n_i; i = 1, \ldots, k, \tag{7.26}$$

and are independent. Let the first stage prior on θ_i and σ_i^2 be such that they are i.i.d. with

$$\theta_i \sim N(\mu_\pi, \sigma_\pi^2), \quad i = 1, \ldots, k;$$
$$\sigma_i^2 \sim \text{inverse Gamma}(a_1, b_1), \quad i = 1, \ldots, k.$$

The second stage prior on μ_π and σ_π^2 is

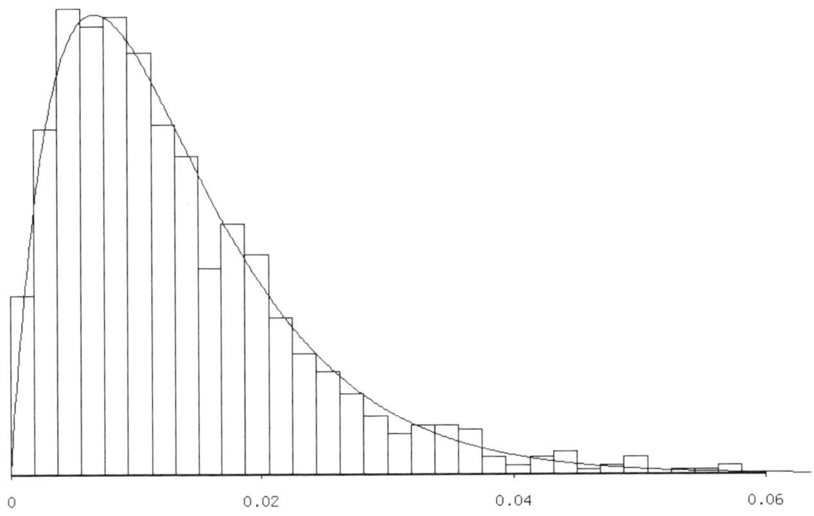

Fig. 7.1. M-H frequency histogram and true posterior density.

$$\mu_\pi \sim N(\mu_0, \sigma_0^2) \quad \text{and} \quad \sigma_\pi^2 \sim \text{inverse Gamma}(a_2, b_2).$$

Here a_1, a_2, b_1, b_2, μ_0, and σ_0^2 are all specified constants. Let us concentrate on computing

$$\boldsymbol{\mu}^\pi(\mathbf{y}) = E^\pi(\boldsymbol{\theta}|\mathbf{y}).$$

Sufficiency reduces this to considering only

$$\bar{Y}_i = \frac{1}{n_i}\sum_{j=1}^{n_i} y_{ij}, i = 1, \ldots, k \text{ and}$$

$$S_i^2 = \sum_{j=1}^{n_i}(y_{ij} - \bar{Y}_i)^2, i = 1, \ldots, k.$$

From normal theory,

$$\bar{Y}_i|\boldsymbol{\theta}, \boldsymbol{\sigma}^2 \sim N(\theta_i, \sigma_i^2/n_i), \quad i = 1, \ldots, k,$$

which are independent and are also independent of
$$S_i^2|\boldsymbol{\theta}, \boldsymbol{\sigma}^2 \sim \sigma_i^2\chi_{n_i-1}^2, \quad i = 1, \ldots, k,$$

which again are independent. To utilize the Gibbs sampler, we need the full conditionals of $\pi(\boldsymbol{\theta}, \boldsymbol{\sigma}^2, \mu_\pi, \sigma_\pi^2|\mathbf{y})$. It can be noted that it is sufficient, and in fact advantageous to consider the conditionals,
(i) $\pi(\boldsymbol{\theta}|\boldsymbol{\sigma}^2, \mu_\pi, \sigma_\pi^2, \mathbf{y})$,
(ii) $\pi(\boldsymbol{\sigma}^2|\boldsymbol{\theta}, \mu_\pi, \sigma_\pi^2, \mathbf{y})$,
(iii) $\pi(\mu_\pi|\sigma_\pi^2, \boldsymbol{\theta}, \boldsymbol{\sigma}^2, \mathbf{y})$,   and
(iv) $\pi(\sigma_\pi^2|\mu_\pi, \boldsymbol{\theta}, \boldsymbol{\sigma}^2, \mathbf{y})$,
rather than considering the set of all univariate full conditionals because of the special structure in this problem. First note that

$$\boldsymbol{\theta}|\mu_\pi, \sigma_\pi^2 \sim N_k(\mu_\pi\mathbf{1}, \sigma_\pi^2 I_k),$$

and hence

$$\boldsymbol{\theta}|\mu_\pi, \sigma_\pi^2, \boldsymbol{\sigma}^2, \mathbf{y} \sim N_k(\boldsymbol{\mu}^{(1)}, \Sigma^{(1)}), \text{where}$$

$$\mu_i^{(1)} = \frac{\sigma_\pi^2}{\sigma_\pi^2 + \sigma_i^2/n_i}\bar{y}_i + \frac{\sigma_i^2/n_i}{\sigma_\pi^2 + \sigma_i^2/n_i}\mu_\pi \text{ and}$$

$$\Sigma^{(1)} \text{ is diagonal with } \Sigma_{ii}^{(1)} = \frac{\sigma_\pi^2\sigma_i^2/n_i}{\sigma_\pi^2 + \sigma_i^2/n_i}, \tag{7.27}$$

which determines (i). Next we note that, given $\boldsymbol{\theta}$, from (7.26), $S_i^{*2} = \sum_{j=1}^{n_i}(y_{ij} - \theta_i)^2$ is sufficient for σ_i^2, and they are independently distributed. Thus we have,

$$S_i^{*2}|\boldsymbol{\sigma}^2, \boldsymbol{\theta} \sim \sigma_i^2\chi_{n_i-1}^2, i = 1, \ldots, k,$$

and are independent, and σ_i^2 are i.i.d. inverse Gamma (a_1, b_1), so that

$$\sigma_i^2|\boldsymbol{\theta},\mu_\pi,\sigma_\pi^2,\mathbf{y} \sim \text{inverse Gamma}(a_1 + \frac{1}{2}n_i, b_1 + \frac{1}{2}S_i^{*2}), \qquad (7.28)$$

and they are independent for $i = 1,\ldots,k$, which specifies (ii). Turning to the full conditional of μ_π, we note from the hierarchical structure that the conditional distribution of $\mu_\pi|\sigma_\pi^2,\boldsymbol{\theta},\boldsymbol{\sigma^2},\mathbf{y}$ is the same as the conditional distribution of $\mu_\pi|\sigma_\pi^2,\boldsymbol{\theta}$. To determine this distribution, note that

$$\theta_i|\mu_\pi,\sigma_\pi^2 \sim N(\mu_\pi,\sigma_\pi^2),$$

for $i = 1,\ldots,k$ and are i.i.d. and $\mu_\pi \sim N(\mu_0,\sigma_0^2)$. Therefore, treating $\boldsymbol{\theta}$ to be a random sample from $N(\mu_\pi,\sigma_\pi^2)$, so that $\bar{\theta} = \sum_{i=1}^{k}\theta_i/k$ is sufficient for μ_π, we have the joint distribution,

$$\bar{\theta}|\mu_\pi,\sigma_\pi^2 \sim N(\mu_\pi,\sigma_\pi^2/k), \text{ and } \mu_\pi \sim N(\mu_0,\sigma_0^2).$$

Thus we obtain,

$$\mu_\pi|\sigma_\pi^2,\boldsymbol{\theta},\boldsymbol{\sigma^2},\mathbf{y} \sim N(\frac{\sigma_0^2}{\sigma_0^2 + \sigma_\pi^2/k}\bar{\theta} + \frac{\sigma_\pi^2/k}{\sigma_0^2 + \sigma_\pi^2/k}\mu_0, \frac{\sigma_0^2\sigma_\pi^2/k}{\sigma_0^2 + \sigma_\pi^2/k}), \quad (7.29)$$

which provides (iii). Just as in the previous case, the conditional distribution of $\sigma_\pi^2|\mu_\pi,\boldsymbol{\theta},\boldsymbol{\sigma^2},\mathbf{y}$ turns out to be the same as the conditional distribution of $\sigma_\pi^2|\mu_\pi,\boldsymbol{\theta}$. To obtain this, note again that

$$\theta_i|\mu_\pi,\sigma_\pi^2 \sim N(\mu_\pi,\sigma_\pi^2),$$

for $i = 1,\ldots,k$ and are i.i.d. so that this time $\sum_{i=1}^{k}(\theta_i - \mu_\pi)^2$ is sufficient for σ_π^2. Further

$$\sum_{i=1}^{k}(\theta_i - \mu_\pi)^2|\sigma_\pi^2 \sim \sigma_\pi^2\chi_k^2 \text{ and } \sigma_\pi^2 \sim \text{inverse Gamma}(a_2,b_2),$$

so that

$$\sigma_\pi^2|\mu_\pi,\boldsymbol{\theta},\boldsymbol{\sigma^2},\mathbf{y} \sim \text{inverse Gamma}(a_2 + \frac{k}{2}, b_2 + \frac{1}{2}\sum_{i=1}^{k}(\theta_i - \mu_\pi)^2). \quad (7.30)$$

This gives us (iv), thus completing the specification of all the required full conditionals. It may be noted that the Gibbs sampler in this problem requires simulations from only the standard normal and the inverse Gamma distributions.

Reversible Jump MCMC

There are situations, especially in model selection problems, where the MCMC procedure should be capable of moving between parameter spaces of different dimensions. The standard M-H algorithm described earlier is incapable of such movements, whereas the *reversible jump* algorithm of Green (1995) is an extension of the standard M-H algorithm to allow exactly this possibility. The basic idea behind this technique as applied to model selection is as follows. Given two models M_1 and M_2 with parameter sets $\boldsymbol{\theta}_1$ and $\boldsymbol{\theta}_2$, which are possibly of different dimensions, fill the difference in the dimensions by supplementing the parameter sets of these models. In other words, find auxiliary variables $\boldsymbol{\gamma}_{12}$ and $\boldsymbol{\gamma}_{21}$ such that $(\boldsymbol{\theta}_1, \boldsymbol{\gamma}_{12})$ and $(\boldsymbol{\theta}_2, \boldsymbol{\gamma}_{21})$ can be mapped with a bijection. Now use the standard M-H algorithm to move between the two models; for moves of the M-H chain within a model, the auxiliary variables are not needed. We sketch this procedure below, but for further details refer to Robert and Casella (1999), Green (1995), Sorensen and Gianola (2002), Waagepetersen and Sorensen (2001), and Brooks et al. (2003).

Consider models $M_1, M_2, \ldots$ where model M_i has a continuous parameter space Θ_i. The parameter space for the model selection problem as a whole may be taken to be

$$\{(M_i, \boldsymbol{\theta}_i) : \boldsymbol{\theta}_i \in \Theta_i, i = 1, 2, \ldots\} .$$

Let $f(\mathbf{x}|M_i, \boldsymbol{\theta}_i)$ be the model density under model M_i, and the prior density be

$$\pi(\boldsymbol{\theta}) = \sum_i \pi_i \pi(\boldsymbol{\theta}_i|M_i) I \left(\boldsymbol{\theta} = \boldsymbol{\theta}_i \in \Theta_i \right),$$

where π_i is the prior probability of model M_i and $\pi(\boldsymbol{\theta}_i|M_i)$ is the prior density conditional on M_i being true. Then the posterior probability of any $B \subset \cup_j \Theta_j$ is

$$\pi(B|\mathbf{x}) = \sum_i \int_{B \cap \Theta_i} \pi(\boldsymbol{\theta}_i|M_i, \mathbf{x}) \, d\boldsymbol{\theta}_i,$$

where

$$\pi(\boldsymbol{\theta}_i|M_i, \mathbf{x}) \propto \pi_i \pi(\boldsymbol{\theta}_i|M_i) f(\mathbf{x}|M_i, \boldsymbol{\theta}_i)$$

is the posterior density restricted to M_i. To compute the Bayes factor of M_k relative to M_l, we will need

$$\frac{P^\pi(M_k|\mathbf{x})}{P^\pi(M_l|\mathbf{x})} \frac{\pi_l}{\pi_k},$$

where

$$P^\pi(M_i|\mathbf{x}) = \frac{\pi_i \int_{\Theta_i} \pi(\boldsymbol{\theta}_i|M_i) f(\mathbf{x}|M_i, \boldsymbol{\theta}_i) \, d\boldsymbol{\theta}_i}{\sum_j \pi_j \int_{\Theta_j} \pi(\boldsymbol{\theta}_j|M_j) f(\mathbf{x}|M_j, \boldsymbol{\theta}_j) \, d\boldsymbol{\theta}_j}$$

is the posterior probability of M_i. Therefore, for the target density $\pi(\boldsymbol{\theta}|\mathbf{x})$, we need a version of the M-H algorithm that will facilitate the above-shown computations. Suppose $\boldsymbol{\theta}_i$ is a vector of length n_i. It suffices to focus on moves between $\boldsymbol{\theta}_i$ in model M_i and $\boldsymbol{\theta}_j$ in model M_j with $n_i < n_j$. The scheme provided by Green (1995) is as follows. If the current state of the chain is $(M_i, \boldsymbol{\theta}_i)$, a new value $(M_j, \boldsymbol{\theta}_j)$ is proposed for the chain from a proposal (transition) distribution $Q(\boldsymbol{\theta}_i, d\boldsymbol{\theta}_j)$, which is then accepted with a certain *acceptance probability*. To move from model M_i to M_j, generate a random vector $\mathbf{V}$ of length $n_j - n_i$ from a proposal density

$$\psi_{ij}(\mathbf{v}) = \prod_{m=1}^{n_j - n_i} \psi(v_m).$$

Identify an appropriate bijection map

$$h_{ij} : \Theta_i \times \mathcal{R}^{n_j - n_i} \longrightarrow \Theta_j,$$

and propose the move from $\boldsymbol{\theta}_i$ to $\boldsymbol{\theta}_j$ using $\boldsymbol{\theta}_j = h_{ij}(\boldsymbol{\theta}_i, \mathbf{V})$. The acceptance probability is then

$$\rho((M_i, \boldsymbol{\theta}_i), (M_j, \boldsymbol{\theta}_j)) = \min\left\{1, \alpha_{ij}(\boldsymbol{\theta}_i, \boldsymbol{\theta}_j)\right\},$$

where

$$\alpha_{ij}(\boldsymbol{\theta}_i, \boldsymbol{\theta}_j) = \frac{\pi(\boldsymbol{\theta}_j | M_j, \mathbf{x}) p_{ji}(\boldsymbol{\theta}_j)}{\pi(\boldsymbol{\theta}_i | M_i, \mathbf{x}) p_{ij}(\boldsymbol{\theta}_i) \psi_{ij}(\mathbf{v})} \left| \frac{\partial h_{ij}(\boldsymbol{\theta}_i, \mathbf{v})}{\partial(\boldsymbol{\theta}_i, \mathbf{v})} \right|,$$

with $p_{ij}(\boldsymbol{\theta}_i)$ denoting the (user-specified) probability that a proposed jump to model M_j is attempted at any step starting from $\boldsymbol{\theta}_i \in \Theta_i$. Note that $\sum_j p_{ij} = 1$.

Example 7.14. For illustration purposes, consider the simple problem of comparing two normal means as in Sorensen and Gianola (2002). Then, the two models to be compared are

$$y_i | M_1, \nu, \sigma^2 \sim N(\nu, \sigma^2), i = 1, 2, \ldots m_1 + m_2 \text{ i.i.d.},$$

$$y_i | M_2, \nu_1, \nu_2, \sigma^2 \sim \begin{cases} N(\nu_1, \sigma^2), i = 1, 2, \ldots m_1; \\ N(\nu_2, \sigma^2), i = m_1 + 1, \ldots m_1 + m_2. \end{cases}$$

To implement the reversible jump M-H algorithm we need the map, h_{12} taking (ν, σ^2, V) to (ν_1, ν_2, σ^2). A reasonable choice for this is the linear map

$$\begin{bmatrix} \nu_1 \\ \nu_2 \\ \sigma^2 \end{bmatrix} = \begin{bmatrix} 1 & 0 & 1 \\ 1 & 0 & -1 \\ 0 & 1 & 0 \end{bmatrix} \begin{bmatrix} \nu \\ \sigma^2 \\ V \end{bmatrix}.$$

7.4.7 Convergence Issues

As we have already seen, Monte Carlo sampling based approaches for inference make use of limit theorems such as the law of large numbers and the central limit theorem to justify their validity. When we add a further dimension to this sampling and adopt MCMC schemes, stronger limit theorems are needed. Ergodic theorems for Markov chains such as those given in equations (7.13) and (7.15) are these useful results. It may appear at first that this procedure necessarily depends on waiting until the Markov chain converges to the target invariant distribution, and sampling from this distribution. In other words, one needs to start a large number of chains beginning with different starting points, and pick the draws after letting these chains run sufficiently long. This is certainly an option, but the law of large numbers for dependent chains, (7.13) says also that this is unnecessary, and one could just use a single long chain. It may, however, be a good idea to use many different chains to ensure that convergence indeed takes place. For details, see Robert and Casella (1999).

There is one important situation, however, where MCMC sampling can lead to absurd inferences. This is where one resorts to MCMC sampling without realizing that the target posterior distribution is not a probability distribution, but an improper one. The following example is similar to the normal problem (see Exercise 13) with lack of identifiability of parameters shown in Carlin and Louis (1996).

Example 7.15. (Example 7.11 continued.) Recall that, in this problem, $(X \mid Y, \theta) \sim \text{binomial}(Y, \theta)$, where $Y \mid \lambda \sim \text{Poisson}(\lambda)$. Earlier, we worked with a known mean λ, but let us now see if it is possible to handle this problem with unknown λ. Because Y is unobservable and only X is observable, there exists an 'identifiability' problem here, as can be seen by noting that $X|\theta \sim \text{Poisson}(\lambda\theta)$. We already have the Beta(α, γ) prior on θ. Suppose $0 < \alpha \leq 1$. Consider an independent prior on λ according to which $\pi(\lambda) \propto I(\lambda > 0)$. Then,

$$\pi(\lambda, \theta|x) \propto \exp(-\lambda\theta)\lambda^x \theta^{x+\alpha-1}(1-\theta)^{\gamma-1}, 0 < \theta < 1, \lambda > 0. \quad (7.31)$$

This joint density is improper because

$$\int_0^\infty \int_0^1 \exp(-\lambda\theta)\lambda^x \theta^{x+\alpha-1}(1-\theta)^{\gamma-1} \, d\lambda \, d\theta$$

$$= \int_0^1 \left(\int_0^\infty \exp(-\lambda\theta)\lambda^x \, d\lambda \right) \theta^{x+\alpha-1}(1-\theta)^{\gamma-1} \, d\theta$$

$$= \int_0^1 \frac{\Gamma(x+1)}{\theta^{x+1}} \theta^{x+\alpha-1}(1-\theta)^{\gamma-1} \, d\theta$$

$$= \Gamma(x+1) \int_0^1 \theta^{\alpha-2}(1-\theta)^{\gamma-1} \, d\theta$$

$$= \infty.$$

In fact, the marginal distributions are also improper. However, it has full conditional distributions that are proper:

$$\lambda|\theta, x \sim \text{Gamma}\,(x+1, \theta) \quad \text{and} \quad \pi(\theta|\lambda, x) \propto \exp(-\lambda\theta)\theta^{x+\alpha-1}(1-\theta)^{\gamma-1}.$$

Thus, for example, the Gibbs sampler can be successfully employed with these proper full conditionals. To generate θ from $\pi(\theta|\lambda, x)$, one may use the independent M-H algorithm described in Example 7.12. Any inference on the marginal posterior distributions derived from this sample, however, will be totally erroneous, whereas inferences can indeed be made on $\lambda\theta$.

In fact, the non-convergence of the chain encountered in the above example is far from being uncommon. Often when we have a hierarchical prior, the prior at the final stage of the hierarchy is an improper objective prior. Then it is not easy to check that the joint posterior is proper. Then none of the theorems on convergence of the chains may apply, but the chain may yet seem to converge. In such cases, inference based on MCMC may be misleading in the sense of what was seen in the example above.

7.5 Exercises

1. (Flury and Zoppè (2000)) A total of $m + n$ lightbulbs are tested in two independent experiments. In the first experiment involving n lightbulbs, the exact lifetimes $y_1, \ldots, y_n$ of all the bulbs are recorded. In the second involving m lightbulbs, the only information available is whether these lightbulbs were still burning at some fixed time $t > 0$. This is known as right-censoring. Assume that the distribution of lifetime is exponential with mean $1/\theta$, and use $\pi(\theta) \propto \theta^{-1}$. Find the posterior mode using the E-M algorithm.

2. (Flury and Zoppè (2000)) In Problem 1, use uniform$(0, \theta)$ instead of exponential for the lifetime distribution, and $\pi(\theta) = I_{(0,\infty)}(\theta)$. Show that the E-M algorithm fails here if used to find the posterior mode.

3. **(Inverse c.d.f. method)** Show that, if the c.d.f. $F(x)$ of a random variable X is continuous and strictly increasing, then $U = F(X) \sim U[0, 1]$, and if $V \sim U[0, 1]$, then $Y \equiv F^{-1}(V)$ has c.d.f F. Using this show that if $U \sim U[0, 1]$, $-\ln U/\beta$ is an exponential random variable with mean β^{-1}.

4. **(Box-Muller transformation method)** Let U_1 and U_2 be a pair of independent Uniform $(0, 1)$ random variables. Consider first a transformation to
$$W \equiv R^2 = -2\log U_1; \quad V = 2\pi U_2,$$
and then let
$$X = R\cos V; \quad Y = R\sin V.$$

Show that X and Y are independent standard normal random variables.

5. Prove that the accept-reject Monte Carlo method given in Example 7.12 indeed generates samples from the target density. Further show that the expected number of draws required from the 'proposal density' per observation is c^{-1}.

6. Using the methods indicated in Exercises 1, 2, and 3 above, or combinations thereof, prove that the standard continuous probability distributions can be simulated.

7. Consider a discrete probability distribution that puts mass p_i on point x_i, $i = 0, 1, \ldots$. Let $U \sim U(0,1)$, and define a new random variable Y as follows.

$$Y = \begin{cases} x_0 \text{ if } U \le p_0; \\ x_i \text{ if } \sum_{i=0}^{i-1} p_j < U \le \sum_{i=0}^{i} p_j, \ i \ge 1. \end{cases}$$

What is the probability distribution of Y?

8. Show that the random sequence generated by the independent M-H algorithm is a Markov chain.

9. (Robert and Casella (1999)) Show that the Gamma distribution with a non-integer shape parameter can be simulated using the accept-reject method or the independent M-H algorithm.

10. **Gibbs Sampling for Multinomial.** Consider the *ABO Blood Group* problem from Rao (1973). The observed counts in the four blood groups, *O*, *A*, *B*, and *AB* are as given in Table 7.3. Assuming that the inheritance of these blood groups is controlled by three alleles, *A*, *B*, and *O*, of which *O* is recessive to *A* and *B*, there are six genotypes *OO*, *AO*, *AA*, *BO*, *BB*, and *AB*, but only four phenotypes. If r, p, and q are the gene frequencies of *O*, *A*, and *B*, respectively (with $p + q + r = 1$), then the probabilities of the four phenotypes assuming Hardy-Weinberg equilibrium are also as shown in Table 7.3. Thus we have here a 4-cell multinomial probability vector that is a function of three parameters p, q, r with $p + q + r = 1$. One may wish to formulate a Dirichlet prior for p, q, r. But it will not be conjugate to the 4-cell multinomial likelihood function in terms of p, q, r from the data, and this makes it difficult to work out the posterior distribution of p, q, r. Although no data are missing in the real sense of the term, it is profitable to split each of the n_A and n_B cells into two: n_A into n_{AA}, n_{AO} with corresponding probabilities $p^2, 2pr$ and n_B into n_{BB}, n_{BO} with corresponding probabilities $q^2, 2qr$, and consider the 6-cell multinomial problem as a *complete* problem with n_{AA}, n_{BB} as 'missing' data.

Table 7.3. *ABO Blood Group* Data

Cell Count	Probability
$n_O = 176$	r^2
$n_A = 182$	$p^2 + 2pr$
$n_B = 60$	$q^2 + 2qr$
$n_{AB} = 17$	$2pq$

Let $N = n_O + n_A + n_B + n_{AB}$, and denote the observed data by $\mathbf{n} = (n_O, n_A, n_B, n_{AB})$. Consider estimation of p, q, r using a Dirichlet prior with parameters α, β, γ with the 'incomplete' observed data $\mathbf{n}$. The likelihood upto a multiplicative constant is

$$L(p, q, r) = r^{2n_O}(p^2 + 2pr)^{n_A}(q^2 + 2qr)^{n_B}(pq)^{n_{AB}}.$$

The posterior density of (p, q, r) given $\mathbf{n}$ is proportional to

$$r^{2n_O + \gamma - 1}(p^2 + 2pr)^{n_A}(q^2 + 2qr)^{n_B}(p)^{n_{AB} + \alpha - 1}(q)^{n_{AB} + \beta - 1}.$$

Let $n_A = n_{AA} + n_{AO}$, $n_B = n_{BB} + n_{BO}$, and write n_{OO} for n_O. Verify that if we have the 'complete' data, $\widetilde{\mathbf{n}} = (n_{OO}, n_{AA}, n_{AO}, n_{BB}, n_{BO}, n_{AB})$, then the likelihood is, upto a multiplicative constant

$$(p^2)^{n_{AA}}(q^2)^{n_{BB}}(r^2)^{n_{OO}}(2pq)^{n_{AB}}(2qr)^{n_{BO}}(2pr)^{n_{AO}}$$

$$= p^{n_A^+}q^{n_B^+}r^{n_O^+},$$

where

$$n_A^+ = n_{AA} + \frac{1}{2}n_{AB} + \frac{1}{2}n_{AO}$$

$$n_B^+ = \frac{1}{2}n_{AB} + n_{BB} + \frac{1}{2}n_{BO}$$

$$n_O^+ = \frac{1}{2}n_{AO} + \frac{1}{2}n_{BO} + n_{OO}.$$

Show that the posterior distribution of (p, q, r) given $\widetilde{\mathbf{n}}$ is Dirichlet with parameters $n_A^+ + \alpha - 1, n_B^+ + \beta - 1, n_O^+ + \gamma - 1$, when the prior is Dirichlet with parameters (α, β, γ).

Show that the conditional distributions of (n_{AA}, n_{BB}) given $\widetilde{\mathbf{n}}$ and (p, q, r) is that of two independent binomials:

$$(n_{AA}|\mathbf{n}, p, q, r) \sim \text{binomial}(n_A, \frac{p^2}{p^2 + 2pr}),$$

$$(n_{BB}|\mathbf{n}, p, q, r) \sim \text{binomial}(n_B, \frac{q^2}{q^2 + 2qr}), \text{ and}$$

$$(p, q, r|\mathbf{n}, n_{AA}, n_{BB}) \sim \text{Dirichlet}(n_A^+ + \alpha - 1, n_B^+ + \beta - 1, n_O^+ + \gamma - 1).$$

Show that the Rao-Blackwellized estimate of (p, q, r) from a Gibbs sample of size m is

$$\frac{1}{m}\sum_{i=1}^{m}(\alpha + n_A^{+i}, \beta + n_B^{+i}, \gamma + n_O^{+i})/(\alpha + \beta + \gamma + N),$$

where the superscript i denotes the ith draw.

11. (**M-H for the Weibull Model:** Robert (2001)). The following twelve observations are from a simulated reliability study:

0.56, 2.26, 1.90, 0.94, 1.40, 1.39, 1.00, 1.45, 2.32, 2.08, 0.89, 1.68.

A Weibull model with the following density form is considered appropriate:

$$f(x|\alpha, \eta) \propto \alpha\eta x^{\alpha - 1}e^{-\eta x^{\alpha}}, 0 < x < \infty,$$

with parameters (α, η). Consider the prior distribution

$$\pi(\alpha, \eta) \propto e^{-\alpha}\eta^{\beta-1}e^{-\xi\eta}.$$

The posterior distribution of (α, η) given the data $(x_1, x_2, \ldots, x_n)$ has density

$$\pi(\alpha, \eta | x_1, x_2, \ldots, x_n) \propto (\alpha\eta)^n \left(\prod_{i=1}^{n} x_i\right)^{\alpha-1} \exp\left\{-\eta \sum_{i=1}^{n} x_i^{\alpha}\right\} \pi(\alpha, \eta).$$

To get a sample from the posterior density, one may use the M–H algorithm with proposal density

$$q(\alpha', \eta' | \alpha, \eta) = \frac{1}{\alpha\eta} \exp\left\{-\frac{\alpha'}{\alpha} - \frac{\eta'}{\eta}\right\},$$

which is a product of two independent exponential distributions with means α, η. Compute the acceptance probability $\rho((\alpha', \eta'), (\alpha^{(t)}, \eta^{(t)}))$ at the tth step of the M-H chain, and explain how the chain is to be generated.

12. Complete the construction of the reversible jump M-H algorithm in Example 7.14. In particular, choose an appropriate prior distribution, proposal distribution and compute the acceptance probabilities.

13. (Carlin and Louis (1996)) Suppose $y_1, y_2, \ldots, y_n$ is an i.i.d. sample with

$$y_i | \theta_1, \theta_2 \sim N(\theta_1 + \theta_2, \sigma^2),$$

where σ^2 is assumed to be known. Independent improper uniform prior distributions are assumed for θ_1 and θ_2.

(a) Show that the posterior density of $(\theta_1, \theta_2 | \mathbf{y})$ is

$$\pi(\theta_1, \theta_2 | \mathbf{y}) \propto \exp(-n(\theta_1 + \theta_2 - \bar{y})^2/(2\sigma^2))I((\theta_1, \theta_2) \in \mathcal{R}^2),$$

which is improper, integrating to ∞ (over $\mathcal{R}^2$)).

(b) Show that the marginal posterior distributions are also improper.

(c) Show that the full conditional distributions of this posterior distribution are proper.

(d) Explain why a sample generated using the Gibbs sampler based on these proper full conditionals will be totally useless for any inference on the marginal posterior distributions, whereas inferences can indeed be made on $\theta_1 + \theta_2$.

14. Suppose $X_1, X_2, \ldots, X_k$ are independent Poisson counts with X_i having mean θ_i. θ_i are *a priori* considered related, but exchangeable, and the prior

$$\pi(\theta_1, \ldots, \theta_k) \propto \left(1 + \sum_{i=1}^{k} \theta_i\right)^{-(k+1)},$$

is assumed.

(a) Show that the prior is a conjugate mixture.

(b) Show how the Gibbs sampler can be employed for inference.

15. Suppose $X_1, X_2, \ldots, X_n$ are i.i.d. random variables with

$$X_i | \lambda_1, \lambda_2 \sim \text{exponential with mean } 1/\lambda_1 \lambda_2,$$

and independent scale-invariant non-informative priors on λ_1 and λ_2 are used. i.e., $\pi(\lambda_1, \lambda_2) \propto (\lambda_1 \lambda_2)^{-1} I(\lambda_1 > 0, \lambda_2 > 0)$.
(a) Show that the marginals of the posterior, $\pi(\lambda_1, \lambda_2 | \mathbf{x})$ are improper, but the full conditionals are standard distributions.
(b) What posterior inferences are possible based on a sample generated from the Gibbs sampler using these full conditionals?

Some Common Problems in Inference

We have already discussed some basic inference problems in the previous chapters. These include the problems involving the normal mean and the binomial proportion. Some other usually encountered problems are discussed in what follows.

8.1 Comparing Two Normal Means

Investigating the difference between two mean values or two proportions is a frequently encountered problem. Examples include agricultural experiments where two different varieties of seeds or fertilizers are employed, or clinical trials involving two different treatments. Comparison of two binomial proportions was considered in Example 4.6 and Problem 8 in Chapter 4. Comparison of two normal means is discussed below.

Suppose the model for the available data is as follows. $Y_{11}, \ldots, Y_{1n_1}$ is a random sample of size n_1 from a normal population, $N(\theta_1, \sigma_1^2)$, whereas $Y_{21}, \ldots, Y_{2n_2}$ is an independent random sample of size n_2 from another normal population, $N(\theta_2, \sigma_2^2)$. All the four parameters θ_1, θ_2, σ_1^2, and σ_2^2 are unknown, but the quantity of inferential interest is $\eta = \theta_1 - \theta_2$.

It is convenient to consider the case, $\sigma_1^2 = \sigma_2^2 = \sigma^2$ separately. In this case, $(\bar{Y}_1, \bar{Y}_2, s^2)$ is jointly sufficient for $(\theta_1, \theta_2, \sigma^2)$ where $s^2 = (\sum_{i=1}^{n_1}(Y_{1i} - \bar{Y}_1)^2 + \sum_{j=1}^{n_2}(Y_{2j} - \bar{Y}_2)^2)/(n_1 + n_2 - 2)$. Further, given $(\theta_1, \theta_2, \sigma^2)$,

$$\bar{Y}_1 \sim N(\theta_1, \frac{\sigma^2}{n_1}), \bar{Y}_2 \sim N(\theta_2, \frac{\sigma^2}{n_2}), \text{ and } (n_1 + n_2 - 2)s^2 \sim \sigma^2 \chi^2_{n_1+n_2-2},$$

and they are independently distributed. Upon utilizing the objective prior, $\pi(\theta_1, \theta_2, \sigma^2) \propto \sigma^{-2}$, one obtains

$$\pi(\theta_1, \theta_2, \sigma^2| \text{ data}) = \pi(\theta_1|\sigma^2, \bar{y}_1)\pi(\theta_2|\sigma^2, \bar{y}_2)\pi(\sigma^2|s^2),$$

and hence

$$\pi(\eta, \sigma^2 |\ \mathrm{data}) = \pi(\eta | \sigma^2, \bar{y}_1, \bar{y}_2) \pi(\sigma^2 | s^2). \tag{8.1}$$

Now, note that

$$\eta | \sigma^2, \bar{y}_1, \bar{y}_2 \sim N\left(\bar{y}_1 - \bar{y}_2, \sigma^2 \left(\frac{1}{n_1} + \frac{1}{n_2}\right)\right).$$

Consequently, integrating out σ^2 from (8.1) yields,

$$\pi(\eta |\ \mathrm{data}) \propto \left\{ 1 + \frac{(\eta - (\bar{y}_1 - \bar{y}_2))^2}{(n_1 + n_2 - 2)s^2(\frac{1}{n_1} + \frac{1}{n_2})} \right\}^{(n_1 + n_2 - 1)/2}, \tag{8.2}$$

or, equivalently

$$\frac{\eta - (\bar{y}_1 - \bar{y}_2)}{s\sqrt{\frac{1}{n_1} + \frac{1}{n_2}}} \Big|\ \mathrm{data} \sim t_{n_1 + n_2 - 2}.$$

In many situations, the assumption that $\sigma_1^2 = \sigma_2^2$ is not tenable. For example, in a clinical trial the populations corresponding with two different treatments may have very different spread. This problem of comparing means when we have unequal and unknown variances is known as the Behrens-Fisher problem, and a frequentist approach to this problem has already been discussed in Problem 17, Chapter 2. We discuss the Bayesian approach now. We have that $(\bar{Y}_1, s_1^2)$ is sufficient for (θ_1, σ_1^2) and $(\bar{Y}_2, s_2^2)$ is sufficient for (θ_2, σ_2^2), where $s_i^2 = \sum_{j=1}^{n_i}(Y_{ij} - \bar{Y}_i)^2/(n_i - 1)$, $i = 1, 2$. Also, given $(\theta_1, \theta_2, \sigma_1^2, \sigma_2^2)$,

$$\bar{Y}_i \sim N\left(\theta_i, \frac{\sigma_i^2}{n_i}\right), \ \ \mathrm{and}\ \ (n_i - 1)s_i^2 \sim \sigma_i^2 \chi_{n_i - 1}^2, i = 1, 2,$$

and further, they are all independently distributed. Now employ the objective prior

$$\pi(\theta_1, \theta_2, \sigma_1^2, \sigma_2^2) \propto \sigma_1^{-2}\sigma_2^{-2},$$

and proceed exactly as in the previous case. It then follows that under the posterior distribution also θ_1 and θ_2 are independent, and that

$$\frac{\sqrt{n_1}(\theta_1 - \bar{y}_1)}{s_1} \Big|\ \mathrm{data} \sim t_{n_1 - 1} \quad \mathrm{and} \quad \frac{\sqrt{n_2}(\theta_2 - \bar{y}_2)}{s_2} \Big|\ \mathrm{data} \sim t_{n_2 - 1}. \tag{8.3}$$

It may be immediately noted that the posterior distribution of $\eta = \theta_1 - \theta_2$, however, is not a standard distribution. Posterior computations are still quite easy to perform because Monte Carlo sampling is totally straightforward. Simply generate independent deviates θ_1 and θ_2 repeatedly from (8.3) and utilize the corresponding $\eta = \theta_1 - \theta_2$ values to investigate its posterior distribution. Problem 4 is expected to apply these results.

Extension to the k-mean problem or one-way ANOVA is straightforward. A hierarchical Bayes approach to this problem and implementation using MCMC have already been discussed in Example 7.13 in Chapter 7.

8.2 Linear Regression

We encountered normal linear regression in Section 5.4 where we discussed prior elicitation issues in the context of the problem of inference on a response variable Y conditional on some predictor variable X. Normal linear models in general, and regression models in particular are very widely used. We have already seen an illustration of this in Example 7.13 in Section 7.4.6. We intend to cover some of the important inference problems related to regression in this section.

Extending the simple linear regression model where $E(Y|\beta_0, \beta_1, X = x) = \beta_0 + \beta_1 x$ to the multiple linear regression case, $E(Y|\boldsymbol{\beta}, \mathbf{X} = \mathbf{x}) = \boldsymbol{\beta}'\mathbf{x}$, yields the linear model

$$\mathbf{y} = X\boldsymbol{\beta} + \boldsymbol{\epsilon}, \tag{8.4}$$

where $\mathbf{y}$ is the n-vector of observations, X the $n \times p$ matrix having the appropriate readings from the predictors, $\boldsymbol{\beta}$ the p-vector of unknown regression coefficients, and $\boldsymbol{\epsilon}$ the n-vector of random errors with mean 0 and constant variance σ^2. The parameter vector then is $(\boldsymbol{\beta}, \sigma^2)$, and most often the statistical inference problem involves estimation of $\boldsymbol{\beta}$ and also testing hypotheses involving the same parameter vector. For convenience, we assume that X has full column rank $p < n$. We also assume that the first column of X is the vector of 1's, so that the first element of $\boldsymbol{\beta}$, namely β_1, is the intercept.

If we assume that the random errors are independent normals, we obtain the likelihood function for $(\boldsymbol{\beta}, \sigma^2)$ as

$$f(\mathbf{y}|\boldsymbol{\beta}, \sigma^2) = \left[\frac{1}{\sqrt{2\pi}\sigma}\right]^n \exp\left\{-\frac{1}{2\sigma^2}(\mathbf{y} - X\boldsymbol{\beta})'(\mathbf{y} - X\boldsymbol{\beta})\right\}$$
$$= \left[\frac{1}{\sqrt{2\pi}\sigma}\right]^n \exp\left\{-\frac{1}{2\sigma^2}\left[(\mathbf{y} - \hat{\mathbf{y}})'(\mathbf{y} - \hat{\mathbf{y}}) + (\boldsymbol{\beta} - \hat{\boldsymbol{\beta}})'X'X(\boldsymbol{\beta} - \hat{\boldsymbol{\beta}})\right]\right\} \tag{8.5}$$

where

$$\hat{\boldsymbol{\beta}} = (X'X)^{-1}X'\mathbf{y}, \text{ and } \hat{\mathbf{y}} = X\hat{\boldsymbol{\beta}}.$$

It then follows that $\hat{\boldsymbol{\beta}}$ is sufficient for $\boldsymbol{\beta}$ if σ^2 is known, and $(\hat{\boldsymbol{\beta}}, (\mathbf{y}-\hat{\mathbf{y}})'(\mathbf{y}-\hat{\mathbf{y}}))$ is jointly sufficient for $(\boldsymbol{\beta}, \sigma^2)$. Further,

$$\hat{\boldsymbol{\beta}}|\sigma^2 \sim N_p(\boldsymbol{\beta}, \sigma^2(X'X)^{-1})$$

and is independent of

$$(\mathbf{y} - \hat{\mathbf{y}})'(\mathbf{y} - \hat{\mathbf{y}}))|\sigma^2 \sim \sigma^2 \chi^2_{n-p}.$$

We take the prior,

$$\pi(\boldsymbol{\beta}, \sigma^2) \propto \frac{1}{\sigma^2}. \tag{8.6}$$

This leads to the posterior,

$$\pi(\boldsymbol{\beta}, \sigma^2 | \mathbf{y}) = \pi(\boldsymbol{\beta} | \hat{\boldsymbol{\beta}}, \sigma^2) \pi(\sigma^2 | (\mathbf{y} - \hat{\mathbf{y}})'(\mathbf{y} - \hat{\mathbf{y}})). \tag{8.7}$$

It can be seen that

$$\boldsymbol{\beta} | \hat{\boldsymbol{\beta}}, \sigma^2 \sim N_p(\hat{\boldsymbol{\beta}}, \sigma^2 (X'X)^{-1})$$

and that the posterior distribution of σ^2 is proportional to an inverse χ^2_{n-p}. Integrating out σ^2 from this joint posterior density yields the multivariate t marginal posterior density for $\boldsymbol{\beta}$, i.e.,

$$\pi(\boldsymbol{\beta} | \mathbf{y})$$

$$= \frac{\Gamma(n/2) |X'X|^{1/2} s^{-p}}{(\Gamma(1/2))^p \Gamma((n-p)/2)(\sqrt{n-p})^p} \left[1 + \frac{(\boldsymbol{\beta} - \hat{\boldsymbol{\beta}})' X'X (\boldsymbol{\beta} - \hat{\boldsymbol{\beta}})}{(n-p)s^2} \right]^{-\frac{n}{2}}, \tag{8.8}$$

where $s^2 = (\mathbf{y} - \hat{\mathbf{y}})'(\mathbf{y} - \hat{\mathbf{y}})/(n-p)$. From this, it can be deduced that the posterior mean of $\boldsymbol{\beta}$ is $\hat{\boldsymbol{\beta}}$ if $n \geq p+2$, and the $100(1-\alpha)\%$ HPD credible region for $\boldsymbol{\beta}$ is given by the ellipsoid

$$\left\{ \boldsymbol{\beta} : (\boldsymbol{\beta} - \hat{\boldsymbol{\beta}})' X'X (\boldsymbol{\beta} - \hat{\boldsymbol{\beta}}) \leq ps^2 F_{p,n-p}(\alpha) \right\}, \tag{8.9}$$

where $F_{p,n-p}(\alpha)$ is the $(1 - \alpha)$ quantile of the $F_{p,n-p}$ distribution. Further, if one is interested in a particular β_j, the fact that the marginal posterior distribution of β_j is given by

$$\frac{\beta_j - \hat{\beta}_j}{s\sqrt{d_{jj}}} | \mathbf{y} \sim t_{n-p}, \tag{8.10}$$

where d_{jj} is the jth diagonal entry of $(X'X)^{-1}$, can be used.

Conjugate priors for the normal regression model are of interest especially if hierarchical prior modeling is desired. This discussion, however, will be deferred to the following chapters where hierarchical Bayesian analysis is discussed.

Example 8.1. Table 8.1 shows the maximum January temperatures (in degrees Fahrenheit), from 1931 to 1960, for 62 cities in the U.S., along with their latitude (degrees), longitude (degrees) and altitude (feet). (See Mosteller and Tukey, 1977.) It is of interest to relate the information supplied by the geographical coordinates to the maximum January temperatures.

The following summary measures are obtained.

$$X'X = \begin{bmatrix} 62.0 & 2365.0 & 5674.0 & 56012.0 \\ 2365.0 & 92955.0 & 217285.0 & 2244586.0 \\ 5674.0 & 217285.0 & 538752.0 & 5685654.0 \\ 56012.0 & 2244586.0 & 5685654.0 & 1.7720873 \times 10^8 \end{bmatrix},$$

Table 8.1. Maximum January Temperatures for U.S. Cities, with Latitude, Longitude, and Altitude

City	Latitude	Longitude	Altitude	Max. Jan. Temp
Mobile, Ala.	30	88	5	61
Montgomery, Ala.	32	86	160	59
Juneau, Alaska	58	134	50	30
Phoenix, Ariz.	33	112	1090	64
Little Rock, Ark.	34	92	286	51
Los Angeles, Calif.	34	118	340	65
San Francisco, Calif.	37	122	65	55
Denver, Col.	39	104	5280	42
New Haven, Conn.	41	72	40	37
Wilmington, Del.	39	75	135	41
Washington, D.C.	38	77	25	44
Jacksonville, Fla.	38	81	20	67
Key West, Fla.	24	81	5	74
Miami, Fla.	25	80	10	76
Atlanta, Ga.	33	84	1050	52
Honolulu, Hawaii	21	157	21	79
Boise, Idaho	43	116	2704	36
Chicago, Ill.	41	87	595	33
Indianapolis, Ind.	39	86	710	37
Des Moines, Iowa	41	93	805	29
Dubuque, Iowa	42	90	620	27
Wichita, Kansas	37	97	1290	42
Louisville, Ky.	38	85	450	44
New Orleans, La.	29	90	5	64
Portland, Maine	43	70	25	32
Baltimore, Md.	39	76	20	44
Boston, Mass.	42	71	21	37
Detroit, Mich.	42	83	585	33
Sault Sainte Marie, Mich.	46	84	650	23
Minneapolis -St. Paul, Minn.	44	93	815	22
St. Louis, Missouri	38	90	455	40
Helena, Montana	46	112	4155	29
Omaha, Neb.	41	95	1040	32
Concord, N.H.	43	71	290	32
Atlantic City, N.J.	39	74	10	43
Albuquerque, N.M.	35	106	4945	46

continues

$$(X'X)^{-1} = 10^{-5} \begin{bmatrix} 94883.1914 & -1342.5011 & -485.0209 & 2.5756 \\ -1342.5011 & 37.8582 & -0.8276 & -0.0286 \\ -485.0209 & -0.8276 & 5.8951 & -0.0254 \\ 2.5756 & -0.0286 & -0.0254 & 0.0009 \end{bmatrix},$$

Table 8.1 continued

City	Latitude	Longitude	Altitude	Max. Jan. Temp
Albany, N.Y.	42	73	20	31
New York, N.Y.	40	73	55	40
Charlotte, N.C.	35	80	720	51
Raleigh, N.C.	35	78	365	52
Bismark, N.D.	46	100	1674	20
Cincinnati, Ohio	39	84	550	41
Cleveland, Ohio	41	81	660	35
Oklahoma City, Okla.	35	97	1195	46
Portland, Ore.	45	122	77	44
Harrisburg, Pa.	40	76	365	39
Philadelphia, Pa.	39	75	100	40
Charlestown, S.C.	32	79	9	61
Rapid City, S.D.	44	103	3230	34
Nashville, Tenn.	36	86	450	49
Amarillo, Tx.	35	101	3685	50
Galveston, Tx.	29	94	5	61
Houston, Tx.	29	95	40	64
Salt Lake City, Utah	40	111	4390	37
Burlington, Vt.	44	73	110	25
Norfolk, Va.	36	76	10	50
Seattle-Tacoma, Wash.	47	122	10	44
Spokane, Wash.	47	117	1890	31
Madison, Wisc.	43	89	860	26
Milwaukee, Wisc.	43	87	635	28
Cheyenne, Wyoming	41	104	6100	37
San Juan, Puerto Rico	18	66	35	81

$$X'\mathbf{y} = \begin{pmatrix} 2739.0 \\ 99168.0 \\ 252007.0 \\ 2158463.0 \end{pmatrix}, \hat{\boldsymbol{\beta}} = \begin{pmatrix} 100.8260 \\ -1.9315 \\ 0.2033 \\ -0.0017 \end{pmatrix}, \text{ and } s = 6.05185.$$

On the basis of the analysis explained above, $\hat{\boldsymbol{\beta}}$ may be taken as the estimate of $\boldsymbol{\beta}$ and the HPD credible region for it can be derived using (8.9). Suppose instead one is interested in the impact of latitude on maximum January temperatures. Then the 95% HPD region for the corresponding regression coefficient β_2 can be obtained using (8.10). This yields the t-interval, $\hat{\beta}_2 \pm s\sqrt{d_{22}}t_{58}(.975)$, or (-2.1623, -1.7007), indicating an expected general drop in maximum temperatures as one moves away from the Equator. If the joint impact of latitude and altitude is also of interest, then one would look at the HPD credible set for (β_2, β_4). This is given by

$$\Big\{ (\beta_2, \beta_4) : (\beta_2 + 1.9315, \beta_4 + 0.0017)C^{-1}(\beta_2 + 1.9315, \beta_4 + 0.0017)' \le 2s^2 F_{2,58}(\alpha) \Big\},$$

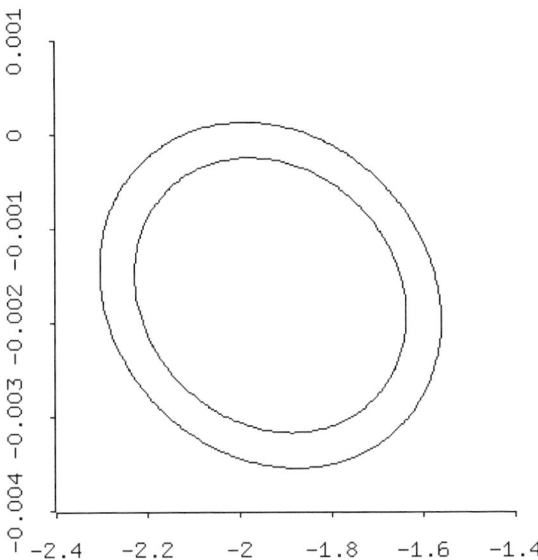

Fig. 8.1. Plot of 95% and 99% HPD credible regions for (β_2, β_4).

where C is the appropriate 2×2 block from $(X'X)^{-1}$,

$$C = 10^{-4} \begin{pmatrix} 3.7858 & -2.8636 \times 10^{-3} \\ -2.8636 \times 10^{-3} & 9.2635 \times 10^{-5} \end{pmatrix}.$$

Plotted in Figure 8.1 are the 95% and 99% HPD credible regions for (β_2, β_4). Impact of altitude on maximum temperatures seems to be very limited for the case under consideration.

Literature on Bayesian approach to linear regression is very large. Some of this material relevant to the discussion given above may be found in Box and Tiao (1973), Leamer (1978), and Gelman et al. (1995).

8.3 Logit Model, Probit Model, and Logistic Regression

We consider a problem here that is related to linear regression but actually belongs to a broad class of generalized linear models. This model is useful for problems involving toxicity tests and bioassay experiments. In such experiments, usually various dose levels of drugs are administered to batches of animals. Most often the responses are dichotomous because what is observed is whether the subject is dead or whether a tumor has appeared. This leads to a setup that can be easily understood in the context of the following example.

Example 8.2. Suppose that k independent random variables $y_1, y_2, \ldots, y_k$ are observed, where y_i has the $\mathrm{B}(n_i, p_i)$ probability distribution, $1 \leq i \leq k$. y_i may be the number of laboratory animals cured of an ailment in an experiment involving n_i such animals. It is certainly possible to make inference on each p_i separately based on the observed y_i (as discussed previously). This, however, is not really useful if we want to predict the results of a similar experiment in future. Suppose that the p_i are related to a covariate or an explanatory variable, such as dosage level in a clinical experiment. Then the natural approach is regression as described in the previous problem, because this allows us to explore and present the relationship between design (explanatory) variables and response variables, and (if needed) predictions of response at desired levels of the explanatory variables. Let t_i be the value of the covariate that corresponds with p_i, $i = 1, 2, \ldots, k$. Because p_i's are probabilities, linking them to the corresponding t_i's through a linear map as was done earlier does not seem appropriate now. Instead it can be made through a link function H such that $p_i = H(\beta_0 + \beta_1 t_i)$. H, here, is a known cumulative distribution function (c.d.f.) and β_0 and β_1 are two unknown parameters. (If H is an invertible function, this is precisely $H^{-1}(p_i) = \beta_0 + \beta_1 t_i$.) If the standard normal c.d.f. is used for H, the model is called the probit model, and if the logistic c.d.f. (i.e., $H(z) = e^{-z}/(1+e^{-z})$) is used, it is called the logit model. The likelihood function for the unknown parameters, β_0 and β_1, is then given by

$$\prod_{i=1}^{k} \binom{n_i}{y_i} H(\beta_0 + \beta_1 t_i)^{y_i} (1 - H(\beta_0 + \beta_1 t_i))^{n_i - y_i} .$$

Suppose $\pi(\beta_0, \beta_1)$ is the prior density on (β_0, β_1) so that the posterior density is

$$\pi(\beta_0, \beta_1 | \text{data}) = \frac{\pi(\beta_0, \beta_1) \prod_{i=1}^{k} H(\beta_0 + \beta_1 t_i)^{y_i} (1 - H(\beta_0 + \beta_1 t_i))^{n_i - y_i}}{\int \pi(a, b) \prod_{i=1}^{k} H(a + bt_i)^{y_i} (1 - H(a + bt_i))^{n_i - y_i} \, da \, db} .$$

It may be noted that the sample size n_i and the covariates t_i (dose level) are treated here as fixed, or equivalently the model conditional on those variables is analyzed (as in the linear regression problem).

More generally, instead of a single covariate t, if we have a set of s covariates represented by $\mathbf{x}$ and the corresponding vector of coefficients $\boldsymbol{\beta}$, the posterior density of $\boldsymbol{\beta}$ will be

$$\pi(\boldsymbol{\beta} | \text{data}) \propto \pi(\boldsymbol{\beta}) \prod_{i=1}^{k} H(\boldsymbol{\beta}' \mathbf{x}_i)^{y_i} (1 - H(\boldsymbol{\beta}' \mathbf{x}_i))^{n_i - y_i} . \tag{8.11}$$

8.3.1 The Logit Model

If we use the logit model whereby $p_i = \exp\{-\boldsymbol{\beta}' \mathbf{x}_i\}/\{1 + \exp\{-\boldsymbol{\beta}' \mathbf{x}_i\}\}$, and hence $-\log(p_i/(1 - p_i)) = \boldsymbol{\beta}' \mathbf{x}_i$, we obtain the likelihood function

$$l(\boldsymbol{\beta}) \propto \prod_{i=1}^{k} \left(\frac{\exp\{-\boldsymbol{\beta}'\mathbf{x}_i\}}{1 + \exp\{-\boldsymbol{\beta}'\mathbf{x}_i\}} \right)^{y_i} (1 + \exp\{-\boldsymbol{\beta}'\mathbf{x}_i\})^{-(n_i - y_i)}$$

$$= \exp\left(-\boldsymbol{\beta}' \sum_{i=1}^{k} y_i \mathbf{x}_i \right) \prod_{i=1}^{k} (1 + \exp(-\boldsymbol{\beta}'\mathbf{x}_i))^{-n_i}, \qquad (8.12)$$

which is largely intractable (but see Problem 7). Usually a flat prior such as $\pi(\boldsymbol{\beta}) \propto 1$ is employed, but because $\boldsymbol{\beta}$ can be considered to be regression coefficients under reparameterization, an approximate conjugate normal prior can also be used. In such a case, a hierarchical prior structure is also meaningful.

To motivate the approximate conjugate hierarchical prior structure, consider the following large sample approximation. For simplicity, let there be only one covariate t. Assume that the n_i are large enough for a satisfactory normal approximation of the binomial model. If we let $\hat{p}_i = y_i/n_i$, then (approximately) these $\hat{p}_i$ are independent $N(p_i, p_i(1 - p_i)/n_i)$ random variates. Now let $\theta_i = -\log(p_i/(1 - p_i))$ and $\hat{\theta}_i = -\log(\hat{p}_i/(1 - \hat{p}_i))$. It can be shown that, approximately, $(\hat{\theta}_i - \theta_i)\sqrt{n_i\hat{p}_i(1 - \hat{p}_i)}$ are independent $N(0,1)$ random variates. Then (again approximately), the likelihood function for (β_0, β_1) is

$$\ell(\beta_0, \beta_1 | \text{data}) = \exp\left(-\frac{1}{2} \sum_{i=1}^{k} w_i(\hat{\theta}_i - (\beta_0 + \beta_1 t_i))^2 \right), \qquad (8.13)$$

where $w_i = n_i\hat{p}_i(1 - \hat{p}_i)$ are known weights. This suggests a bivariate normal prior on (β_0, β_1) as the first level in the hierarchical structure. Now the problem is very similar to that discussed in Section 5.4. Further, there is also another important use for the approximate likelihood in (8.13). Its product with the conjugate normal prior discussed above can be used as a natural proposal density for the M-H algorithm in the computations required for inferential purposes (see Problem 9). If instead a flat prior on $\boldsymbol{\beta}$ is to be employed, then (8.13) itself (up to a constant) can be used as the proposal density.

Example 8.3. (**Example 8.2 continued**). The data given in Table 8.2 is from Finney (1971) (which originally appeared in Martin (1942)) where results of spraying rotenone of different concentrations on the chrysanthemum aphids in batches of about fifty are presented. The concentration is in milligrams per liter of the solution and the dosage x is measured on the logarithmic scale. The median lethal dose LD50, the dose at which 50% of the subjects will expire, is one of the quantities of inferential interest.

The plot of $\hat{p} = y/n$ against x shown in Figure 8.2 is S-shaped, so a linear fit for $\hat{p}$ based on x is unsatisfactory. Instead, as suggested by Figure 8.3, the logistic regression is more appropriate here. Suppose that a flat prior on (β_0, β_1) is to be used. Then the implementation of M-H algorithm as explained above using the bivariate normal proposal density is straightforward. A scatter plot of 1000 values of (β_0, β_1) so obtained is shown in Figure 8.4.

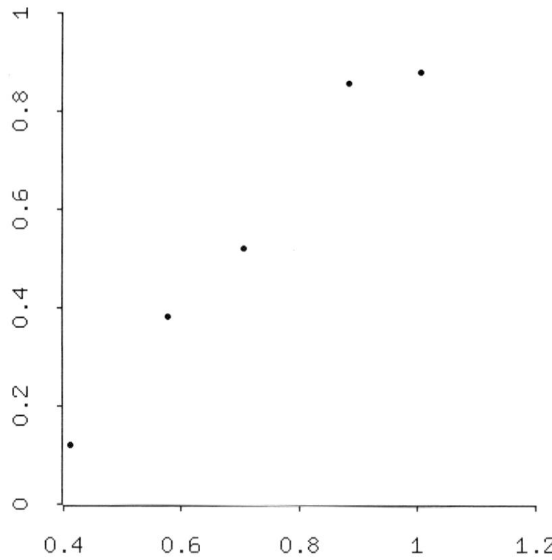

Fig. 8.2. Plot of proportion of deaths against dosage level.

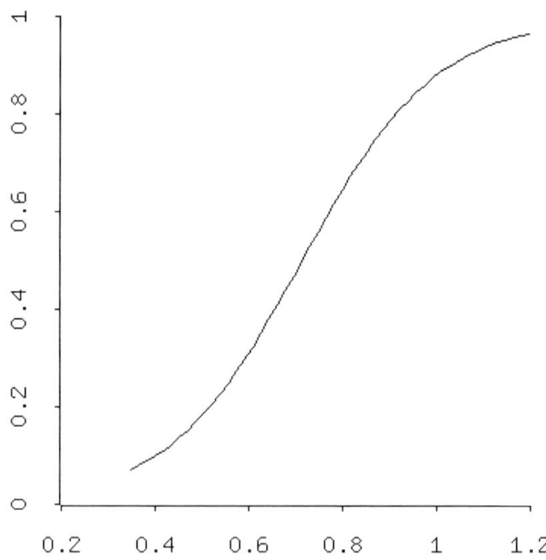

Fig. 8.3. Plot of logistic response function: $e^{-5+7x}/(1 + e^{-5+7x})$.

Table 8.2. Toxicity of Rotenone

Concentration (c_i)	Dose $(x_i = \log(c_i))$	Batch Size (n_i)	Deaths (y_i)
2.6	0.4150	50	6
3.8	0.5797	48	16
5.1	0.7076	46	24
7.7	0.8865	49	42
10.2	1.0086	50	44

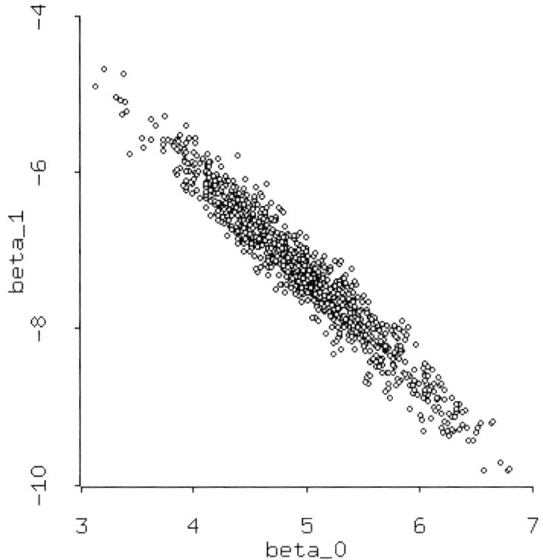

Fig. 8.4. Scatter plot of 1000 (β_0, β_1) values sampled using M-H algorithm.

A comparison of the estimates of β_0 and β_1 obtained using MLE and posterior means are shown in Table 8.3.

Histograms of the M-H samples of β_0 and β_1 are shown in Figure 8.5 and Figure 8.6, respectively. They seem to be skewed and hence the posterior estimates seem more appropriate.

Let us consider the estimation of LD50 next. Note that for the logit model LD50 is that dosage level t_0 for which $E(y_i/n_i|t_i = t_0) = \exp(-(\beta_0 + \beta_1 t_0))/(1 + \exp(-(\beta_0 + \beta_1 t_0))) = 0.5$. This means that LD50 $= -\beta_0/\beta_1$. We

Table 8.3. Estimates of β_0 and β_1 from Different Methods

Method	β_0	s.e.(β_0)	β_1	s.e.(β_1)
MLE from logistic regression	4.826		-7.065	
Posterior estimates	4.9727	0.6312	-7.266	0.8859

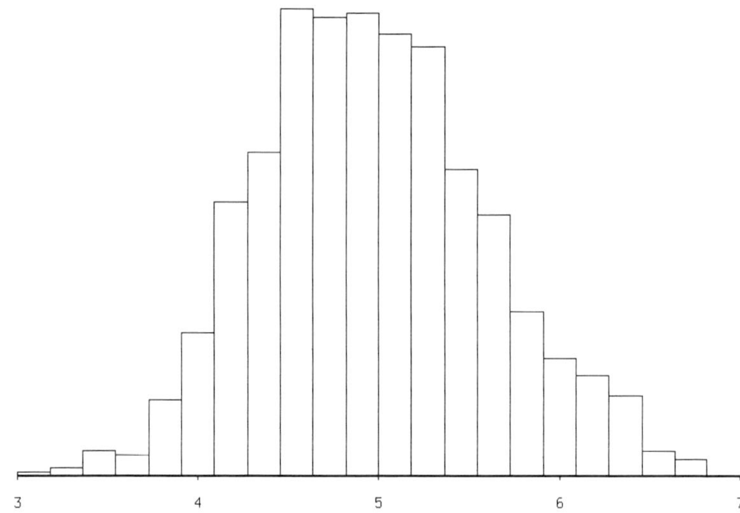

Fig. 8.5. Histogram of 1000 β_0 values sampled using M-H algorithm.

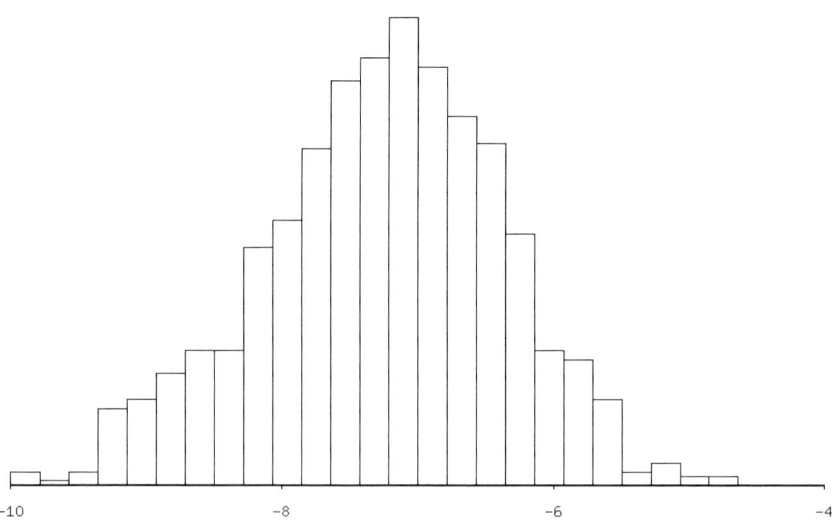

Fig. 8.6. Histogram of 1000 β_1 values sampled using M-H algorithm.

can easily estimate this using our M-H sample, and we obtain an estimate of
0.6843 (in the logarithmic scale) with a standard error of 0.022.

8.3.2 The Probit Model

As mentioned previously, if the standard normal c.d.f., Φ is used for the link
function H above, we obtain the probit model. Then, assuming that $\pi(\boldsymbol{\beta})$ is
the prior density on $\boldsymbol{\beta}$ the posterior density is obtained as

$$\pi(\boldsymbol{\beta}|\text{data}) \propto \pi(\boldsymbol{\beta}) \prod_{i=1}^{k} \Phi(\boldsymbol{\beta}'\mathbf{x}_i)^{y_i} \left(1 - \Phi(\boldsymbol{\beta}'\mathbf{x}_i)\right)^{n_i - y_i}. \tag{8.14}$$

Analytically, this is even less tractable than the posterior density for the logit
model. However, the following computational scheme developed by Albert and
Chib (1993) based on the Gibbs sampler provides a convenient strategy.

First consider the simpler case involving Bernoulli y_i's, i.e., $n_i \equiv 1$. Then,

$$\pi(\boldsymbol{\beta}|\text{data}) \propto \pi(\boldsymbol{\beta}) \prod_{i=1}^{k} \Phi(\boldsymbol{\beta}'\mathbf{x}_i)^{y_i} \left(1 - \Phi(\boldsymbol{\beta}'\mathbf{x}_i)\right)^{1 - y_i}.$$

The computational scheme then proceeds by introducing k independent latent
variables $Z_1, Z_2, \ldots, Z_k$, where $Z_i \sim N(\boldsymbol{\beta}'\mathbf{x}_i, 1)$. If we let $Y_i = I(Z_i > 0)$,
then $Y_1, \ldots, Y_k$ are independent Bernoulli with $p_i = P(Y_i = 1) = \Phi(\boldsymbol{\beta}'\mathbf{x}_i)$.
Now note that the posterior density of $\boldsymbol{\beta}$ and $\mathbf{Z} = (Z_1, \ldots, Z_k)$ given $\mathbf{y} = (y_1, \ldots, y_k)$ is

$$\pi(\boldsymbol{\beta}, \mathbf{Z}|\mathbf{y}) \propto \pi(\boldsymbol{\beta}) \prod_{i=1}^{k} \left\{I(Z_i > 0)I(y_i = 1) + I(Z_i \leq 0)I(y_i = 0)\right\} \phi(Z_i - \boldsymbol{\beta}'\mathbf{x}_i),$$
$$\tag{8.15}$$

where ϕ is the standard normal p.d.f. Even though (8.15) is not a joint density
which allows sampling from it directly, it allows Gibbs sampler to handle it
since $\pi(\boldsymbol{\beta}|\mathbf{Z}, \mathbf{y})$ and $\pi(\mathbf{Z}|\boldsymbol{\beta}, \mathbf{y})$ allow sampling from them. It is clear that

$$\pi(\boldsymbol{\beta}|\mathbf{Z}, \mathbf{y}) \propto \pi(\boldsymbol{\beta}) \prod_{i=1}^{k} \phi(Z_i - \boldsymbol{\beta}'\mathbf{x}_i), \tag{8.16}$$

whereas

$$Z_i|\boldsymbol{\beta}, \mathbf{y} \sim \begin{cases} N(\boldsymbol{\beta}'\mathbf{x}_i, 1) & \text{truncated at the left by 0 if } y_i = 1; \\ N(\boldsymbol{\beta}'\mathbf{x}_i, 1) & \text{truncated at the right by 0 if } y_i = 0. \end{cases} \tag{8.17}$$

Sampling $\mathbf{Z}$ from (8.17) is straightforward. On the other hand, (8.16) is simply
the posterior density for the regression parameters in the normal linear model
with error variance 1. Therefore, if a flat noninformative prior on $\boldsymbol{\beta}$ is used,
then

Table 8.4. Lethality of Chloracetic Acid

Dose	Fatalities	Dose	Fatalities
.0794	1	.1778	4
.1000	2	.1995	6
.1259	1	.2239	4
.1413	0	.2512	5
.1500	1	.2818	5
.1558	2	.3162	8

$$\boldsymbol{\beta}|\mathbf{Z}, \mathbf{y} \sim N_s(\hat{\boldsymbol{\beta}}_{\mathbf{Z}}, (\mathbf{X}'\mathbf{X})^{-1}),$$

where $\mathbf{X} = (\mathbf{x}_1', \ldots, \mathbf{x}_k')'$ and $\hat{\boldsymbol{\beta}}_{\mathbf{Z}} = (\mathbf{X}'\mathbf{X})^{-1}\mathbf{X}'\mathbf{Z}$. If a proper normal prior is used a different normal posterior will emerge. In either case, it is easy to sample $\boldsymbol{\beta}$ from this posterior conditional on $\mathbf{Z}$.

Extension of this scheme to binomial counts $Y_1, Y_2, \ldots, Y_k$ is straightforward. We let $Y_i = \sum_{j=1}^{n_i} Y_{ij}$ where $Y_{ij} = I(Z_{ij} > 0)$, with $Z_{ij} \sim N(\boldsymbol{\beta}'\mathbf{x}_i, 1)$ are independent, $j = 1, 2, \ldots, n_i$, $i = 1, 2, \ldots, k$. We then proceed exactly as above but at each Gibbs step we will need to generate $\sum_{i=1}^{k} n_i$ many (truncated) normals Z_{ij}. This procedure is employed in the following example.

Example 8.4. (**Example 8.2 continued**). Consider the data given in Table 8.4, taken from Woodward et al. (1941) where several data sets on toxicity of certain acids were reported. This particular data set examines the relationship between exposure to chloracetic acid and the death of mice. At each of the dose levels (measured in grams per kilogram of body weight), ten mice were exposed. The median lethal dose LD50 is again one of the quantities of inferential interest.

The Gibbs sampler as explained above is employed to generate a sample from the posterior distribution of (β_0, β_1) given the data. A scatter plot of 1000 points so generated is shown in Figure 8.7. From this sample we have the estimate of (-1.4426, 5.9224) for (β_0, β_1) along with standard errors of 0.4451 and 2.0997, respectively.

To estimate the LD50, note that for the probit model LD50 is the dosage level t_0 for which $E(y_i/n_i|t_i = t_0) = \Phi(\beta_0 + \beta_1 t_0) = 0.5$. As before, this implies that LD50 $= -\beta_0/\beta_1$. Using the sample provided by the Gibbs sampler we estimate this to be 0.248.

8.4 Exercises

1. Show how a random deviate from the Student's t is to be generated.
2. Construct the 95% HPD credible set for $\theta_1 - \theta_2$ for the two-sample problem in Section 8.1 assuming $\sigma_1^2 = \sigma_2^2$.

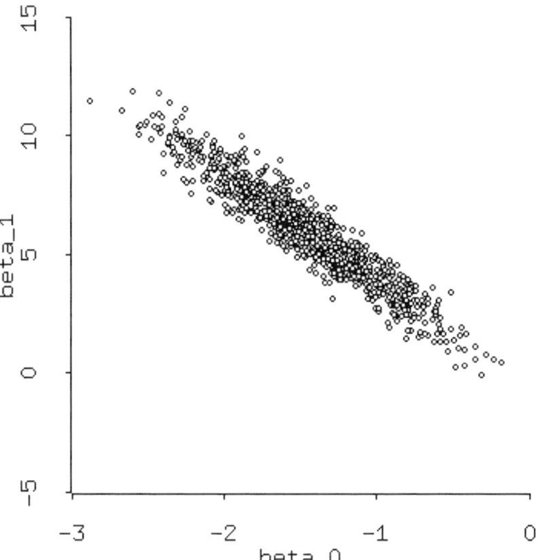

Fig. 8.7. Scatter plot of 1000 (β_0, β_1) values sampled using Gibbs sampler.

3. Show that Student's t can be expressed as scale mixtures of normals. Using this fact, explain how the 95% HPD credible set for $\theta_1 - \theta_2$ can be constructed for the case given in (8.3).
4. Consider the data in Table 8.5 from a clinical trial conducted by Mr. S. Sahu, a medical student at Bangalore Medical College, Bangalore, India (personal communication). The objective of the study was to compare two treatments, surgical and non-surgical medical, for short-term management of benign prostatic hyperplasia (enlargement of prostate). The random observable of interest is the 'improvement score' recorded for each of the patients by the physician, which we assume to be normally distributed. There were 15 patients in the non-surgical group and 14 in the surgical group.

Table 8.5. Improvement Scores

Surgical	15	9	12	16	14	15	18	13	12	11	15	9	16	9	
Non-surgical	6	8	7	4	4	6	8	3	7	8	9	6	3	6	4

Apply the results from Problems 2 and 3 above to make inferences about the difference in mean improvement in this problem.
5. Consider the linear regression model (8.4).
   (a) Show that $(\hat{\boldsymbol{\beta}}, (\mathbf{y} - \hat{\mathbf{y}})'(\mathbf{y} - \hat{\mathbf{y}}))$ is jointly sufficient for $(\boldsymbol{\beta}, \sigma^2)$.

(b) Show that $\hat{\boldsymbol{\beta}}|\sigma^2 \sim N_p(\boldsymbol{\beta}, \sigma^2(X'X)^{-1})$, $(\mathbf{y}-\hat{\mathbf{y}})'(\mathbf{y}-\hat{\mathbf{y}}))|\sigma^2 \sim \sigma^2\chi^2_{n-p}$, and they are independently distributed.

(c) Under the default prior (8.6), show that $\boldsymbol{\beta}|\mathbf{y}$ has the multivariate t distribution having density (8.8).

6. Construct 95% HPD credible set for (β_2, β_3) in Example 8.1.

7. Consider the model given in (8.12).

(a) What is a sufficient statistic for $\boldsymbol{\beta}$?

(b) Show that the likelihood equations for deriving MLE of $\boldsymbol{\beta}$ must satisfy

$$\sum_{i=1}^{k} \frac{n_i \exp\{-\boldsymbol{\beta}'\mathbf{x}_i\}}{1 + \exp\{-\boldsymbol{\beta}'\mathbf{x}_i\}} x_{ij} = \sum_{i=1}^{k} y_i \mathbf{x}_{ij}, \quad j = 1, 2, \ldots$$

8. Justify the approximate likelihood given in (8.13).

9. Consider a multivariate normal prior on $\boldsymbol{\beta}$ for Problem 7.

(a) Explain how the M-H algorithm can be used for computing the posterior quantities.

(b) Compare the above scheme with an importance sampling strategy where the importance function is proportional to the product of the normal prior and the approximate normal likelihood given in (8.13).

10. Apply the results from Problem 9 to Example 8.3 with some appropriate choice for the hyperparameters.

11. Justify (8.16) and (8.17), and explain how Gibbs sampler handles (8.15).

12. Analyze the problem in Example 8.4 with an additional data point having 9 fatalities at the dosage level of 0.3400.

13. Analyze the problem in Example 8.4 using logistic regression. Compare the results with those obtained in Section 8.3.2 using the probit model.

9

High-dimensional Problems

Rather than begin by defining what is meant by high-dimensional, we begin with a couple of examples.

Example 9.1. (Stein's example) Let $N(\boldsymbol{\mu}_{p\times 1}, \Sigma_{p\times p} \equiv \sigma^2 I_{p\times p})$ be a p-variate normal population and $\boldsymbol{X}_i = (X_{i1}, \ldots, X_{ip})$, $i = 1, \ldots, n$ be n i.i.d. p-variate samples. Because $\Sigma = \sigma^2 I$, we may alternatively think of the data as p independent samples of size n from p univariate normal populations $N(\mu_j, \sigma^2)$, $j = 1, \ldots, p$. The parameters of interest are the μ_j's. For convenience, we initially assume σ^2 is known. Usually, the number of parameters, p, is large and the sample size n is small compared with p. These have been called problems with large p, small n. Note that n in Stein's example is the sample size, if we think of the data as a p-variate sample of size n. However, we could also think of the data as univariate samples of size n from each of p univariate populations. Then the total sample size would be np. The second interpretation leads to a class of similar examples. Note that the observations are not exchangeable except in subgroups, in this sense one may call them partially exchangeable.

Example 9.2. Let $f(x|\mu_j), j = 1, \ldots, p$, denote the densities for p populations, and $X_{ij}, i = 1, \ldots, n, j = 1, \ldots, p$ denote p samples of size n from these p populations. As in Example 9.1, $f(x|\mu_j)$ may contain additional common parameters. The object is to make inference about the μ_j's.

In several path-breaking papers Stein (1955), James and Stein (1960), Stein (1981), Robbins (1955, 1964), Efron and Morris (1971, 1972, 1973a, 1975) have shown classical objective Bayes or classical frequentist methods, e.g., maximum likelihood estimates, will usually be inappropriate here. See also Kiefer and Wolfowitz (1956) for applications to examples like those of Neyman and Scott (1948). These approaches are discussed in Sections 9.1 through 9.4, with stress on the parametric empirical Bayes (PEB) approach of Efron and Morris, as extended in Morris (1983).

It turns out that exchangeability of $\mu_1, \ldots, \mu_p$ plays a fundamental role in all these approaches. Under this assumption, there is a simple and natural Bayesian solution of the problem based on a hierarchical prior and MCMC. Much of the popularity of Bayesian methods is due to the fact that many new examples of this kind could be treated in a unified way.

Because of the fundamental role of exchangeability of μ_j's and the simplicity, at least in principle, of the Bayesian approach, we begin with these two topics in Section 9.1. This leads in a natural way to the PEB approach in Sections 9.2 and 9.3 and the frequentist approach in Section 9.4.

All the above sections deal with point or interval estimation. In Section 9.6 we deal with testing and multiple testing in high-dimensional problems, with an application to microarrays. High-dimensional inference is closely related to model selection in high-dimensional problems. A brief overview is presented in Sections 9.7 and 9.8. We discuss several general issues in Sections 9.5 and 9.9.

9.1 Exchangeability, Hierarchical Priors, Approximation to Posterior for Large p, and MCMC

We follow the notations of Example 9.1 and Example 9.2, i.e., we consider p similar but not identical populations with densities $f(x|\mu_1), \ldots, f(x|\mu_p)$. There is a sample of size n, $X_{1j}, \ldots, X_{nj}$, from the jth population. These p populations may correspond with p adjacent small areas with unknown per capita income $\mu_1, \ldots, \mu_p$, as in small area estimation, Ghosh and Meeden (1997, Chapters 4, 5). They could also correspond with p clinical trials in a particular hospital and μ_j, $j = 1, \ldots, p$, are the mean effects of the drug being tested. In all these examples, the different studied populations are related to each other. In Morris (1983), which we closely follow in Section 9.2, the p populations correspond to p baseball players and μ_j's are average scores. Other such studies are listed in Morris and Christiansen (1996).

In order to assign a prior distribution for the μ_j's, we model them as exchangeable rather than i.i.d. or just independent. An exchangeable, dependent structure is consistent with the assumption that the studies are similar in a broad sense, so they share many common elements.

On the other hand, independence may be unnecessarily restrictive and somewhat unintuitive in the sense that one would expect to have separate analysis for each sample if the μ_j's were independent and hence unrelated. However, to justify exchangeability one would need a particular kind of dependence. For example, Morris (1983) points out that the baseball players in his study were all hitters; his analysis would have been hard to justify if he had considered both hitters and pitchers.

Using a standard way of generating exchangeable random variables, we introduce a vector of hyperparameters $\boldsymbol{\eta}$ and assume μ_j's are i.i.d. $\pi(\mu|\boldsymbol{\eta})$ given $\boldsymbol{\eta}$. Typically, if $f(x|\mu)$ belongs to an exponential family, it is convenient to take $\pi(\mu|\boldsymbol{\eta})$ to be a conjugate prior. It can be shown that for even moderately large p – in the baseball example of Morris (1983), $p = 18$ – there is a lot of information in the data on $\boldsymbol{\eta}$. Hence the choice of a prior for $\boldsymbol{\eta}$ does not have much influence on inference about μ_j's. It is customary to choose a uniform or one of the other objective priors (vide Chapter 5) for $\boldsymbol{\eta}$.

We illustrate these ideas by exploring in detail Example 9.1.

Example 9.3. (Example 9.1, continued) Let $f(x|\mu_j)$ be the density of $N(\mu_j, \sigma^2)$ where we initially assume σ^2 is known. We relax this assumption in Subsection 9.1.1.

The prior for μ_j is taken to be $N(\eta_1, \eta_2)$ where η_1 is the prior guess about the μ_j's and η_2 is a measure of the prior uncertainty about this guess, vide Example 2.1, Chapter 2. The prior for η_1, η_2 is $\pi(\eta_1, \eta_2)$, which we specify a bit later.

Because $(\bar{X}_j = \sum_{i=1}^n X_{ij}/n, j = 1, \ldots, p)$ form a sufficient statistic and $\bar{X}_j$'s are independent, the Bayes estimate for μ_j under squared error loss is

$$E(\mu_j|\boldsymbol{X}) = E(\mu_j|\bar{\boldsymbol{X}}) = \int E(\mu_j|\bar{\boldsymbol{X}}, \boldsymbol{\eta})\pi(\boldsymbol{\eta}|\bar{\boldsymbol{X}})d\boldsymbol{\eta}.$$

where $\boldsymbol{X} = (X_{ij}, i = 1, \ldots, n, j = 1, \ldots, p)$, $\bar{\boldsymbol{X}} = (\bar{X}_1, \ldots, \bar{X}_p)$ and (vide Example 2.1)

$$E(\mu_j|\boldsymbol{X}, \boldsymbol{\eta}) = E(\mu_j|\bar{X}_j, \boldsymbol{\eta}) = \frac{\eta_2\bar{X}_j + (\sigma^2/n)\eta_1}{\eta_2 + (\sigma^2/n)} = (1 - B)\bar{X}_j + B\eta_1, \quad (9.1)$$

with $B = (\sigma^2/n)/(\eta_2 + \sigma^2/n)$, depends on data only through $\bar{X}_j$.

The Bayes estimate of μ_j, on the other hand, depends on $\bar{X}_j$ as above and also on $(\bar{X}_1, \ldots, \bar{X}_p)$ because the posterior distribution of $\boldsymbol{\eta}$ depends on all the $\bar{X}_j$'s. Thus the Bayes estimate learns from the full sufficient statistic justifying simultaneous estimation of all the μ_j's. This learning process is sometimes referred to as borrowing strength. If the modeling of μ_j's is realistic, we would expect the Bayes estimates to perform better than the $\bar{X}_j$'s. This is what is strikingly new in the case of large p, small n and follows from the exchangeability of μ_j's.

The posterior density $\pi(\boldsymbol{\eta}|\boldsymbol{X})$ is also easy to calculate in principle. For known σ^2, one can get it explicitly.

Integrating out the μ_j's and holding $\boldsymbol{\eta}$ fixed, we get $\bar{X}_j$'s are independent and

$$\bar{X}_j|\boldsymbol{\eta} \sim N(\eta_1, \eta_2 + \sigma^2/n). \quad (9.2)$$

Let the prior density of (η_1, η_2) be constant on $\mathcal{R} \times \mathcal{R}^+$. (See Problem 1 and Gelman et al. (1995) to find out why some other choices like $\pi(\eta_1, \eta_2) = 1/\eta_2$ are not suitable here.)

Given (9.2) and $\pi(\eta_1, \eta_2)$ as above,

$$\pi(\boldsymbol{\eta}|\boldsymbol{X}) \propto \left\{2\pi(\eta_2 + \frac{\sigma^2}{n})\right\}^{-p/2} \exp\left\{-\frac{1}{2(\eta_2 + \frac{\sigma^2}{n})}\sum_{j=1}^{p}(\bar{X}_j - \eta_1)^2\right\}\pi(\boldsymbol{\eta})$$

$$\propto \left\{2\pi(\eta_2 + \frac{\sigma^2}{n})\right\}^{-1/2} \exp\left\{-\frac{p}{2(\eta_2 + \frac{\sigma^2}{n})}(\eta_1 - \bar{X})^2\right\}$$

$$\times \left(\eta_2 + \frac{\sigma^2}{n}\right)^{-(p-1)/2} \exp\left\{-\frac{1}{2(\eta_2 + \frac{\sigma^2}{n})}S\right\}, \tag{9.3}$$

where $\bar{X} = \frac{1}{p}\sum_{j=1}^{p}\bar{X}_j$ and $S = \sum_{j=1}^{p}(\bar{X}_j - \bar{X})^2$.

In a similar way,

$$\pi(\boldsymbol{\mu}|\boldsymbol{X}) = \int \prod_{j=1}^{p} \pi(\mu_j|\bar{X}_j, \boldsymbol{\eta})\pi(\boldsymbol{\eta}|\bar{\boldsymbol{X}})\, d\boldsymbol{\eta}, \tag{9.4}$$

where $(\mu_j|\bar{X}_j, \boldsymbol{\eta})$ are independent normal with

$$\text{mean as in (9.1) and variance } \frac{\eta_2\sigma^2/n}{\eta_2 + \sigma^2/n} \tag{9.5}$$

and

$$\pi(\boldsymbol{\eta}|\bar{\boldsymbol{X}}) = \pi(\eta_1|\bar{\boldsymbol{X}}, \eta_2)\pi(\eta_2|\bar{\boldsymbol{X}}) \tag{9.6}$$

is the product of a normal and inverse-Gamma (as given in (9.3)).

Construction of a credible interval for μ_j is, in principle, simple. Consider $\pi(\mu_j|\bar{\boldsymbol{X}})$ and fix $0 < \alpha < 1$. Calculate the posterior quantiles $\underline{\mu}_j(\bar{\boldsymbol{X}}), \bar{\mu}_j(\bar{\boldsymbol{X}})$ of orders $100\alpha/2$ and $100(1 - \alpha/2)$ for given data. Then

$$P\{\underline{\mu}_j(\bar{\boldsymbol{X}}) \le \mu_j \le \bar{\mu}_j(\bar{\boldsymbol{X}})|\bar{\boldsymbol{X}}\} = 1 - \alpha.$$

In general, to calculate $\underline{\mu}_j$ and $\bar{\mu}_j$ one would have to resort to MCMC which is explained in Subsection 9.1.1.

For large p, good approximations are available. To do this we anticipate to some extent Section 9.2.

Because p is large, we can invoke the theorem on posterior normality (Chapter 4). Thus the posterior for $\boldsymbol{\eta}$ is nearly normal with mean $\hat{\boldsymbol{\eta}}$ and variances of order $O(1/p)$, $\hat{\boldsymbol{\eta}}$ being the MLE of $\boldsymbol{\eta}$ based on the "likelihood"

$$\prod_{j=1}^{p} f(\bar{x}_j|\boldsymbol{\eta}).$$

Hence, $\pi(\boldsymbol{\eta}|\bar{\boldsymbol{x}})$ is approximately (in the sense of convergence in distribution) degenerate at $\hat{\boldsymbol{\eta}}$. This implies

$$\pi(\mu_j|\bar{\boldsymbol{X}}) = \int \pi(\mu_j|\bar{X}_j,\boldsymbol{\eta})\pi(\boldsymbol{\eta}|\bar{\boldsymbol{X}})\,d\boldsymbol{\eta}$$

$$= \pi(\mu_j|\bar{X}_j,\hat{\boldsymbol{\eta}}) \text{ (approximately) .} \tag{9.7}$$

This in turn implies the Bayes estimate of μ_j is

$$E(\mu_j|\bar{\boldsymbol{X}}) = E(\mu_j|\bar{X}_j,\hat{\boldsymbol{\eta}}) \text{ (approximately)} \tag{9.8}$$

which, by (9.1), is a shrinkage estimate that shrinks $\bar{X}_j$ towards $\hat{\eta}_1$, and which depends on all the sample means.

The approximation (9.8) is correct up to $O(1/p)$. A similar argument provides an approximation to the posterior s.d. but the accuracy is only $O(1/\sqrt{p})$.

Simulations indicate the approximation for the Bayes estimate is quite good but that for the posterior s.d. is much less accurate. It is known, vide Morris (1983), that the approximation is also inadequate for credible intervals.

As a final application of this approximation we indicate it is possible to check whether the prior $\pi(\mu_j|\boldsymbol{\eta})$ is consistent with data. More precisely, we check the consistency of $f(\bar{x}_j|\boldsymbol{\eta})$ with data, but a check for $f(\bar{x}_j|\boldsymbol{\eta})$ is indirectly a check for $\pi(\mu_j|\boldsymbol{\eta})$.

In the light of the data, $\boldsymbol{\eta} = \hat{\boldsymbol{\eta}}$ is the most likely value of the hyperparameter $\boldsymbol{\eta}$. Under $\hat{\boldsymbol{\eta}}$, $\bar{X}_j$'s are i.i.d normal with mean and variance given by (9.2) with $\boldsymbol{\eta}$ replaced by $\hat{\boldsymbol{\eta}}$. We can now check the fit of this model to the empirical distribution via the Q-Q plot. For each $0 < p < 1$, we plot the $100p$th quantiles for the theoretical and empirical distributions as $(x(p),y(p))$. If the fit is good, the resulting curve $\{(x(p),y(p)),0 < p < 1\}$ should scatter around an equiangular line passing through the origin.

9.1.1 MCMC and E-M Algorithm

We begin with p exponential densities of the same form, namely,

$$\exp\left\{nc(\boldsymbol{\theta}_j) + \sum_{i=1}^{k} t_{ji}(\boldsymbol{x}_j)\theta_{ji}\right\} h(\boldsymbol{x}_j), \quad j = 1,\ldots,p. \tag{9.9}$$

The conjugate prior density for the jth model is proportional to

$$\exp\{\eta_0 c(\boldsymbol{\theta}_j) + \sum_{i=1}^{k} \eta_i\theta_{ji}\}, \quad j = 1,\ldots,p. \tag{9.10}$$

Note that the hyperparameters $(\eta_0,\eta_1,\ldots,\eta_k)$ are the same for all j. Finally, consider a prior $\pi(\boldsymbol{\eta})$ for the hyperparameters.

Let $\boldsymbol{X} = (\boldsymbol{X_1},\ldots,\boldsymbol{X_p})$ and $\boldsymbol{\theta} = (\boldsymbol{\theta}_1,\ldots,\boldsymbol{\theta}_p)$. The conditional density of $\boldsymbol{\theta}$ given $\boldsymbol{X},\boldsymbol{\eta}$ is

$$\pi(\boldsymbol{\theta}|\boldsymbol{X},\boldsymbol{\eta}) \propto \prod_{j=1}^{p} \exp\{(\eta_0 + n)c(\boldsymbol{\theta}_j) + \sum_{i=1}^{k}(t_{ji}(\boldsymbol{x}_j) + \eta_i)\theta_{ji}\} \tag{9.11}$$

which shows conditionally $\boldsymbol{\theta}_j$'s remain independent. Also

$$\pi(\boldsymbol{\eta}|\boldsymbol{X},\boldsymbol{\theta}) \propto \exp\{pd(\boldsymbol{\eta}) + (\eta_0 + n)\sum_{j=1}^{p} c(\boldsymbol{\theta}_j) + \sum_{j=1}^{p}\sum_{i=1}^{k}(\eta_i + t_{ji}(\boldsymbol{x}_j))\theta_{ji}\}\pi(\boldsymbol{\eta})$$

$$(9.12)$$

where $\exp(d(\boldsymbol{\eta}))$ is the normalizing constant of the expression in (9.10).

By (9.12), the Bayes formula and cancellation of some common terms in the numerator and denominator of the Bayes formula,

$$\pi(\boldsymbol{\eta}|\boldsymbol{X},\boldsymbol{\theta}) \propto \exp\{pd(\boldsymbol{\eta}) + \eta_0\sum_{j=1}^{p} c(\boldsymbol{\theta}_j) + \sum_{j=1}^{p}\sum_{i=1}^{k}\eta_i\theta_{ji}\}\pi(\boldsymbol{\eta}).$$

If $d(\boldsymbol{\eta})$ has an explicit form, as is often the case, one can apply Gibbs sampling to draw samples from the joint posterior of $\boldsymbol{\theta}$ and $\boldsymbol{\eta}$ using the conditionals $\pi(\boldsymbol{\theta}|\boldsymbol{X},\boldsymbol{\eta})$ and $\pi(\boldsymbol{\eta}|\boldsymbol{X},\boldsymbol{\theta})$. Otherwise one can apply Metropolis-Hastings.

In general, the approximations based on $\hat{\boldsymbol{\eta}}$ are still valid but computation of $\hat{\boldsymbol{\eta}}$ is non-trivial. It turns out that the E-M algorithm can be applied, vide Gelman et al. (1995, Chapter 9).

We illustrate the algorithms for MCMC and E-M in the case of $N(\mu_j, \sigma^2)$, $j = 1,\ldots,p$, with $(\mu_1,\ldots,\mu_p)$ and σ^2 unknown. MCMC is quite straightforward here. Recall Example 7.13 from Chapter 7. The hierarchical Bayesian analysis of the usual one-way ANOVA was discussed there. With minimal modification, the same algorithm fits here to allow Gibbs sampling. Application of the E-M algorithm is also easy, as discussed in Gelman et al. (1995). We assume as before that μ_j are i.i.d. $N(\eta_1, \eta_2)$, and further take $\pi(\eta_1, \sigma^2, \eta_2) = 1/\sigma^2$. Then, recall from Section 7.2 that we need to apply the E and M steps to

$$\log\pi(\boldsymbol{\mu},\eta_1,\sigma^2,\eta_2|\boldsymbol{X}) = -(\frac{n}{2}+1)\log\sigma^2 - \frac{p}{2}\log\eta_2 - \frac{1}{2\eta_2}\sum_{j=1}^{p}(\mu_j - \eta_1)^2$$

$$-\frac{1}{2\sigma^2}\sum_{j=1}^{p}\sum_{i=1}^{n}(X_{ij} - \mu_j)^2 + \text{ constants} . \qquad (9.13)$$

The E-step requires the conditional distribution of $(\boldsymbol{\mu}, \sigma^2)$ given $\boldsymbol{X}$ and the current value $(\eta_1^{(c)}, \eta_2^{(c)})$ of (η_1, η_2). This is just the normal, inverse Gamma model. In the M-step we need to maximize $E^{(c)}(\log\pi(\boldsymbol{\mu},\eta_1,\sigma^2,\eta_2|\boldsymbol{X}))$ as a function of (η_1, η_2) which is straightforward.

9.2 Parametric Empirical Bayes

To explain the basic ideas, we consider once more the special case of $N(\mu_j, \sigma^2)$, σ^2 known. Explicit formulas are available in this special case for comparison

with the estimates of Stein. Another interesting special case is discussed in
Carlin and Louis (1996, Chapter 3).

The PEB approach was introduced by Efron and Morris in a series of
papers including Efron and Morris (1971, 1972, 1973a, 1973b, 1975, 1976).
In this section we generally follow Morris (1983).

Efron and Morris tried to take an intermediate position between a fully
Bayes and a fully frequentist approach by treating the likelihood as given by
$f(\bar{x}_j|\boldsymbol{\eta})$ obtained by integrating out the μ_j's as in (9.2). The $\boldsymbol{\eta}$'s are treated
as unknown parameters as in frequentist analysis whereas the μ_j's are treated
as random variables as in Bayesian analysis. This leads to a reduction of a
high-dimensional frequentist problem about μ_j's to a low-dimensional semi-
frequentist problem about $\boldsymbol{\eta}$, about which there is a lot of information in the
data. The fully Bayesian and the PEB approach differ in that no prior is as-
signed to $\boldsymbol{\eta}$, and $\boldsymbol{\eta}$ is estimated by MLE or by a suitable unbiased estimate.
So one may, if one wishes, think of the PEB approach as an approximation
to the hierarchical Bayes approach of Section 9.1. A disadvantage of PEB is
that accounting for the uncertainty about $\boldsymbol{\eta}$ is more difficult than in hierarchical Bayes – a point that would be discussed again in subsection 9.2.1. An
advantage is that one gets an explicit estimate for μ_j, namely, (9.1) with $\boldsymbol{\eta}$
replaced by an estimate of $\boldsymbol{\eta}$.

Note that under the likelihood (9.2), the complete sufficient statistic is the
pair

$$(\bar{X} = \frac{1}{p}\sum_{j=1}^{p}\bar{X}_j, \quad S = \sum_{j=1}^{p}(\bar{X}_j - \bar{X})^2).\tag{9.14}$$

Also, $\bar{X}$ and

$$\hat{B} = (p-3)\frac{\sigma^2/n}{S}\tag{9.15}$$

are unbiased estimates of η_1 and

$$B = \frac{\sigma^2/n}{\sigma^2/n + \eta_2}.\tag{9.16}$$

Then the best unbiased predictor of the RHS of (9.1) is

$$\hat{\mu}_j = (1 - \hat{B})\bar{X}_j + \hat{B}\bar{X}\tag{9.17}$$

which is the famous James-Stein-Lindley estimate of μ_j. It shrinks $\bar{X}_j$ towards
the overall mean $\bar{X}$.

The amount of shrinkage is determined by $\hat{B}$, which is close to 1 if $S/(p-3)$
is close to σ^2/n and close to zero if $S/(p-3)$ is much larger than σ^2/n. Note
that if $S/(p-3)$ is small, as in the first case, then the $\bar{X}_j$'s are close to $\bar{X}$
indicating μ_j's are close to each other. This justifies a fairly strong shrinkage
towards the grand mean. On the other hand, a large $S/(p-3)$ indicates
heterogeneity among the μ_j's, suggesting relatively large weight for $\bar{X}_j$.

Because

$$E(S/(p-1)) = \frac{\sigma^2}{n} + \eta_2, \tag{9.18}$$

an unbiased estimate of η_2 is $\hat{\eta}_2 = S/(p-1) - \sigma^2/n$. Because $\eta_2 \geq 0$, a more plausible estimate is $\hat{\eta}_2^+ = \max(0, \hat{\eta}_2)$, the positive part of the unbiased estimate. This suggests changing the estimate of B to

$$\widetilde{B} = \frac{(p-3)}{(p-1)} \frac{\sigma^2}{n} \Big/ \Big(\frac{\sigma^2}{n} + \hat{\eta}_2^+ \Big), \tag{9.19}$$

which is the James-Stein-Lindley positive part estimate.

If we take $\eta_1 = 0$, i.e., μ_j's are i.i.d $N(0, \eta_2)$, then the two estimates are of the form

$$\hat{\mu}_j = (1 - \hat{B})\bar{X}_j \text{ and } \widetilde{\mu}_j = (1 - \widetilde{B})\bar{X}_j. \tag{9.20}$$

These are the James-Stein and James-Stein positive part estimates. They shrink the estimate towards an arbitrary point zero and so do not seem attractive in the exchangeable case. But they have turned out to be quite useful in estimating coefficients in an orthogonal expansion of an unknown function with white noise as error, vide Cai et al. (2000). We study frequentist properties of these two estimates in Section 9.4.

9.2.1 PEB and HB Interval Estimates

Morris defines a random confidence interval $(\underline{\mu}_j(\bar{\boldsymbol{X}}), \bar{\mu}_j(\bar{\boldsymbol{X}}))$ for μ_j to have PEB confidence coefficient $(1 - \alpha)$ if

$$P\boldsymbol{\eta}\{\underline{\mu}_j \leq \mu_j \leq \bar{\mu}_j\} \geq 1 - \alpha. \tag{9.21}$$

Let $S_j^2 = [1 - ((p-1)/p)\hat{B}]\sigma^2/n + (2/(p-3))\hat{B}^2(\bar{X}_j - \bar{X})^2$. Morris has conjectured

$$\bar{X}_j \pm z_{\alpha/2} S_j \tag{9.22}$$

is a PEB confidence interval with confidence coefficient $1 - \alpha$.

It is shown in Basu et al. (2003) that the conjecture is not true but the violations are so rare and so small in magnitude that it hardly matters. Basu et al. (2003) suggest an adjustment that would make (9.22) true up to $O(p^{-2})$. It is also shown there that asymptotically the adjusted interval is equivalent to a PEB interval proposed by Carlin and Louis (1996, Chapter 3).

A trouble with Morris's interval is that it is somewhat ad hoc. We are not told how exactly it is derived. It seems he puts a noninformative prior on η_1, η_2 and adjusts somewhat the HB credible interval to get a conservative frequentist coverage probability.

There is a natural alternative that does not require additional adjustment and ensures (9.21) with the inequality replaced by an equality up to $O(p^{-2})$. To do this, one has to choose a prior for $\boldsymbol{\eta}$ that is probability matching in the sense of

$$P_{\boldsymbol{\eta}}\{\underline{\mu}_j \leq \mu_j \leq \bar{\mu}_j\} = 1 - \alpha + O(p^{-2}), \tag{9.23}$$

where

$$P\{\mu_j > \bar{\mu}_j | \bar{\boldsymbol{X}}\} = \alpha/2,$$
$$P\{\mu_j < \underline{\mu}_j | \bar{\boldsymbol{X}}\} = \alpha/2, \tag{9.24}$$

and the probabilities in (9.24) are the posterior probabilities under the prior for $\boldsymbol{\eta}$. This leads to probability matching differential equations. A solution is

$$\pi(\boldsymbol{\eta}) = \frac{\sigma^2/n}{\eta_2 + \sigma^2/n}, \tag{9.25}$$

vide Datta, Ghosh, and Mukerjee (2000).

9.3 Linear Models for High-dimensional Parameters

We can extend the HB and PEB approach to a more general setup using covariates and linear models. The parameters are no longer exchangeable but are affected by a common set of low-dimensional hyperparameters assuming the role of $\boldsymbol{\eta}$. The model in Sections 9.1 and 9.2 is a special case of the linear model below.

Following Morris (1983), we change our notations slightly

$$Y_j | \theta_j \sim N(\theta_j, V), \quad j = 1, \ldots, p \quad \text{independent}, \tag{9.26}$$

and given $\boldsymbol{\beta}, A$,

$$\boldsymbol{\theta}_{p \times 1} = Z_{p \times r} \boldsymbol{\beta}_{r \times 1} + \boldsymbol{\epsilon}_{p \times 1}, \tag{9.27}$$

ϵ_j's are i.i.d $N(0, A)$. Here p is at least moderately large, r is relatively small. Morris allows the variance of ϵ_j to depend on j, which is often a more realistic assumption. Keeping the same variance A for all j simplifies the algebra considerably.

In the HB analysis we need to put a further prior on $\boldsymbol{\beta}$. A conjugate prior for $\boldsymbol{\beta}$ given A is

$$\boldsymbol{\beta} \sim N(\gamma_1, \gamma_2(Z'Z)^{-1}). \tag{9.28}$$

Finally, A is given an inverse Gamma or a uniform or the standard prior $1/A$ for scale parameters. Assuming V is known, our specification of priors is complete.

To do MCMC we partition the parameters and (random) hyperparameters into three sets $(\boldsymbol{\theta}, \boldsymbol{\beta}, A)$. The conditionals are as follows.
(1) Given $\boldsymbol{\beta}, A$ (and $\boldsymbol{Y}$), $\boldsymbol{\theta}$ is multivariate normal.
(2) Given $\boldsymbol{\theta}, A$ (and $\boldsymbol{Y}$), $\boldsymbol{\beta}$ is multivariate normal.
(3) Given $\boldsymbol{\theta}, \boldsymbol{\beta}$ (and $\boldsymbol{Y}$), A has an inverse Gamma distribution. You are asked to find the parameters of these conditionals in Problem 6.

In the PEB approach, one first notes

$$\theta_i|Y_i, \boldsymbol{\beta}, A \sim N(\theta_i^*, V(1 - B)), \qquad (9.29)$$

where

$$\theta_i^* = (1 - B)Y_i + BZ_i'\boldsymbol{\beta} \qquad (9.30)$$

with $B = V/(V + A)$. Here Z_i is the ith column vector of Z. This is the shrinkage estimate corresponding with (9.1) of Section 9.1.

In the PEB approach one has to estimate $\boldsymbol{\beta}$ and B either by maximizing the likelihood of the independent Y_i's with

$$Y_i|\boldsymbol{\beta}, A \sim N(Z_i'\boldsymbol{\beta}, V + A) \qquad (9.31)$$

or by finding a suitable unbiased estimate as in (9.18). Let

$$\hat{\boldsymbol{\beta}} = (Z'Z)^{-1}(Z'\mathbf{Y}).$$

The statistic $\hat{\boldsymbol{\beta}}$ and

$$S = (\mathbf{Y} - Z\hat{\boldsymbol{\beta}})'(\mathbf{Y} - Z\hat{\boldsymbol{\beta}})$$

are independent, complete sufficient statistics for $(\boldsymbol{\beta}, A)$. Hence the best unbiased estimates for $\boldsymbol{\beta}$ and B are $\hat{\boldsymbol{\beta}}$ and

$$\hat{B} = (p - r - 2)V/S$$

(vide Problem 10). Substituting in the shrinkage estimate θ_i^* for θ_i, one gets

$$\hat{\theta}_i = (1 - \hat{B})Y_i + \hat{B}Z'\hat{\boldsymbol{\beta}}.$$

This is the analogue of James-Stein-Lindley estimate under the regression model.

In Problem 8, you are asked to show that the PEB risk of $\hat{\theta}_i$, namely $E_{\boldsymbol{\beta},A}(\hat{\theta}_i - \theta_i)^2$ is smaller than the PEB risk of Y_i, namely, $E_{\boldsymbol{\beta},A}(Y_i - \theta_i)^2$. The relative strength of the PEB estimate comes through the use of $\hat{\boldsymbol{\beta}}$, which is based on the full data set $\mathbf{Y}$.

In Section 8.3, linear regression is discussed as a common statistical problem where an objective Bayesian analysis is done. You may wish to explore how similar ideas are used in this section to model a high-dimensional problem.

9.4 Stein's Frequentist Approach to a High-dimensional Problem

Once again we study Example 9.1. Let $\bar{X}_j$'s be independent, $\bar{X}_j \sim N(\mu_j, \sigma^2/n)$. Classical criteria like maximum likelihood, minimaxity or minimizing variance among unbiased estimates, all lead to $(\bar{X}_1, \ldots, \bar{X}_p)$ as estimate of $(\mu_1, \ldots, \mu_p)$. Let $p \geq 3$. Stein, in a series of papers, Stein (1956), James and Stein (1960), Stein (1981), showed that if we define a loss function

$$L(\bar{\boldsymbol{X}}, \boldsymbol{\mu}) = \sum_{j=1}^{p} (\bar{X}_j - \mu_j)^2 \tag{9.32}$$

and generally

$$L(\boldsymbol{T}, \boldsymbol{\mu}) = \sum_{j=1}^{p} (T_j(\bar{\boldsymbol{X}}) - \mu_j)^2 \tag{9.33}$$

for a general estimate $\boldsymbol{T}$, it is possible to choose a $\boldsymbol{T}$ that is better than $\bar{\boldsymbol{X}}$ in the sense

$$E_{\boldsymbol{\mu}}(L(\boldsymbol{T}, \boldsymbol{\mu})) < E_{\boldsymbol{\mu}}(L(\bar{\boldsymbol{X}}, \boldsymbol{\mu})) \text{ for all } \boldsymbol{\mu}. \tag{9.34}$$

Stein (1956) provides a heuristic motivation that suggests $\bar{\boldsymbol{X}}$ is too large in a certain sense explained below. To see this compare the expectation of the squared norm of $\bar{\boldsymbol{X}}$ with the squared norm of $\boldsymbol{\mu}$.

$$E_{\boldsymbol{\mu}}(\|\bar{\boldsymbol{X}}\|^2) = \|\boldsymbol{\mu}\|^2 + p\sigma^2/n > \|\boldsymbol{\mu}\|^2. \tag{9.35}$$

The larger the p the bigger the deviation between the LHS and RHS. So one would expect at least for sufficiently large p, one can get a better estimate by shrinking each coordinate of $\bar{\boldsymbol{X}}$ suitably towards zero. We present below two of the most well-known shrinkage estimates, namely, the James-Stein and the positive part James-Stein estimate. Both have already appeared in Section 9.2 as PEB estimates. It seems to us that the PEB approach provides the most insight about Stein's estimates, even though the PEB interpretation came much later.

Morris points out that there is no exchangeability or prior in Stein's approach but summing the individual losses produces a similar effect. Moreover, pooling the individual losses would be a natural thing to do only when the different μ_j's are related in some way. If they are totally unrelated, Stein's result would be merely a curious fact with no practical significance, not a profound new data analytic tool. It is in the case of exchangeable high-dimensional problems that it provides substantial improvement.

We present two approaches to proving that the Stein-James estimate is superior to the classical estimate. One is based on Stein (1981) with details as in Ibragimov and Has'minskii (1981). The other is an interesting variation on this due to Schervish (1995).

Stein's Identity. *Let $X \sim N(\mu, \sigma^2)$ and $\phi(x)$ be a real valued function differentiable on $\mathcal{R}$ with $\int_a^x \phi'(u)du = \phi(x) - \phi(a)$. Then*

$$\sigma^2 E(\phi'(X)) = E((X - \mu)\phi(X)).$$

Proof. Integrating by parts or changing the order of integration

$$E(\phi'(X)) = \frac{1}{\sqrt{2\pi}\sigma} \int_{-\infty}^{\infty} \phi'(x) \exp\left\{\frac{-(x-\mu)^2}{2\sigma^2}\right\} dx$$

$$= -\frac{1}{\sqrt{2\pi}\sigma} \int_{-\infty}^{\infty} \phi(x) \frac{d}{dx} \exp\left\{\frac{-(x-\mu)^2}{2\sigma^2}\right\} dx$$

$$= \sigma^{-2} E(\phi(X)(X-\mu)).\square \tag{9.36}$$

For more details see the proof in Stein (1981).

As a corollary we have the following result.

Corollary. *Let* $(X_1, X_2, \ldots, X_p)$ *be a random vector* $\sim N_p(\boldsymbol{\mu}, \frac{\sigma^2}{n}I)$. *Let* $\phi = (\phi_1, \phi_2, \ldots, \phi_p) : \mathcal{R}^p \to \mathcal{R}^p$ *be differentiable,* $E|\frac{\partial \phi_j}{\partial X_j}| < \infty$,
$\phi_j(x_1, \ldots, x_{j-1}, x, x_{j+1}, \ldots, x_p) = \int_a^x \frac{\partial \phi_j}{\partial x_j} dx_j$ *and assume that*
$\phi_j(x_1, \ldots, x_{j-1}, x, x_{j+1}, \ldots) \exp\{\frac{-(x-\mu_j)^2}{2\sigma^2/n}\} \to 0$ *as* $|x| \to \infty$. *Then*

$$E\left\{\sigma^2 \frac{\partial \phi_j}{\partial X_j}\right\} = E((X_j - \mu_j)\phi_j). \tag{9.37}$$

We now return to Example 9.1. The classical estimate for $\boldsymbol{\mu}$ is $\bar{\boldsymbol{X}} = (\bar{X}_1, \bar{X}_2, \ldots, \bar{X}_p)$. Consider an alternative estimate of the form

$$\widetilde{\boldsymbol{\mu}} = \bar{\boldsymbol{X}} + n^{-1}\sigma^2 g(\bar{\boldsymbol{X}}), \tag{9.38}$$

where $g(\boldsymbol{x}) = (g_1, g, \ldots, g_p) : \mathcal{R}^p \to \mathcal{R}^p$ with g satisfying the conditions in the corollary.

Then, by the corollary,

$$E_{\boldsymbol{\mu}}\|\bar{\boldsymbol{X}} - \boldsymbol{\mu}\|^2 - E_{\boldsymbol{\mu}}\|\bar{\boldsymbol{X}} + n^{-1}\sigma^2 g(\bar{\boldsymbol{X}}) - \boldsymbol{\mu}\|^2$$
$$= -2n^{-1}\sigma^2 E_{\boldsymbol{\mu}}\{(\bar{\boldsymbol{X}} - \boldsymbol{\mu})'g(\bar{\boldsymbol{X}})\} - n^{-2}\sigma^4 E_{\boldsymbol{\mu}}\|g(\bar{\boldsymbol{X}})\|^2$$
$$= -2n^{-2}\sigma^4 E_{\boldsymbol{\mu}}\{\sum_1^p \frac{\partial g_j}{\partial X_j}\} - n^{-2}\sigma^4 E_{\boldsymbol{\mu}}\|g(\bar{\boldsymbol{X}})\|^2. \tag{9.39}$$

Now suppose $g(\boldsymbol{x}) = \text{grad}(\log \phi(\boldsymbol{x}))$, where ϕ is a twice continuously differentiable function from $\mathcal{R}^p$ into $\mathcal{R}$. Then

$$\sum_1^p \frac{\partial g_j}{\partial x_j} = -\|g\|^2 + \frac{1}{\phi}\Delta\phi \tag{9.40}$$

where $\Delta = \sum_1^p \frac{\partial^2}{\partial x_j^2}$. Hence

$$E_{\boldsymbol{\mu}}\|\bar{\boldsymbol{X}} - \boldsymbol{\mu}\|^2 - E_{\boldsymbol{\mu}}\|\widetilde{\boldsymbol{\mu}} - \boldsymbol{\mu}\|^2 = n^{-2}\sigma^4 E_{\boldsymbol{\mu}}\|g\|^2 - n^{-2}\sigma^4 E_{\boldsymbol{\mu}}\left\{\frac{1}{\phi(\bar{\boldsymbol{X}})}\Delta\phi(\bar{\boldsymbol{X}})\right\} \tag{9.41}$$

which is positive if $\phi(x)$ is a positive non-constant superharmonic function, i.e,

$$\Delta\phi \leq 0. \tag{9.42}$$

Thus $\widetilde{\mu}$ is superior to $\bar{X}$ if (9.42) holds. It is known that such functions exist if and only if $p \geq 3$.

To show the superiority of the James-Stein positive part estimate for $p \geq 3$, take

$$\phi(x) = \begin{cases} \|x\|^{-(p-2)} & \text{if } \|x\| \geq \sqrt{p-2} \\ (p-2)^{-(p-2)/2} \exp\left\{\frac{1}{2}(p-2) - \|x\|^2)\right\} & \text{otherwise.} \end{cases} \tag{9.43}$$

Then $\mathrm{grad}(\log \phi)$ is easily verified to be the James-Stein positive part estimate.

To show the superiority of the James-Stein estimate, take

$$\phi(x) = \|x\|^{p-2}. \tag{9.44}$$

We observed earlier that shrinking towards zero is natural if one modeled μ_j's as exchangeable with common mean equal to zero. We expect substantial improvement if $\mu = 0$.

Calculation shows

$$E_{\mu}\|\widetilde{\mu} - \mu\|^2 = \frac{2}{p} E_{\mu}\|\bar{X} - \mu\|^2 = 2 \tag{9.45}$$

if $\mu = 0, \sigma^2 = 1, n = 1$.

It appears that borrowing strength in the frequentist formulation is possible because Stein's loss adds up the losses of the component decision problems. Such addition would make sense only when the different problems are connected in a natural way, in which case the exchangeability assumption and the PEB or hierarchical models are also likely to hold. It is natural to ask how good are the James-Stein estimates in the frequentist sense. They are certainly minimax since they dominate minimax estimates. Are they admissible? Are they Bayes (not just PEB)? For the James-Stein positive part estimate the answer to both questions is no, see Berger (1985a, pp. 542, 543). On the other hand, Strawderman (1971) constructs a proper Bayes minimax estimate for $p \geq 5$. Berger (1985a, pp. 364, 365) discusses the question of which among the various minimax estimates to choose. Note that the PEB approach leads in a natural way to James-Stein positive part estimate, suggesting that it can't be substantially improved even though it is not Bayes. See in this connection Robert (1994, p. 66). There is a huge literature on Stein estimates as well as questions of admissibility in multidimensional problems. Berger (1985a) and Robert (1994) provide excellent reviews of the literature. There are intriguing connections between admissibility and recurrence of suitably constructed Markov processes, see Brown (1971), Srinivasan (1981), and Eaton (1992, 1997, 2004).

When extreme μ's may occur, the Stein estimates do not offer much improvement. Stein (1981) and Berger and Dey (1983) suggest how this problem can be solved by suitably truncating the sample means. For Stein type results for general ridge regression estimates see Strawderman (1978) where several other references are given.

Of course, instead of zero we could shrink towards an arbitrary $\boldsymbol{\mu}_0$. Then a substantial improvement will occur near $\boldsymbol{\mu}_0$. Exactly similar results hold for the James-Stein-Lindley estimate and its positive part estimate if $p \geq 4$.

For the James-Stein estimate, Schervish (1995, pp. 163–165) uses Stein's identity as well as (9.40) but then shows directly (with $\sigma^2 = 1, n = 1$)

$$\|g\|^2 + 2 \sum_{j=1}^{p} \frac{\partial}{\partial x_j} g_j = \frac{-(p-2)^2}{\sum_1^p x_j^2} < 0.$$

Clearly for $\widetilde{\boldsymbol{\mu}} =$ James-Stein estimate,

$$E_{\boldsymbol{\mu}} \|\widetilde{\boldsymbol{\mu}} - \boldsymbol{\mu}\|^2 = p - E_{\boldsymbol{\mu}} \left\{ \frac{(p-2)^2}{\sum \bar{X}_j^2} \right\},$$

which shows how the risk can be evaluated by simulating a noncentral χ^2-distribution.

9.5 Comparison of High-dimensional and Low-dimensional Problems

In the low-dimensional case, where n is large or moderate and p small, the prior is washed away by the data, the likelihood influences the posterior more than the prior. This is not so when p is much larger than n – the so-called high-dimensional case. The prior is important, so elicitation, if possible, is important. Checking the prior against data is possible and should be explored. We discuss this below.

In the high-dimensional cases examined in Sections 9.2 and 9.3 some aspects of the prior, namely $\pi(\mu_j|\hat{\boldsymbol{\eta}})$, can be checked against the empirical distribution. We have discussed this earlier mathematically, but one can approach this from a more intuitive point of view. Because we have many μ_j's as sample from $\pi(\mu_j|\hat{\boldsymbol{\eta}})$ and $\bar{X}_j$'s provide approximate estimates of μ_j's, the empirical distribution of the $\bar{X}_j$'s should provide a check on the appropriateness of $\pi(\mu_j|\hat{\boldsymbol{\eta}})$.

Thus there is a curious dichotomy. In the low-dimensional case, the data provide a lot of information about the parameters but not much information about their distribution, i.e., the prior. The opposite is true in high-dimensional problems. The data don't tell us much about the parameters but there is information about the prior.

This general fact suggests that the smoothed empirical distribution of estimates could be used to generate a tentative prior if the likelihood is not exponential and so conjugate priors cannot be used. Adding a location-scale hyperparameter η could provide a family of priors as a starting point of objective high-dimensional Bayesian analysis.

Bernardo (1979) has shown that at least for Example 9.1 a sensible Bayesian analysis can be based on a reference prior with a suitable reparameterization. It does seem very likely that this example is not an exception but a general theory of the right reparameterization needs to be developed.

9.6 High-dimensional Multiple Testing (PEB)

Multiple tests have become very popular because of application in many areas including microarrays where one searches for genes that have been expressed. We provide a minimal amount of modeling that covers a variety of such applications arising in bioinformatics, statistical genetics, biology, etc. Microarrays are discussed in Appendix D. Whereas PEB or HB high-dimensional estimation has been around for some time, PEB or HB high-dimensional multiple testing is of fairly recent origin, e.g., Efron et al. (2001a), Newton et al. (2003), etc.

We have p samples, each of size n, from p normal populations. In the simplest case we assume the populations are homoscedastic. Let σ^2 be the common unknown variance, and the means $\mu_1, \ldots, \mu_p$.

For μ_j, consider the hypotheses $H_{0j} : \mu_j = 0, H_{1j} : \mu_j \sim N(\eta_1, \eta_2), j = 1, \ldots, p$. The data are $X_{ij}, i = 1, \ldots n, j = 1, \ldots, p$. In the gene expression problem, $X_{ij}, i = 1, \ldots n$ are n i.i.d. observations on the expression of the jth gene. The value of $|X_{ij}|$ may be taken as a measure of observed intensity of expression. If one accepts H_{0j}, it amounts to saying the jth gene is not expressed in this experiment. On the other hand, accepting H_{1j} is to assert that the jth gene has been expressed. Roughly speaking, a gene is said to be expressed when the gene has some function in the cell or cells being studied, which could be a malignant tumor. For more details, see the appendix. In addition to H_{0j} and H_{1j}, the model involves $\pi_0 =$ probability that H_{0j} is true and $\pi_1 = 1 - \pi_0 =$ probability that H_{1j} is true. If

$$I_j = \begin{cases} 1 & \text{if } H_{1j} \text{ is true;} \\ 0 & \text{if } H_{0j} \text{ is true,} \end{cases}$$

then we assume $I_1, \ldots, I_p$ are i.i.d. $\sim B(1, \pi_1)$.

The interpretation of π_1 has a subjective and a frequentist aspect. It represents our uncertainty about expression of each particular gene as well as approximate proportion of expression among p genes.

If $\sigma^2, \pi_1, \eta_1, \eta_2$ are all known, $\bar{X}_j$ is sufficient for μ_j and a Bayes test is available for each j. Calculate the posterior probability of H_{1j}:

$$\pi_{1j} = \frac{\pi_1 f_1(\bar{X}_j)}{\pi_1 f_1(\bar{X}_j) + \pi_0 f_0(\bar{X}_j)}$$

which is a function of $\bar{X}_j$ only. Here f_0 and f_1 are densities of $\bar{X}_j$ under H_{0j} and H_{1j}.

$$\text{If } \pi_{1j} > \frac{1}{2} \quad \text{accept } H_{1j} \quad \text{and}$$

$$\text{if } \pi_{1j} < \frac{1}{2} \quad \text{accept } H_{0j}.$$

This test is based only on the data for the jth gene.

In practice, we do not know π_1, η_1, η_2. In PEB testing, we have to estimate all three. In HB testing, we have to put a prior on (π_1, η_1, η_2). To us a natural prior would be a uniform for π_1 on some range $(0, \delta), \delta$ being upper bound to π_1, uniform prior for η_1 on $\mathcal{R}$ and uniform or some other objective prior for η_2.

In the PEB approach, we have to estimate π_1, η_1, η_2. If σ^2 is also unknown, we have to put a prior on σ^2 also or estimate it from data. An estimate of σ^2 is $\sum_i \sum_j (X_{ij} - \bar{X}_j)^2 / \{p(n-1)\}$.

For fixed π_1, we can estimate η_1 and η_2 by the method of moments using the equations,

$$\bar{X} \equiv \frac{1}{p} \sum \bar{X}_j = \pi_1 \eta_1, \tag{9.46}$$

$$\frac{1}{p} \sum (\bar{X}_j - \bar{X})^2 = \frac{\sigma^2}{n} + \pi_1 \eta_2 + \pi_1(1 - \pi_1)\eta_1^2, \tag{9.47}$$

from which it follows that

$$\hat{\eta}_1 = \frac{1}{\pi_1} \bar{X}, \tag{9.48}$$

$$\hat{\eta}_2 = \frac{1}{\pi_1} \left\{ \frac{1}{p} \sum (\bar{X}_j - \bar{X})^2 - \frac{\sigma^2}{n} - \frac{1 - \pi_1}{\pi_1} (\bar{X})^2 \right\}^+. \tag{9.49}$$

Alternatively, if it is felt that $\eta_1 = 0$, then the estimate for η_2 is given by

$$\hat{\eta}_2 = \frac{1}{\pi_1} \left\{ \frac{1}{p} \sum (\bar{X}_j - \bar{X})^2 - \frac{\sigma^2}{n} \right\}^+. \tag{9.50}$$

Now we may maximize the joint likelihood of $\bar{X}_j$'s with respect to π_1.

Using these estimates, we can carry out the Bayes test for each j, provided we know π_1 or put a prior on π_1. We do not know of good PEB estimates of π_1.

Scott and Berger (2005) provide a very illuminating fully Bayesian analysis for microarrays.

9.6.1 Nonparametric Empirical Bayes Multiple Testing

Nonparametric empirical Bayes (NPEB) solutions were introduced by Robbins (1951, 1955, 1964). It is a Bayes solution based on a nonparametric estimate of the prior. Robbins applied these ideas in an ingenious way in several problems. It was regarded as a breakthrough, but the method never became popular because the nonparametric methods did not perform well even in moderately large samples and were somewhat unstable.

Recently Efron et al. (2001a, b) have made a successful application to a microarray with p equal to several thousands. The data are massive enough for NPEB to be stable and perform well.

After some reductions the testing problem takes the following form. For $j = 1, 2, \ldots, p$, we have random variables Z_j. $Z_j \sim f_0(z)$ under H_{0j} and $Z_j \sim f_1(z)$ under H_{1j} where f_0 is completely specified but $f_1(z) \neq f_0(z)$ is completely unknown. This is what makes the problem nonparametric. Finally, as in the case of parametric empirical Bayes, the indicator of H_{1j} is $I_j = 1$ with probability π_1 and $= 0$ with probability $\pi_0 = 1 - \pi_1$. If π_1 and f_1 were known we could use the Bayes test of H_{0j} based on the posterior probability of H_{1j}

$$P(H_{1j}|z_j) = \frac{\pi_1 f_1(z_j)}{\pi_1 f_1(z_j) + (1 - \pi_1)f_0(z_j)}.$$

Let $f(z) = \pi_1 f_1(z) + (1 - \pi_1)f_0(z)$. We know $f_0(z)$. Also we can estimate $f(z)$ using any standard method – kernel, spline, nonparametric Bayes, vide Ghosh and Ramamoorthi (2003) – from the empirical distribution of the z_j's. But since π_1 and f_1 are both unknown, there is an identifiability problem and hence estimation of π_1, f_1 is difficult. The two papers, Efron et al. (2001a, b), provide several methods for bounding π_1.

One bound follows from

$$\pi_0 \leq \min_z [f(z)/f_0(z)],$$

$$\pi_1 \geq 1 - \min_z [f(z)/f_0(z)].$$

So the posterior probability of H_{1j} is

$$P\{H_{1j}|z_j\} = 1 - \frac{\pi_0 f_0(z_j)}{f(z_j)} \geq 1 - \left\{ \min_z \frac{f(z)}{f_0(z)} \right\} \frac{f_0(z_j)}{f(z_j)}$$

which is estimated by $1 - \left\{ \min_z \frac{\hat{f}(z)}{f_0(z)} \right\} \frac{f_0(z_j)}{\hat{f}(z_j)}$, where $\hat{f}$ is an estimate of f as mentioned above. The minimization will usually be made over observed values of z.

Another bound is given by

$$\pi_0 \leq \frac{\int_A f(z)dz}{\int_A f_0(z)dz}.$$

Now minimize the RHS over different choices of A. Intuition suggests a good choice would be an interval centered at the mode of $f_0(z)$, which will usually be at zero. A fully Bayesian nonparametric approach is yet to be worked out. Other related papers are Efron (2003, 2004). For an interesting discussion of microarrays and the application of nonparametric empirical Bayes methodology, see Young and Smith (2005).

9.6.2 False Discovery Rate (FDR)

The false discovery rate (FDR) was introduced by Benjamini and Hochberg (1995). Controlling it has become an important frequentist concept and method in multiple testing, specially in high-dimensional problems. We provide a brief review, because it has interesting similarities with NPEB, as noted, e.g., in Efron et al. (2001a, b). We consider the multiple testing scenario introduced earlier in this section. Consider a fixed test. The (random) FDR for the test is defined as $\frac{U(\boldsymbol{z})}{V(\boldsymbol{z})} I_{\{V(\boldsymbol{z})>0\}}$, where $U =$ total number of false discoveries, i.e., number of true H_{0j}'s that are rejected by the test for a $\boldsymbol{z}$, and $V =$ total number of discoveries, i.e., number of H_{0j}'s that are rejected by a test. The (expected) FDR is

$$FDR = E_{\boldsymbol{\mu}} \left(\frac{U}{V} I_{\{V>0\}} \right).$$

To fix ideas suppose all H_{0j}'s are true, i.e., all μ_j's are zero, then $U = V$ and so

$$\frac{U}{V} I_{\{V>0\}} = I_{\{V>0\}}$$

and

$$FDR = P_{\boldsymbol{\mu}=0}(\text{ at least one } H_{0j} \text{ is rejected })$$
$$= \text{ Type 1 error probability under the full null.}$$

This is usually called family wise error rate (FWER). The Benjamini-Hochberg (BH) algorithm (see Benjamini and Hochberg (1995)) for controlling FDR is to define

$$j_0 = \max\{j : P_{(j)} \leq \frac{j}{p}\alpha\}$$

where $P_j =$ the P-value corresponding with the test for jth null and $P_{(j)} =$ jth order statistic among the P-values with $P_{(1)} =$ the smallest, etc.

The algorithm requires rejecting all H_{0j} for which $P_j \leq P_{(j_0)}$. Benjamini and Hochberg (1995) showed this ensures

$$E_{\boldsymbol{\mu}} \left(\frac{U}{V} I_{\{V>0\}} \right) \leq \frac{p_0}{p+1}\alpha \leq \alpha \; \forall \boldsymbol{\mu}$$

where p_0 is the number of true H_{0j}'s. It is a remarkable result because it is valid for all $\boldsymbol{\mu}$. This exact result has been generalized by Sarkar (2003).

Benjamini and Liu (1999) have provided another algorithm. See also Benjamini and Yekutieli (2001). Genovese and Wasserman (2001) provide a test based on an asymptotic evaluation of j_0 and a less conservative rejection rule. An asymptotic evaluation is also available in Genovese and Wasserman (2002). See also Storey (2002) and Donoho and Jin (2004). Scott and Berger (2005) discuss FDR from a Bayesian point of view.

Controlling FDR leads to better performance under alternatives than controlling FWER. Many successful practical applications of FDR control are known. On the other hand, from a decision theoretic point of view it seems more reasonable to control the sum of false discoveries and false negatives rather than FDR and proportion of false negatives.

9.7 Testing of a High-dimensional Null as a Model Selection Problem[1]

Selection from among nested models is one way of handling testing problems as we have seen in Chapter 6. Parsimony is taken care of to some extent by the prior on the additional parameters of the more complex model. As in estimation or multiple testing, consider samples of size r from p normal populations $N(\mu_i, \sigma^2)$. For simplicity σ^2 is assumed known. Usually σ^2 will be unknown. Because $S^2 = \sum_i \sum_j (X_{ij} - \bar{X}_i)^2/p(r-1)$ is an unbiased estimate of σ^2 with lots of degrees of freedom, it does not matter much whether we put one of the usual objective priors for σ^2 or pretend that σ^2 is known to be S^2.

We wish to test $H_0 : \mu_i = 0 \ \forall i$ versus H_1: at least one $\mu \neq 0$. This is sometimes called Stone's problem, Berger et al. (2003), Stone (1979). We may treat this as a model selection problem with $M_0 \equiv H_0 : \mu_i = 0 \ \forall i$ and $M_1 = H_0 \cup H_1$, i.e., $M_1 : \boldsymbol{\mu} \in \mathcal{R}^p$. In this formulation, $M_0 \subset M_1$ whereas H_0 and H_1 are disjoint. On grounds of parsimony, H_0 is favored if both M_0 and M_1 are equally plausible.

To test a null or select a model, we have to define a prior $\pi(\boldsymbol{\mu})$ under M_1 and calculate the Bayes factor

$$B_{01} = \frac{\prod_{i=1}^{p} f_0(\boldsymbol{X}_i)}{\int_{\mathcal{R}^p} \prod_{i=1}^{p} f_1(\boldsymbol{X}_i|\mu_i)\pi(\boldsymbol{\mu})d\boldsymbol{\mu}}.$$

There is no well developed theory of objective priors, specially for testing problems. However as in estimation it appears natural to treat μ_j's as exchangeable rather than independent. A popular prior in this context is the Zellner and Siow (1980) multivariate Cauchy prior

$$\pi(\boldsymbol{\mu}) = \frac{\Gamma(\frac{(p+1)}{2})}{\pi^{\frac{p+1}{2}}\sigma^p}(1 + \frac{\boldsymbol{\mu}'\boldsymbol{\mu}}{\sigma^2})^{-\frac{(p+1)}{2}}$$

---

[1] Section 9.7 may be omitted at first reading.

$$= \int_0^\infty \frac{t^{\frac{p}{2}}}{(2\pi)^{\frac{p}{2}}\sigma^p} e^{-\frac{t}{2\sigma^2}\mu'\mu} \frac{1}{\sqrt{2\pi}} e^{-\frac{t}{2}} t^{-\frac{1}{2}} dt.$$

$$(9.51)$$

Another plausible prior is the smooth Cauchy prior given by

$$\pi_{sc}(\mu) = \frac{\Gamma(\frac{p+1}{2})}{\Gamma(\frac{p+2}{2})\Gamma(\frac{1}{2})(2\pi\sigma^2)^{\frac{p}{2}}} e^{-\frac{\mu'\mu}{2\sigma^2}} M(\frac{1}{2}, \frac{p+2}{2}, \frac{\mu'\mu}{2\sigma^2})$$

$$= \int_0^1 \frac{t^{\frac{p}{2}}}{(2\pi)^{\frac{p}{2}}\sigma^p} e^{-\frac{t}{2\sigma^2}\mu'\mu} \frac{dt}{\pi\sqrt{t(1-t)}},$$

where $M(\frac{1}{2}, \frac{p+2}{2}, \frac{\mu'\mu}{2\sigma^2})$ is the hypergeometric $_1F_1$ function of Abramowitz and Stegun (1970).

It is tempting to use the difference (between the two models) of BIC as an approximation to the logarithm of Bayes factor (BF) even though it was developed by Schwarz for low-dimensional problems. Stone was the first to point out that the use of BIC is problematic in high-dimensional problems. Berger et al. (2003) have developed a generalization of BIC called GBIC, which provides a good approximation to the integrated likelihood for priors like the above Cauchy priors which are obtained by integrating the scale parameter for $N(\mu_i, \sigma^2)$. In Stone's problem one has the normal linear model setup

$$X_{ij} = \mu_i + \epsilon_{ij}; \ i = 1, \ldots, p; \ j = 1, \ldots, r; \ n = pr. \qquad (9.52)$$

It is assumed that as $n \to \infty$, $p \to \infty$ and r is fixed. Under these assumptions, Berger et al. (2003) provide a Laplace approximation and a GBIC. The GBIC also approximates the BIC for low-dimensional problems. The formula for ΔGBIC (the difference of GBIC for the comparison of M_1 and M_0) is given by

$$\Delta\text{GBIC} = (\frac{r}{2}\bar{\mathbf{X}}'\bar{\mathbf{X}} - \frac{p}{2}\log(rc_p) - \frac{p}{2})^+ - \frac{\log p}{2}, \qquad (9.53)$$

where $c_p = \frac{1}{p}\sum_{i=1}^p \bar{X}_i^2$. Table 9.1, taken from Berger et al. (2003) provides some idea of the accuracy of BIC, GBIC and Laplace approximation. One has $p = 50$ and $r = 2$ for these calculations and the multivariate Cauchy prior was used.

Substantial new results appear in Liang et al. (2005). They propose a mixture of Zellner's (Zellner (1986)) popular g-prior. In Zellner's form, the prior looks like $\mu|M_1 \sim N(\mathbf{0}, \frac{g}{\sigma^2}(\mathbf{Z}'\mathbf{Z})^{-1})$ where $\mathbf{Z}$ is the design matrix (in our problem only composed of 0's and 1's). This g is usually elicited through an empirical Bayes method. The above authors consider a family of mixtures of g-priors (under which the Zellner-Siow Cauchy prior is a special case) and use those for model selection. They propose Laplace approximations to the

Table 9.1. Comparison of the Performance of GBIC and Laplace Approximation with BIC

c_p	True Log Bayes Factor	ΔBIC	$\Delta GBIC$	ΔLaplace Approx
0.1	-8.5348	-110.129	-1.956	-8.5776
0.5	-3.8251	-90.129	-1.956	-3.9083
1.0	6.0388	-65.129	5.715	5.9236
1.5	20.8203	-40.129	20.579	20.7564
2.0	38.4814	-15.129	38.387	38.4408
10.0	397.369	384.871	398.151	397.369

marginal likelihood under these general priors and show that the models thus selected are generally correct asymptotically if the complex model is true. Under the null model, this type of consistency still holds under the Zellner-Siow prior.

Further generalizations to non-normal problems appear in Berger (2005) and Chakrabarti and Ghosh (2005a). Both papers provide generalizations of BIC when the observations come from an exponential family of distributions in high-dimensional problems. In Table 9.2, using simulation results reported in Chakrabarti and Ghosh (2005a), the performance of GBIC and the Laplace approximation ($\log \hat{m}_2$) with BIC are compared in approximating the integrated likelihood under the more complex model (denoted by m_2) when the more complex model is actually true and observations come from Bernoulli, exponential, and Poisson distributions. In this case one has p groups of observations, each group having a (potentially) different parameter value and each group has r observations. Under the simpler model, these different groups are assumed to have the same (specified) parameter value, while for the more complex model the parameter vector is assumed to belong to $\mathcal{R}^p$. See the paper for details on the priors used.

In principle, the same methods apply to any two nested models
$$M_0 : \mu_i = 0, 1 \le i \le p_1, p_1 < p \quad \text{versus} \quad M_1 : \boldsymbol{\mu} \in \mathcal{R}^p.$$

Table 9.2. Approximation to Integrated Likelihood in the Exponential Family

Distribution	p	r	$\log m_2$	$\log \hat{m}_2$	BIC	$GBIC$
Bernoulli	50	10	-327.45	-327.684	-349.577	-327.863
Bernoulli	50	200	-4018.026	-4018.072	-4052.757	-4018.587
Exponential	50	10	-662.526	-661.979	-640.320	-660.384
Exponential	50	200	-22186.199	-22186.100	-22178.759	-22186.117
Poisson	50	10	-671.504	-670.775	-683.383	-671.374
Poisson	50	200	-15704.585	-15704.618	-15713.139	-15705.010

9.8 High-dimensional Estimation and Prediction Based on Model Selection or Model Averaging[2]

Given a set of data from an experiment or observational study done on a given population, a statistician is asked the following three questions quite frequently. First, which among a given set of possible statistical models seems to be the correct model describing the underlying mechanism producing the data? Second, what will be the predicted value of a future observation, if the experimental conditions are kept at predetermined levels? Third, what is the estimate of a single parameter or a vector (may be infinite dimensional) of parameters? We will focus in this section on some Bayesian approaches to answer the last two types of questions. But before going into the details, we will explain briefly in the next paragraph how one would pose the above three questions from a decision theoretic point of view and what is the basic difference in the Bayesian approaches in tackling such questions.

Bayesian approaches to such questions are basically dictated by the goal of obtaining decision theoretic optimality, and hence the solutions are also heavily dependent upon the type of loss functions being used. The loss function, on the other hand, is mostly determined by the goal of the statistician or practitioner. The goal of the statistician in the first problem above is to select the correct model (which is assumed to be one in the list of models considered). The loss function often used in this problem is the 0-1 loss function. In the Bayesian approach to model selection, the statistician would put prior probabilities on the set of candidate models and a simple argument shows that for this loss, the optimum Bayesian model would be the posterior mode, i.e., the model that has the maximum posterior probability. As explained in the earlier section, BIC and GBIC can be used to select a model using the Bayesian paradigm with 0-1 loss if the sample size is large, in appropriate situations, as they approximate the integrated likelihood and hence can be used to find the model with highest posterior probability. On the other hand, if one is interested in answering the second or third question above (i.e., if one is interested in prediction or estimation of a parameter), the problem can be approached in two different ways. First, one might be interested in finding a particular model that does the best job of prediction (in some appropriate sense). Secondly, one might only want a predicted value, not a particular model for repeated future use in prediction. In either case, the most popular loss function is the squared prediction error loss, i.e., the square of the difference between the predicted/estimated value and the value being predicted/estimated. The best predictor/estimator turns out to be the Bayesian model averaging estimate (to be explained later) and the best predictive model is the one which minimizes the expected posterior predictive loss.

We now consider the problem of optimal prediction from a Bayesian approach. We use the ideas, notations, and results of Barbieri and Berger (2004)

---

[2] Section 9.8 may be omitted at first reading.

for this part. Consider the canonical model

$$\mathbf{y} = \mathbf{X}\boldsymbol{\beta} + \boldsymbol{\epsilon}, \tag{9.54}$$

where $\mathbf{y}$ is an $n \times 1$ vector of observations, $\mathbf{X}$ is the $n \times k$ full rank design matrix, $\boldsymbol{\beta}$ is the unknown $k \times 1$ vector of regression coefficients and $\boldsymbol{\epsilon}$ is the $n \times 1$ vector of random errors, which are i.i.d. $N(0, \sigma^2)$, σ^2 being known or unknown. Our goal is to predict a future observation y^*, given by

$$y^* = \mathbf{x}^*\boldsymbol{\beta} + \epsilon, \tag{9.55}$$

where $\mathbf{x}^* = (x_1^*, \ldots, x_k^*)$ is the value of the covariate vector for which the prediction is to be made. We consider the loss in predicting y^* by $\hat{y}^*$ as

$$L(\hat{y}^*, y^*) = (\hat{y}^* - y^*)^2; \tag{9.56}$$

i.e., the squared error prediction loss. Assume that we have submodels

$$M_l : \mathbf{y} = \mathbf{X}_l\boldsymbol{\beta}_l + \boldsymbol{\epsilon}, \tag{9.57}$$

where $\mathbf{l} = (l_1, \ldots, l_k)$ with $l_i = 1$ or 0 according as the ith covariate is in the model M_l or not, $\mathbf{X}_l$ is a matrix containing columns of $\mathbf{X}$ corresponding with the nonzero coordinates of $\mathbf{l}$ and $\boldsymbol{\beta}_l$ is the corresponding vector of regression coefficients. Let k_l denote the number of covariates included in the model; then $\mathbf{X}_l$ is of dimension $(n \times k_l)$ and $\boldsymbol{\beta}_l$ is a $(k_l \times 1)$ vector.

We put prior probabilities $P(M_l)$ to each model M_l included in the model space such that $\sum_l P(M_l) = 1$, and given model M_l, a prior $\pi_l(\boldsymbol{\beta}_l, \sigma)$ is assumed on the parameters $(\boldsymbol{\beta}_l, \sigma)$ included in model M_l. Using standard posterior calculations, one obtains the quantities (a) $p_l = P(M_l|\mathbf{y})$, the posterior probability of model M_l and (b) $\pi_l(\boldsymbol{\beta}_l, \sigma|\mathbf{y})$, the posterior distribution of the unknown parameters in M_l. With this setup in mind, we shall now discuss two optimal prediction strategies, as described below.

First note that the best predictor of y^* for a given value of $\mathbf{x}^*$ comes out as $\bar{y}^* = E(y^*|\mathbf{y})$, where the expectation is taken with respect to the posterior/predictive distribution of y^* given $\mathbf{y}$. This follows by noting that

$$E[(y^* - \hat{y}^*)^2] = E^{\mathbf{y}} E[(y^* - \hat{y}^*)^2|\mathbf{y}], \tag{9.58}$$

where the expectation inside is taken with respect to the posterior distribution of y^* given $\mathbf{y}$. But note that

$$\bar{y}^* = E(y^*|\mathbf{y}) = \sum_l p_l E(y^*|\mathbf{y}, M_l) = \mathbf{x}^* \sum_l p_l H_l \tilde{\boldsymbol{\beta}}_l, \tag{9.59}$$

where H_l is a $(k \times k_l)$ matrix such that $\mathbf{x}^* H_l$ is the subvector of $\mathbf{x}^*$ corresponding to the nonzero coordinates of $\mathbf{l}$ and $\tilde{\boldsymbol{\beta}}_l$ is the posterior mean of $\boldsymbol{\beta}_l$ with respect to $\pi_l(\boldsymbol{\beta}_l, \sigma|\mathbf{y})$. Noting that if we knew that M_l were the true model, then the optimal predictor of y^* for $\mathbf{x}$ fixed at $\mathbf{x}^*$ would be given by

$$\hat{y}_1^* = \mathbf{x}^* H_1 \tilde{\boldsymbol{\beta}}_1, \text{ we have} \tag{9.60}$$

$$\bar{y}^* = E(y^*|\mathbf{y}) = \mathbf{x}^* \bar{\boldsymbol{\beta}} \equiv \mathbf{x}^* \sum_1 p_1 H_1 \tilde{\boldsymbol{\beta}}_1 = \sum_1 p_1 \hat{y}_1^*. \tag{9.61}$$

$\bar{y}^*$ is called the Bayesian model averaging estimate, in that it is a weighted average of the optimal Bayesian predictors under each individual model, the weights being the posterior probabilities of each model. Many authors have argued the use of the model averaging estimate as an appropriate predictive estimate. They justify this by saying that in using model selection to choose the best model and then making inference based on the assumption that the selected model is true, does not take into account the fact that there is uncertainty about the model itself. As a result, one might underestimate the uncertainty about the quantity of interest. See, for example, Madigan and Raftery (1994), Raftery, Madigan, and Hoeting (1997), Hoeting, Madigan, Raftery, and Volinsky (1999), and Clyde (1999); just to name a few, for detailed discussion on this point of view. However if the number of models in the model space is very large (e.g., in case all subsets of parameters are allowed in the model space, as will happen in high or even moderately high dimensions), the task of computing the Bayesian model averaging estimate exactly might be virtually impossible. Moreover, it is not prudent to keep in the model average those models that have small posterior probability indicating relative incompatibility with observed data. There are some proposals to get around this difficulty, as discussed in the literature cited above. Two of them are based on the 'Occam's window' method of Madigan and Raftery (1994) and the Markov chain Monte Carlo approach of Madigan and York (1995).

In the first approach, the averaging is done over a small set of appropriately selected models, which are parsimonious and supported by data. In the second approach, one constructs a Markov chain with state space same as the model space and equilibrium distribution $\{P(M_1|\mathbf{y})\}$ where M_1 varies over the model space. Upon simulation from this chain, the Bayesian model averaging estimator is approximated by taking average value of the posterior expectations under each model visited in the chain. But it must be commented that Bayesian model averaging (BMA) has its limitations in high-dimensional problems. Each approach addresses both issues but it is unclear how well.

Although BMA is the optimal predictive estimation procedure, often a single model is desired for prediction. For example, choice of a single model will require observing only the covariates included in the model. Also, as noted earlier, in high dimensions, BMA has its problems. We will assume now that the future predictions will be made for covariates $\mathbf{x}^*$ such that

$$Q = E(\mathbf{x}^{*\prime}\mathbf{x}^*)$$

exists and is positive definite. A frequent choice of Q is $Q = \mathbf{X}'\mathbf{X}$, i.e., the future covariates will be like the ones observed in the past. In general, the best single model will depend on $\mathbf{x}^*$, but we present here some general characterizations which give the optimal predictive model without this dependence. In general, the optimal predictive model is not the one with the highest posterior probability. However, there are interesting exceptions. If there are only two models, it is easy to show the posterior mode with shrinkage estimate is optimal for prediction (Berger (1997) and Mukhopadhyay (2000)). This also holds sometimes in the context of variable selection for linear models with orthogonal design matrix, as in Clyde and Parmigiani (1996). As Berger (1997) notes, it is easy to see that if one is considering only two models, say M_1 and M_2 with prior probabilities $\frac{1}{2}$ each and proper priors are assigned to the unknown parameters under each model, the best predictive model turns out to be M_1 or M_2 according as the Bayes factor of M_1 to M_2 is greater than one or not, and hence the best predictive model is the one with the highest posterior probability. The characterizations we will describe here are in terms of what is called the 'median probability model.' If it exists, the median probability model $M_{\mathbf{l}*}$ is defined to be the model consisting of those variables only whose posterior inclusion probabilities are at least $\frac{1}{2}$. The posterior inclusion probability for variable i is

$$p_i = \sum_{\mathbf{l}:l_i=1} P(M_{\mathbf{l}}|\mathbf{y}). \tag{9.62}$$

So, $\mathbf{l}^*$ is defined coordinatewise as $l_i = 1$ if $p_i \geq \frac{1}{2}$ and $l_i = 0$ otherwise. It is possible that the median probability model does not exist, in that the variables included according to the definition of $\mathbf{l}^*$ do not correspond with any model under consideration. But in the variable selection problem, if we are allowed to include or exclude any variable in the possible models, i.e., all possible values of $\mathbf{l}$ are allowed, then the median probability model will obviously exist. Another important class of models is a class of models with 'graphical model structure' for which the median probability model will always exist (this fact follows directly from the definition below).

Definition 9.4. *Suppose that for each variable index i, there is a corresponding index set $I(i)$ of other variables. A subclass of linear models is said to have 'graphical model structure' if it consists of all models satisfying the condition 'for each i, if variable x_i is in the model, then variables x_j with $j \in I(i)$ are in the model.'*

The class of models with 'graphical model structure' includes the class of models with all possible subsets of variables and sequences of nested models, $M_{\mathbf{l}(j)}$, $j = 0, 1, \ldots, k$, where $\mathbf{l}(j) = (1, \ldots, 1, 0, \ldots, 0)$ with j ones and $k - j$ zeros. For the all subsets scenario, $I(i)$ is the null set while in the nested case $I(i) = \{j : 1 \leq j < i\}$ for $i \geq 2$ and $I(i)$ is the null set for $i = 0$ or 1. The latter are natural in many examples including polynomial regression models, where j refers to the degree of polynomial used. Another example of nested models is provided by nonparametric regression (vide Chapter 10,

Sections 10.2, 10.3). The unknown function is approximated by partial sums of its Fourier expansion, with all coefficients after stage j assumed to be zero. Note that in this situation, the median probability model has a simple description; one calculates the cumulative sum of posterior model probabilities beginning from the smallest model, and the median probability model is the first model for which this sum equals or exceeds $\frac{1}{2}$. Mathematically, the median probability model is $M_{\mathbf{1}(j^*)}$, where

$$\sum_{i=0}^{j^*-1} P(M_{\mathbf{1}(i)}|\mathbf{y}) < \frac{1}{2} \text{ and } \sum_{i=0}^{j^*} P(M_{\mathbf{1}(i)}|\mathbf{y}) \geq \frac{1}{2}. \tag{9.63}$$

We present some results on the optimality of the posterior median model in prediction. The best predictive model is found as follows. Once a model is selected, the best Bayesian predictor assuming that model is true is obtained. In the next stage, one finds the model such that the expected prediction loss (this expectation does not assume any particular model is true, but is an over-all expectation) using this Bayesian predictor is minimized. The minimizer is the best predictive model. There are some situations where the median probability model and the highest posterior probability are the same. Obviously, if there is one model with posterior probability greater than $\frac{1}{2}$, this will be trivially true. Barbieri and Berger (2004) observe that when the highest posterior probability model has substantially larger probability than the other models, it will typically also be the median probability model. We describe another such situation later in the corollary to Theorem 9.8.

We state and prove two simple lemmas.

Lemma 9.5. *(Barbieri and Berger, 2004) Assume Q exists and is positive definite. The optimal model for predicting y^* under the squared error loss, is the unique model minimizing*

$$R(M_{\mathbf{1}}) \equiv (H_{\mathbf{1}}\tilde{\boldsymbol{\beta}}_{\mathbf{1}} - \bar{\boldsymbol{\beta}})'Q(H_{\mathbf{1}}\tilde{\boldsymbol{\beta}}_{\mathbf{1}} - \bar{\boldsymbol{\beta}}), \tag{9.64}$$

where $\bar{\boldsymbol{\beta}}$ is defined in (9.61).

Proof. As noted earlier, $\hat{y}_{\mathbf{1}}^*$ is the optimal Bayesian predictor assuming $M_{\mathbf{1}}$ is the true model. The optimal predictive model is found by minimizing with respect to $\mathbf{1}$, where $\mathbf{1}$ belongs to the space of models under consideration, the quantity $E(y^* - \hat{y}_{\mathbf{1}}^*)^2$. Minimizing this is equivalent to minimizing for each $\mathbf{y}$ the quantity $E[(y^* - \hat{y}_{\mathbf{1}}^*)^2|\mathbf{y}]$. It is easy to see that for a fixed $\mathbf{x}^*$,

$$E[(y^* - \hat{y}_{\mathbf{1}}^*)^2|\mathbf{y}] = C + (\bar{y}^* - \hat{y}_{\mathbf{1}}^*)^2, \tag{9.65}$$

where the symbols have been defined earlier and C is a quantity independent of $\mathbf{1}$. The expectation above is taken with respect to the predictive distribution of y^* given $\mathbf{y}$ and $\mathbf{x}^*$. So the optimal predictive model will be found by finding the minimizer of the expression obtained by taking a further expectation over $\mathbf{x}^*$ on the second quantity on the right hand side of (9.65). By plugging in the values of $\hat{y}_1^*$ and $\bar{y}^*$, we immediately get

$$(\bar{y}^* - \hat{y}_1^*)^2 = (H_1\tilde{\beta}_1 - \bar{\beta})'\mathbf{x}^{*'}\mathbf{x}^*(H_1\tilde{\beta}_1 - \bar{\beta}). \tag{9.66}$$

The lemma follows. The uniqueness follows from the fact that Q is positive definite. $\square$

Lemma 9.6. *(Barbieri and Berger, 2004) If Q is diagonal with diagonal elements $q_i > 0$, and the posterior means $\tilde{\beta}_1$ satisfy $\tilde{\beta}_1 = \mathbf{H}_1'\tilde{\beta}$ (where $\tilde{\beta}$ is the posterior mean under the full model as in (9.54)) then*

$$R(M_1) = \sum_{i=1}^{k} \tilde{\beta}_i^2 q_i (l_i - p_i)^2. \tag{9.67}$$

Proof. From the fact $\tilde{\beta}_1 = \mathbf{H}_1'\tilde{\beta}$, it follows that

$$\bar{\beta} = \sum_l p_l \mathbf{H}_1\tilde{\beta}_1 = \sum_l p_l \mathbf{H}_1\mathbf{H}_1'\tilde{\beta} = D(\mathbf{p})\tilde{\beta}, \tag{9.68}$$

where $D(\mathbf{p})$ is the diagonal matrix with diagonal elements p_i, by noting that $H_1(i,j) = 1$ if $l_i = 1$ and $j = \sum_{r=1}^{i} l_r$ and $H_1(i,j) = 0$ otherwise. Similarly,

$$R(M_1) = (\mathbf{H}_1\mathbf{H}_1'\tilde{\beta} - D(\mathbf{p})\tilde{\beta})'Q(\mathbf{H}_1\mathbf{H}_1'\tilde{\beta} - D(\mathbf{p})\tilde{\beta})$$
$$= \tilde{\beta}'(D(\mathbf{1}) - D(\mathbf{p}))Q(D(\mathbf{1}) - D(\mathbf{p}))\tilde{\beta}, \tag{9.69}$$

from where the result follows. $\square$

Remark 9.7. The condition $\tilde{\beta}_1 = \mathbf{H}_1'\tilde{\beta}$, simply means that the posterior mean of $\tilde{\beta}_1$ is found by taking the relevant coordinates of the posterior mean in the full model as in (9.54). As Barbieri and Berger (2004) comment, this will happen in two important cases. Assume $X'X$ is diagonal. In the first case, if one uses the reference prior $\pi_1(\beta_1, \sigma) = 1/\sigma$ or a constant prior if σ is known, the LSE becomes same as the posterior means and the diagonality of $(\mathbf{X}'\mathbf{X})$ implies that the above condition will hold. Secondly, suppose in the full model $\pi(\beta, \sigma) = N_k(\mu, \sigma^2\Delta)$ where Δ is a known diagonal matrix, and for the submodels the natural corresponding prior $N_{k_1}(\mathbf{H}_1'\mu, \sigma^2\mathbf{H}_1'\Delta\mathbf{H}_1)$. Then it is easy to see that for any prior on σ^2 or if σ^2 is known, the above will hold.

We now state the first theorem.

Theorem 9.8. *(Barbieri and Berger, 2004) If Q is diagonal with $q_i > 0$ and $\tilde{\beta}_1 = \mathbf{H}_1'\tilde{\beta}$, and the models have graphical model structure, then the median probability model is the best predictive model.*

Proof. Because $q_i > 0$, $\tilde{\beta_i}^2 \geq 0$ for each i and p_i (defined in (9.62)) does not depend on $\mathbf{l}$, to minimize $R(M_{\mathbf{l}})$ among all possible models, it suffices to minimize $(l_i - p_i)^2$ for each individual i and that is achieved by choosing $l_i = 1$ if $p_i \geq \frac{1}{2}$ and $l_i = 0$ if $p_i < \frac{1}{2}$, whence $\mathbf{l}$ as defined will be the median probability model. The graphical model structure ensures that this model is among the class of models under consideration. $\square$

Remark 9.9. The above theorem obviously holds if we consider all submodels, this class having graphical model structure; provided the conditions of the theorem hold. By the same token, the result will hold under the situation where the models under consideration are nested.

Corollary 9.10. *(Barbieri and Berger, 2004) If the conditions of the above theorem hold, all submodels of the full model are allowed, σ^2 is known, $\mathbf{X'X}$ is diagonal and β_i's have $N(\mu_i, \lambda_i \sigma^2)$ distributions and*

$$P(M_{\mathbf{l}}) = \prod_{i=1}^{k} (p_i^0)^{l_i} (1 - p_i^0)^{(1-l_i)}, \tag{9.70}$$

where p_i^0 is the prior probability that variable x_i is in the model, then the optimal predictive model is the model with highest posterior probability which is also the median probability model.

Proof. Let $\hat{\beta}_i$ be the least squares estimate of β_i under the full model. Because $X'X$ is diagonal, $\hat{\beta}_i$'s are independent and the likelihood under M_l factors as

$$L(M_{\mathbf{l}}) \propto \prod_{i=1}^{k} (\lambda_i^0)^{l_i} (\lambda_i')^{1-l_i}$$

where λ_i^0 depends only on $\hat{\beta}_i$ and β_i, λ_i' depends only on $\hat{\beta}_i$ and the constant of proportionality here and below depend an $\mathbf{Y}$ and $\hat{\beta}_i$'s.

Also, the conditional prior distribution of β_i's given M_l has a factorization

$$\pi(\boldsymbol{\beta}|M_{\mathbf{l}}) = \prod_{i=1}^{k} [N(\mu_i, \lambda_i \sigma^2)]^{l_i} [\delta\{0\}]^{1-l_i}$$

where $\delta\{0\} = $ degenerate distribution with all mass at zero.

It follows from (9.70) and the above two factorizations that the posterior probability of M_l has a factorization

$$P(M_{\mathbf{l}}|\mathbf{Y}) \alpha \prod_{i=1}^{k} \{p_i^0 \int_{-\infty}^{\infty} \lambda_i^0 N(\mu_i, \lambda_i \sigma^2) d\beta\}^{l_i} \{(1 - p_i^0)\lambda_i' \delta\{0\}\}^{1-l_i}$$

which in turn implies that the marginal posterior of including or not including ith variable is proportional to the two terms respectively in the ith factor. This completes the proof, vide Problem 21. (The integral can be evaluated as in Chapter 2.) $\square$

We have noted before that if the conditions in Theorem 9.8 are satisfied and the models are nested, then the best predictive model is the median probability model. Interestingly even if Q is not necessarily diagonal, the best predictive model turns out to be the median probability model under some mild assumptions, in the nested model scenario. Consider

Assumption 1: $Q = \gamma \mathbf{X}'\mathbf{X}$ for some $\gamma > 0$, i.e., the prediction will be made at covariates that are similar to the ones already observed in the past.

Assumption 2: $\tilde{\boldsymbol{\beta}}_1 = b\hat{\boldsymbol{\beta}}_1$, where $b > 0$, i.e, the posterior means are proportional to the least squares estimates.

Remark 9.11. Barbieri and Berger (2004) list two situations when the second assumption will be satisfied. First, if one uses the reference prior $\pi_1(\boldsymbol{\beta}_1, \sigma) = 1/\sigma$, whereby the posterior means will be the LSE's. It will also be satisfied with $b = c/(1 + c)$, if one uses g-type normal priors of Zellner (1986), where $\pi_1(\boldsymbol{\beta}_1|\sigma) \sim N_{k_1}(\mathbf{0}, c\sigma^2(\mathbf{X}_1'\mathbf{X}_1)^{-1})$ and the prior on σ is arbitrary.

Theorem 9.12. *For a sequence of nested models for which the above two conditions hold, the best predictive model is the median probability model.*

Proof. See Barbieri and Berger (2004). $\square$

Barbieri and Berger(2004, Section 5) present a geometric formulation for identification of the optimal predictive model. They also establish conditions under which the median probability model and the maximum posterior probability model coincides; and that it is typically not enough to know only the posterior probabilities of each model to determine the optimal predictive model.

Till now we have concentrated on some Bayesian approaches to the prediction problem. It turns out that model selection based on the classical Akaike information criterion (AIC) also plays an important role in Bayesian prediction and estimation for linear models and function estimation. Optimality results for AIC in classical statistics are due to Shibata (1981, 1983), Li (1987), and Shao (1997).

The first Bayesian result about AIC is taken from Mukhopadhyay (2000). Here one has observations $\{y_{ij} : i = 1, \ldots, p, \ j = 1, \ldots, r, \ n = pr\}$ given by

$$y_{ij} = \mu_i + \epsilon_{ij}, \tag{9.71}$$

where ϵ_{ij} are i.i.d. $N(0, \sigma^2)$ with σ^2 known. The models are $M_1 : \mu_i = 0$ for all i and $M_2 : \eta^2 = \lim_{p->\infty} \frac{1}{p} \sum_{i=1}^{p} \mu_i^2 > 0$. Under M_2, we assume a $N(0, \tau^2 I_p)$ prior on μ where τ^2 is to be estimated from data using an empirical Bayes method. It is further assumed that $p \to \infty$ as $n \to \infty$. The goal is to predict a future set of observations $\{z_{ij}\}$ independent of $\{y_{ij}\}$ using the usual prediction error loss, with the 'constraint' that once a model is selected, least squares estimates have to be used to make the predictions. Theorem 9.13 shows that the constrained empirical Bayes rule is equivalent to AIC asymptotically. A weaker result is given as Problem 17.

Theorem 9.13. *(Mukhopadhyay, 2000) Suppose M_2 is true, then asymptotically the constrained empirical Bayes rule and AIC select the same model. Under M_1, AIC and the constrained empirical Bayes rule choose M_1 with probability tending to 1. Also under M_1, the constrained empirical Bayes rule chooses M_1 whenever AIC does so.*

The result is extended to general nested problems in Mukhopadhyay and Ghosh (2004a). It is however also shown in the above reference that if one uses Bayes estimates instead of least squares estimates, then the unconstrained Bayes rule does better than AIC asymptotically. The performance of AIC in the PEB setup of George and Foster (2000) is studied in Mukhopadhyay and Ghosh (2004a).

As one would expect from this, AIC also performs well in nonparametric regression which can be formulated as an infinite dimensional linear problem. It is shown in Chakrabarti and Ghosh (2005b) that AIC attains the optimal rate of convergence in an asymptotically equivalent problem and is also adaptive in the sense that it makes no assumption about the degree of smoothness. Because this result is somewhat technical, we only present some numerical results for the problem of nonparametric regression.

In the nonparametric regression problem

$$Y_i = f(\frac{i}{n}) + \epsilon_i, \ i = 1, \dots, n, \tag{9.72}$$

one has to estimate the unknown smooth function f. In Table 9.3, we consider $n = 100$ and $f(x) = (\sin(2\pi x))^3$, $(\cos(\pi x))^4$, $7 + \cos(2\pi x)$, and $e^{\sin(2\pi x)}$, the loss function $L(f, \hat{f}) \equiv \int_0^1 (f(x) - \hat{f}(x))^2 dx$, and report the average loss of modified James-Stein estimator of Cai et al. (2000), AIC, and the kernel method with Epanechnikov kernel in 50 simulations. To use the first two methods, we express f in its (partial sum) Fourier expansion with respect to the usual sine-cosine Fourier basis of $[0, 1]$ and then estimate the Fourier coefficients by the regression coefficients. Some simple but basic insight about the AIC may be obtained from Problems 15–17. It is also worth remembering that AIC was expected by Akaike to perform well in high-dimensional estimation or prediction problem when the true model is too complex to be in the model space.

9.9 Discussion

Bayesian model selection is passing through a stage of rapid growth, especially in the context of bioinformatics and variable selection. The two previous sections provide an overview of some of the literature. See also the review by Ghosh and Samanta (2001). For a very clear and systematic approach to different aspects of model selection, see Bernardo and Smith (1994).

Model selection based on AIC is used in many real-life problems by Burnham and Anderson (2002). However, its use for testing problems with 0-1

Table 9.3. Comparison of Simulation Performance of Various Estimation Methods in Nonparametric Regression

$Function$	Modified James-Stein	AIC	Kernel Method
$[Sin(2\pi x)]^3$	0.2165	0.0793	0.0691
$[Cos(\pi x)]^4$	0.2235	0.078	0.091
$7 + Cos(2\pi x)$	0.2576	0.0529	0.5380
$e^{Sin(2\pi x)}$	0.2618	0.0850	0.082

loss is questionable vide Problem 16. A very promising new model selection criterion due to Spiegelhalter et al. (2002) may also be interpreted as a generalization of AIC, see, e.g., Chakrabarti and Ghosh (2005a). In the latter paper, GBIC is also interpreted from the information theoretic point of view of Rissanen (1987).

We believe the Bayesian approach provides a unified approach to model selection and helps us see classical rules like BIC and AIC as still important but by no means the last word in any sense. We end this section with two final comments.

One important application of model selection is to examine model fit. Gelfand and Ghosh (1998) (see also Gelfand and Dey (1994)) use leave-k-out cross-validation to compare each collection of k data points and their predictive distribution based on the remaining observations. Based on the predictive distributions, one may calculate predicted values and some measure of deviation from the k observations that are left out. An average of the deviation over all sets of k left out observations provides some idea of goodness of fit. Gelfand and Ghosh (1998) use these for model selection. Presumably, the average distance for a model can be used for model check also. An interesting work of this kind is Bhattacharya (2005).

Another important problem is computation of the Bayes factor. Gelfand and Dey (1994) and Chib (1995) show how one can use MCMC calculations by relating the marginal likelihood of data to the posterior via $P(y) = L(\theta|y)P(\theta)/P(\theta|y)$. Other relevant papers are Carlin and Chib (1995), Chib and Greenberg (1998), and Basu and Chib (2003). There are interesting suggestions also in Gelman et al (1995).

9.10 Exercises

1. Show that $\pi(\eta_2|\mathbf{X})$ is an improper density if we take $\pi(\eta_1, \eta_2) = 1/\eta_2$ in Example 9.3.
2. Justify (9.2) and (9.3).
3. Complete the details to implement Gibbs sampling and E-M algorithm in Example 9.3 when μ and σ^2 are unknown. Take $\pi(\eta_1, \sigma^2, \eta_2) = 1/\sigma^2$.
4. Let X_i's be independent with density $f(x|\theta_i)$, $i = 1, 2, \ldots, p$, $\theta_i \in \mathcal{R}$. Consider the problem of estimating $\boldsymbol{\theta} = (\theta_1, \ldots, \theta_p)'$ with loss function

$$L(\boldsymbol{\theta}, \boldsymbol{a}) = \sum_{i=1}^{p} L(\theta_i, a_i) = \sum_{i=1}^{p} (\theta_i - a_i)^2, \quad \boldsymbol{\theta}, \boldsymbol{a} \in \mathcal{R}^p.$$

i.e., the total loss is the sum of the losses in estimating θ_i by a_i. An estimator for $\boldsymbol{\theta}$ is the vector $(T_1(\boldsymbol{X}), T_2(\boldsymbol{X}), \ldots, T_p(\boldsymbol{X}))$. We call this a compound decision problem with p components.

(a) Suppose $\sup_\delta f(x|\delta) = f(x|T(x))$, i.e., $T(x)$ is the MLE (of θ_j in $f(x|\theta_j)$). Show that $(T(X_1), T(X_2), \ldots T(X_p))$ is the MLE of $\boldsymbol{\theta}$.

(b) Suppose $T(X)$ (not necessarily the $T(X)$ of (a)) satisfies the sufficient condition for a minimax estimate given at the end of Section 1.5. Is $(T(X_1), T(X_2), \ldots, T(X_p))$ minimax for $\boldsymbol{\theta}$ in the compound decision problem?

(c) Suppose $T(X)$ is the Bayes estimate with respect to squared error loss for estimating θ of $f(x|\theta)$. Is $(T(X_1), \ldots, T(X_p))$ a Bayes estimate for $\boldsymbol{\theta}$?

(d) Suppose $\mathbf{T} = (T_1(X_1), \ldots, T_p(X_p))$ and $T_j(X_i)$ is admissible in the jth component decision problem. Is $\mathbf{T}$ admissible?

5. Verify the claim of the best unbiased predictor (9.17).

6. Given the hierarchical prior of Section 9.3 for Morris's regression setup, calculate the posterior and the Bayes estimate as explicitly as possible. Find the full conditionals of the posterior distribution in order to implement MCMC.

7. Prove the claims of superiority made in Section 9.4 for the James-Stein-Lindley estimate and the James-Stein positive part estimate using Stein's identity.

8. Under the setup of Section 9.3, show that the PEB risk of $\hat{\theta}_i$ is smaller than the PEB risk of Y_i.

9. Refer to Sections 9.3 and 9.4. Compare the PEB risk of $\hat{\theta}_i$ and Stein's frequentist risk of $\hat{\boldsymbol{\theta}}$ and show that the two risks are of the same form but one has $E(\hat{B})$ and the other $E_\theta(\hat{B})$. (Hint: See equations (1.17) and (1.18) of Morris (1983)).

10. Consider the setup of Section 9.3. Show that $\hat{B}$ is the best unbiased estimate of B.

11. (Disease mapping) (See Section 10.1 for more details on the setup.) Suppose that the area to be mapped is divided into N regions. Let O_i and E_i be respectively the observed and expected number of cases of a disease in the ith region, $i = 1, 2, \ldots, N$. The unknown parameters of interest are θ_i, the relative risk in the ith region, $i = 1, 2, \ldots, N$. The traditional model for O_i is the Poisson model, which states that given $(\theta_1, \ldots, \theta_N)$, O_i's are independent and

$$O_i|\theta_i \sim \text{ Poisson } (E_i\theta_i).$$

Let $\theta_1, \theta_2, \ldots, \theta_N$ be i.i.d. $\sim$ Gamma(a, b). Find the PEB estimates of $\theta_1, \theta_2, \ldots, \theta_N$. In Section 10.1, we will consider hierarchical Bayes analysis for this problem.

12. Let Y_i be i.i.d $N(\theta_i, V)$, $i = 1, 2, \ldots, p$. Stein's heuristics (Section 9.4) shows $\|\boldsymbol{Y}\|^2$ is too large in a frequentist sense. Verify by a similar argument that if θ_i are i.i.d uniform on $\mathcal{R}$ then $\|\boldsymbol{Y}\|^2$ is too small in an improper Bayesian sense, i.e., there is extreme divergence between frequentist probability and naive objective Bayes probability in a high-dimensional case.

13. (Berger (1985a, p. 542)) Consider a multiparameter exponential family $f(\boldsymbol{x}|\boldsymbol{\theta}) = c(\boldsymbol{\theta}) \exp(\boldsymbol{\theta}' T(\boldsymbol{x})) h(\boldsymbol{x})$, where $\boldsymbol{x}$ and $\boldsymbol{\theta}$ are vectors of the same dimension. Assuming Stein's loss, show that (under suitable conditions) the Bayes estimate can be written as $\mathrm{gradient}(\log m(\boldsymbol{x})) - \mathrm{gradient}(\log h(\boldsymbol{x}))$ where $m(\boldsymbol{x})$ is the marginal density of $\boldsymbol{x}$ obtained by integrating out $\boldsymbol{\theta}$.

14. Simulate data according to the model in Example 9.3, Section 9.1.
(a) Examine how well the model can be checked from the data X_{ij}, $i = 1, 2, \ldots n$, $j = 1, 2, \ldots p$.
(b) Suppose one uses the empirical distribution of $\bar{X}_j$'s as a surrogate prior for μ_j's. Compare critically the Bayes estimate of $\boldsymbol{\mu}$ for this prior with the PEB estimate.

15. (Stone's problem) Let
$Y_{ij} = \alpha + \mu_i + \epsilon_{ij}$, $\epsilon_{ij} \sim N(0, \sigma^2)$, $i = 1, 2, \ldots, p$, $j = 1, 2, \ldots, r$, $n = pr$ with σ^2 assumed known or estimated by $S^2 = \sum_{i=1}^{p} \sum_{j=1}^{r} (Y_{ij} - \bar{Y}_i)^2 / p(r - 1)$.
The two models are

$$M_1 : \mu_i = 0 \forall i \text{ and } M_2 : \boldsymbol{\mu} \in \mathcal{R}^p.$$

Suppose $n \to \infty$, $p \log n / n \to \infty$ and $\sum_{i=1}^{p} (\mu_i - \bar{\mu})^2 / (p - 1) \to \tau^2 > 0$.
(a) Show that even though M_2 is true, BIC will select M_1 with probability tending to 1. Also show that AIC will choose the right model M_2 with probability tending to one.
(b) As a Bayesian how important do you think is this notion of consistency?
(c) Explore the relation between AIC and selection of model based on estimation of residual sum of squares by leave-one-out cross validation.

16. Consider an extremely simple testing problem. $X \sim N(\mu, 1)$. You have to test $H_0 : \mu = 0$ versus $H_1 : \mu \neq 0$. Is AIC appropriate for this? Compare AIC, BIC, and the usual likelihood ratio test, keeping in mind the conflict between P-values and posterior probability of the sharp null hypothesis.

17. Consider two nested models and an empirical Bayes model selection rule with the evaluation based on the more complex model. Though you know the more complex model is true, you may be better off predicting with the simpler model.
Let $Y_{ij} = \mu_i + \epsilon_{ij}$, ϵ_{ij} i.i.d $N(0, \sigma^2)$, $i = 1, 2, \ldots, p$, $j = 1, 2, \ldots, r$ with known σ^2. The models are

$$M_1 : \boldsymbol{\mu} = 0$$
$$M_2 : \boldsymbol{\mu} \in \mathcal{R}^p, \boldsymbol{\mu} \sim N_p(0, \tau^2 I_p), \tau^2 > 0.$$

(a) Assume that in PEB evaluation under M_2 you estimate τ^2 by the moment estimate:

$$\hat{\tau}^2 = \left[\frac{1}{p}\sum_{i=1}^{p}\bar{Y}_i^2 - \frac{\sigma^2}{r}\right]^+.$$

Show with PEB evaluation of risk under M_2 and M_1, $\bar{Y}$ is preferred if and only if AIC selects M_2.

(b) Why is it desirable to have large p in this problem?

(c) How will you try to justify in an intuitive way occasional choice of the simple but false model?

(d) Use (a) to motivate how the penalty coefficient 2 arises in AIC.

(This problem is based on a result in Mukhopadhyay (2001)).

18. Burnham and Anderson (2002) generated data to mimic a real-life experiment of Stromberg et al. (1998). Select a suitable model from among the 9 models considered by Ghosh and Samanta (2001). The main issue is computation of the integrated likelihood under each model. You can try Laplace approximation, the method based on MCMC suggested at the end of Section 9.9, and importance sampling. All methods are difficult, but they give very close answers in this problem. The data and the models can be obtained from the Web page

`http://www.isical.ac.in/~tapas/book`

19. Let $X_i \sim N(\mu, 1), i = 1, \ldots, n$ and $\mu \sim N(\eta_1, \eta_2)$. Find the PEB estimate of η_1 and η_2 and examine its implications for the inadequacy of the PEB approach in low-dimensional problems.

20. Consider NPEB multiple testing (Section 9.6.1) with known π_1 and an estimate $\hat{f}$ of $(1-\pi_1)f_0 + \pi_1 f_1$. Suppose for each i, you reject $H_{0i} : \mu_i = 0$ if

$$f_0(x_i) \leq \hat{f}(x_i)\alpha, \text{ where } o < \alpha < 1.$$

Examine whether this test provides any control on the (frequentist) FDR. Define a Bayesian FDR and examine if, for small π_1, this is also controlled by the test. Suggest a test that would make the Bayesian FDR approximately equal to α. (The idea of controlling a Bayesian FDR is due to Storey (2003). The simple rules in this problem are due to Bogdan, Ghosh, and Tokdar (personal communication).)

21. For all subsets variable selection models show that the posterior median model and the posterior mode model are the same if

$$P(M_l|X) = \prod_{i=1}^{p} p_i^{l_i}(1 - p_i)^{1-l_i}$$

where $l_i = 1$ if the ith variable is included in M_l and $l_i = 0$ otherwise.

10

Some Applications

The popularity of Bayesian methods in recent times is mainly due to their successful applications to complex high-dimensional real-life problems in diverse areas such as epidemiology, microarrays, pattern recognition, signal processing, and survival analysis. This chapter presents a few such applications together with the required methodology. We describe the method without going into the details of the critical issues involved, for which references are given. This is followed by an application involving real or simulated data.

We begin with a hierarchical Bayesian modeling of spatial data in Section 10.1. This is in the context of disease mapping, an area of epidemiological interest. The next two sections, 10.2 and 10.3, present nonparametric estimation of regression function using wavelets and Dirichlet multinomial allocation. They may also be treated as applications involving Bayesian data smoothing. For several recent advances in Bayesian nonparametrics, see Dey et al. (1998) and Ghosh and Ramamoorthi (2003).

10.1 Disease Mapping

Our first application is from the area of epidemiology and involves hierarchical Bayesian spatial modeling. Disease mapping provides a geographical distribution of a disease displaying some index such as the relative risk of the disease in each subregion of the area to be mapped. Suppose that the area to be mapped is divided into N regions. Let O_i and E_i be respectively the observed and expected number of cases of a disease in the ith region, $i = 1, 2, \ldots, N$. The unknown parameters of interest are θ_i, the relative risk in the ith region, $i = 1, 2, \ldots, N$. Here E_i is a simple-minded expectation assuming all regions have the same disease rate (at least after adjustment for age), vide Banerjee et al. (2004, p. 158). The relative risk θ_i is the regional effect in a multiplicative model of expected number of cases: $E(O_i) = E_i\theta_i$. If $\theta_i = 1$, we have $E(O_i) = E_i$. The objective is to make inference about θ_i's across regions. Among other things, this helps epidemiologists and public health professionals to identify regions or cluster of regions having high relative risks and hence needing attention and also to identify covariates causing high relative risk. The traditional model for O_i is the Poisson model, which states that given $(\theta_1, \ldots, \theta_N)$, O_i's are independent and

$$O_i|\theta_i \sim \text{ Poisson } (E_i\theta_i). \tag{10.1}$$

Under this model E_i's are assumed fixed. The classical maximum likelihood estimate of θ_i is $\hat{\theta}_i = O_i/E_i$, known as the standardized mortality ratio (SMR) for region i and $\text{Var}(\hat{\theta}_i) = \theta_i/E_i$, which may be estimated as $\hat{\theta}_i/E_i$. However, it was noted in Chapter 9 that the classical estimates may not be appropriate here for simultaneous estimation of the parameters $\theta_1, \theta_2, \ldots, \theta_N$.

As mentioned in Chapter 9, because of the assumption of exchangeability of $\theta_1, \ldots, \theta_N$, there is a natural Bayesian solution to the problem. A Bayesian modeling involves specification of prior distribution of $(\theta_1, \ldots \theta_N)$. Clayton and Kaldor (1987) followed the empirical Bayes approach using a model that assumes

$$\theta_1, \theta_2, \ldots, \theta_N \text{ i.i.d. } \sim \text{ Gamma } (a, b) \tag{10.2}$$

and estimating the hyperparameters a and b from the marginal density of $\{O_i\}$ given a, b (see Section 9.2). Here we present a full Bayesian approach adopting a prior model that allows for spatial correlation among the θ_i's. A natural extension of (10.2) could be a multivariate Gamma distribution for $(\theta_1, \ldots, \theta_N)$. We, however, assume a multivariate normal distribution for the log-relative risks $\log \theta_i$, $i = 1, \ldots, N$. The model may also be extended to allow for explanatory covariates $\boldsymbol{x}_i$ which may affect the relative risk. Thus we consider the following hierarchical Bayesian model

$$O_i|\theta_i \text{ are independent } \sim \text{ Poisson } (E_i\theta_i) \tag{10.3}$$
$$\text{where } \log \theta_i = \boldsymbol{x}_i'\boldsymbol{\beta} + \phi_i, \ i = 1, \ldots, N.$$

The usual prior for $\boldsymbol{\phi} = (\phi_1, \ldots, \phi_N)$ is given by the conditionally autoregressive (CAR) model (Besag, 1974), which is briefly described below. For details see, e.g., Besag (1974) and Banerjee et al. (2004, pp. 79–83, 163, 164). Suppose the full conditionals are specified as

$$\phi_i|\phi_j, j \neq i \sim N(\sum_{j \neq i} a_{ij}\phi_j, \sigma_i^2), \ i = 1, 2, \ldots, N. \tag{10.4}$$

These will lead to a joint distribution having density proportional to

$$\exp\left\{-\frac{1}{2}\boldsymbol{\phi}'D^{-1}(I - A)\boldsymbol{\phi}\right\} \tag{10.5}$$

where $D = \text{Diag}(\sigma_1^2, \ldots, \sigma_N^2)$ and $A = (a_{ij})_{N \times N}$. We look for a model that allows for spatial correlation and so consider a model where correlation depends on geographical proximity. A proximity matrix $W = (w_{ij})$ is an $N \times N$ matrix where w_{ij} spatially connects regions i and j in some manner. We consider here binary choices. We set $w_{ii} = 0$ for all i, and for $i \neq j$, $w_{ij} = 1$ if i is a neighbor of j, i.e., i and j share some common boundary and $w_{ij} = 0$ otherwise. Also, w_{ij}'s in each row may be standardized as $\widetilde{w}_{ij} = w_{ij}/w_{io}$ where $w_{io} = \sum_j w_{ij}$ is the number of neighbors of region i. Returning to our model (10.5), we now set $a_{ij} = \alpha w_{ij}/w_{io}$ and $\sigma_i^2 = \lambda/w_{io}$. Then (10.5) becomes

$$\exp\left\{-\frac{1}{2\lambda}\phi'(D_w - \alpha W)\phi\right\}$$

where $D_w = \mathrm{Diag}(w_{10}, w_{20}, \ldots, w_{N0})$. This also ensures that $D^{-1}(I - A) = \frac{1}{\lambda}(D_w - \alpha W)$ is symmetric.

Thus the prior for ϕ is multivariate normal

$$\phi \sim N(\mathbf{0}, \Sigma) \text{ with } \Sigma = \lambda(D_w - \alpha W)^{-1}. \tag{10.6}$$

We take $0 < \alpha < 1$, which ensures propriety of the prior and positive spatial correlation; only the values of α close to 1 give enough spatial similarity. For $\alpha = 1$ we have the standard improper CAR model. One may use the improper CAR prior because it is known that the posterior will typically emerge as proper. For this and other relative issues, see Banerjee et al. (2004).

Having specified priors for all the unknown parameters including the spatial variance parameter λ and propriety parameter α $(0 < \alpha < 1)$, one can now do Bayesian analysis using MCMC techniques. We illustrate through an example.

Example 10.1. Table 10.1 presents data from Clayton and Kaldor (1987) on observed (O_i) and expected (E_i) cases of lip cancer during the period 1975–1980 for $N = 56$ counties of Scotland. Also available are x_i, values of a covariate, the percentage of the population engaged in agriculture, fishing, and forestry (AFF), for the 56 counties. The log-relative risk is modeled as

$$\log \theta_i = \beta_0 + \beta_1 x_i + \phi_i, \quad i = 1, \ldots, N \tag{10.7}$$

where the prior for $(\phi_1, \ldots, \phi_N)$ is as specified in (10.6). We use vague priors for β_0 and β_1 and a prior having high concentration near 1 for the parameter α. The data may be analyzed using WinBUGS. A WinBUGS code for this example is put in the web page of Samanta. A part of the results – the Bayes estimates $\widetilde{\theta}_i$ of the relative risks for the 56 counties – are presented in Table 10.1. The θ_i's are smoothed by pooling the neighboring values in an automatic adaptive way as suggested in Chapter 9. The estimates of β_0 and β_1 are obtained as $\widehat{\beta}_0 = -0.2923$ and $\widehat{\beta}_1 = 0.3748$ with estimates of posterior s.d. equal to 0.3426 and 0.1325, respectively.

Table 10.1. Lip Cancer Incidence in Scotland by County: Observed Numbers (O_i), Expected Numbers (E_i), Values of the Covariate AFF (x_i), and Bayes Estimates of the Relative Risk ($\widetilde{\theta}_i$).

County	O_i	E_i	x_i	$\widetilde{\theta}_i$	County	O_i	E_i	x_i	$\widetilde{\theta}_i$
1	9	1.4	16	4.705	29	16	14.4	10	1.222
2	39	8.7	16	4.347	30	11	10.2	10	0.895
3	11	3.0	10	3.287	31	5	4.8	7	0.860
4	9	2.5	24	2.981	32	3	2.9	24	1.476
5	15	4.3	10	3.145	33	7	7.0	10	0.966
6	8	2.4	24	3.775	34	8	8.5	7	0.770
7	26	8.1	10	2.917	35	11	12.3	7	0.852
8	7	2.3	7	2.793	36	9	10.1	0	0.762
9	6	2.0	7	2.143	37	11	12.7	10	0.886
10	20	6.6	16	2.902	38	8	9.4	1	0.601
11	13	4.4	7	2.779	39	6	7.2	16	1.008
12	5	1.8	16	3.265	40	4	5.3	0	0.569
13	3	1.1	10	2.563	41	10	18.8	1	0.532
14	8	3.3	24	2.049	42	8	15.8	16	0.747
15	17	7.8	7	1.809	43	2	4.3	16	0.928
16	9	4.6	16	2.070	44	6	14.6	0	0.467
17	2	1.1	10	1.997	45	19	50.7	1	0.431
18	7	4.2	7	1.178	46	3	8.2	7	0.587
19	9	5.5	7	1.912	47	2	5.6	1	0.470
20	7	4.4	10	1.395	48	3	9.3	1	0.433
21	16	10.5	7	1.377	49	28	88.7	0	0.357
22	31	22.7	16	1.442	50	6	19.6	1	0.507
23	11	8.8	10	1.185	51	1	3.4	1	0.481
24	7	5.6	7	0.837	52	1	3.6	0	0.447
25	19	15.5	1	1.188	53	1	5.7	1	0.399
26	15	12.5	1	1.007	54	1	7.0	1	0.406
27	7	6.0	7	0.946	55	0	4.2	16	0.865
28	10	9.0	7	1.047	56	0	1.8	10	0.773

10.2 Bayesian Nonparametric Regression Using Wavelets

Let us recall the nonparametric regression problem that was stated in Example 6.1. In this problem, it is of interest to fit a general regression function to a set of observations. It is assumed that the observations arise from a real-valued regression function defined on an interval on the real line. Specifically, we have

$$y_i = g(x_i) + \varepsilon_i, \quad i = 1, \ldots, n, \text{ and } x_i \in \mathcal{T}, \tag{10.8}$$

where ε_i are i.i.d. $N(0, \sigma^2)$ errors with unknown error variance σ^2, and g is a function defined on some interval $\mathcal{T} \subset \mathcal{R}^1$.

It can be immediately noted that a Bayesian solution to this problem involves specifying a prior distribution on a large class of regression functions. In general, this is a rather difficult task. A simple approach that has been successful is to decompose the regression function g into a linear combination of a set of basis functions and to specify a prior distribution on the regression coefficients. In our discussion here, we use the (orthonormal) wavelet basis. We provide a very brief non-technical overview of wavelets including multi-resolution analysis (MRA) here, but for a complete and thorough discussion refer to Ogden (1997), Daubechies (1992), Hernández and Weiss (1996), Müller and Vidakovic (1999), and Vidakovic (1999).

10.2.1 A Brief Overview of Wavelets

Consider the function

$$\psi(x) = \begin{cases} 1 & 0 \le x < 1/2; \\ -1 & 1/2 \le x \le 1; \\ 0 & \text{otherwise.} \end{cases} \tag{10.9}$$

which is known as the Haar wavelet, simplest of the wavelets. Note that its dyadic dilations along with integer translations, namely,

$$\psi_{j,k}(x) = 2^{j/2}\psi(2^j x - k), \quad j,k \in \mathcal{Z}, \tag{10.10}$$

provide a complete orthonormal system for $\mathcal{L}^2(\mathcal{R})$. This says that any $f \in \mathcal{L}^2(\mathcal{R})$ can be approximated arbitrarily well using step functions that are simply linear combinations of wavelets $\psi_{j,k}(x)$. What is more interesting and important is how a finer approximation for f can be written as an orthogonal sum of a coarser approximation and a detail function. In other words, for $j \in \mathcal{Z}$, let

$$V_j = \Big\{ f \in \mathcal{L}^2(\mathcal{R}) : \ f \text{ is piecewise constant on intervals}$$

$$[k2^{-j}, (k+1)2^{-j}), k \in \mathcal{Z} \Big\}. \tag{10.11}$$

Now suppose $P^j f$ is the projection of $f \in \mathcal{L}^2(\mathcal{R})$ onto V_j. Then note that

$$\begin{aligned} P^j f &= P^{j-1} f + g^{j-1} \\ &= P^{j-1} f + \sum_{k \in \mathcal{Z}} < f, \psi_{j-1,k} > \psi_{j-1,k}, \end{aligned} \tag{10.12}$$

with g^{j-1} being the detail function as shown, so that

$$V_j = V_{j-1} \oplus W_{j-1}, \tag{10.13}$$

where $W_j = \text{span}\,\{\psi_{j,k}, k \in \mathcal{Z}\}$. Also, corresponding with the 'mother' wavelet ψ (Haar wavelet in this case), there is a father wavelet or scaling function

$\phi = I_{[0,1]}$ such that $V_j = \text{span}\{\phi_{j,k}, k \in \mathcal{Z}\}$, where $\phi_{j,k}$ is the dilation and translation of ϕ similar to the definition (10.10), i.e.,

$$\phi_{j,k}(x) = 2^{j/2}\phi(2^j x - k), \quad j, k \in \mathcal{Z}, \tag{10.14}$$

In fact, the sequence of subspaces $\{V_j\}$ has the following properties:

1. $\cdots \subset V_{-2} \subset V_{-1} \subset V_0 \subset V_1 \subset V_2 \subset \cdots$.
2. $\cap_{j \in \mathcal{Z}} V_j = \{0\}, \overline{\cup_{j \in \mathcal{Z}} V_j} = \mathcal{L}^2(\mathcal{R})$.
3. $f \in V_j$ iff $f(2.) \in V_{j+1}$.
4. $f \in V_0$ implies $f(. - k) \in V_0$ for all $k \in \mathcal{Z}$.
5. There exists $\phi \in V_0$ such that $\quad \text{span}\{\phi_{0,k} = \phi(. - k), k \in \mathcal{Z}\} = V_0$.

Given this ϕ, the corresponding ψ can be easily derived (see Ogden (1997) or Vidakovic (1999)). What is interesting and useful to us is that there exist scaling functions ϕ with desirable features other than the Haar function. Especially important are Daubechies wavelets that are compactly supported and each having a different degree of smoothness.

Definition: Closed subspaces $\{V_j\}_{j \in \mathcal{Z}}$ satisfying properties 1–5 are said to form a multi-resolution analysis (MRA) of $\mathcal{L}^2(\mathcal{R})$. If $V_j = \text{span}\{\phi_{j,k}, k \in \mathcal{Z}\}$ form an MRA of $\mathcal{L}^2(\mathcal{R})$, then the corresponding ϕ is also said to generate this MRA.

In statistical inference, we deal with finite data sets, so wavelets with compact support are desirable. Further, the regression functions (or density functions) that we need to estimate are expected to have certain degree of smoothness. Therefore, the wavelets used here should have some smoothness also. The Haar wavelet does have compact support but is not very smooth. In the application discussed later, we use wavelets from the family of compactly supported smooth wavelets introduced by Daubechies (1992). These, however, cannot be expressed in closed form. A sketch of their construction is as follows.

Because, from property 5 above of MRA, $\phi \in V_0 \subset V_1$, we have

$$\phi(x) = \sum_{k \in \mathcal{Z}} h_k \phi_{1,k}(x), \tag{10.15}$$

where the 'filter' coefficients h_k are given by

$$h_k = <\phi, \phi_{1,k}> = \sqrt{2} \int \phi(x)\phi(2x - k)\, dx. \tag{10.16}$$

For compactly supported wavelets ϕ, only finitely many h_k's will be non-zero. Define the 2π-periodic trigonometric polynomial

$$m_o(\omega) = \frac{1}{\sqrt{2}} \sum_{k \in \mathcal{Z}} h_k e^{-ik\omega} \tag{10.17}$$

associated with $\{h_k\}$. The Fourier transforms of ϕ and ψ can be shown to be of the form

$$\hat{\phi}(\omega) = \frac{1}{\sqrt{2}} \prod_{j=1}^{\infty} m_0(2^{-j}\omega), \tag{10.18}$$

$$\hat{\psi}(\omega) = -e^{-i\omega/2}\overline{m_0(\frac{\omega}{2} + \pi)}\hat{\phi}(\frac{\omega}{2}). \tag{10.19}$$

Depending on the number of non-zero elements in the filter $\{h_k\}$, wavelets of different degree of smoothness emerge.

It is natural to wonder what is special about MRA. Smoothing techniques such as linear regression, splines, and Fourier series all try to represent a signal in terms of component functions. At the same time, wavelet-based MRA studies the detail signals or differences in the approximations made at adjacent resolution levels. This way, local changes can be picked up much more easily than with other smoothing techniques.

With this short introduction to wavelets, we return to the nonparametric regression problem in (10.8). Much of the following discussion closely follows Angers and Delampady (2001). We begin with a compactly supported wavelet function $\psi \in \mathcal{C}^s$, the set of real-valued functions with continuous derivatives up to order s. We note that then g has the wavelet decomposition

$$g(x) = \sum_{|k|\leq K_0} \alpha_k\phi_k(x) + \sum_{j\geq 0} \sum_{|k|\leq K_j} \beta_{jk}\psi_{j,k}(x), \tag{10.20}$$

with

$$\phi_k(x) = \phi(x - k), \text{ and}$$
$$\psi_{j,k}(x) = 2^{j/2}\psi(2^j x - k),$$

where K_j is such that $\phi_k(x)$ and $\psi_{j,k}(x)$ vanish on $\mathcal{T}$ whenever $|k| > K_j$, and ϕ is the scaling function ('father wavelet') corresponding with the 'mother wavelet' ψ. Such K_j's exist (and are finite) because the wavelet function that we have chosen has compact support. For any specified resolution level J, we have

$$g(x) = \sum_{|k|\leq K_0} \alpha_k\phi_k(x) + \sum_{j=0}^{J} \sum_{|k|\leq K_j} \beta_{jk}\psi_{j,k}(x) + \sum_{j=J+1}^{\infty} \sum_{|k|\leq K_j} \beta_{jk}\psi_{j,k}(x)$$
$$= g_J(x) + R_J(x), \tag{10.21}$$

where

$$g_J(x) = \sum_{|k|\leq K_0} \alpha_k\phi_k(x) + \sum_{j\geq 0} \sum_{|k|\leq K_j} \beta_{jk}\psi_{j,k}(x), \text{ and}$$

$$R_J(x) = \sum_{j=J+1}^{\infty} \sum_{|k|\leq K_j} \beta_{jk}\psi_{j,k}(x). \tag{10.22}$$

In the representation (10.22), we note that the ϕ functions appearing in the first part detect the global features of g, and subsequently the ψ functions in the second part check for local details.

To proceed further, many standard wavelet based procedures apply the 'discrete wavelet transform' to the data and work with the resulting wavelet coefficients (see Vidakovic (1999), Müller and Vidakovic (1999)). We, however, use the familiar hierarchical Bayesian approach to specify the prior model for g in (10.8). At the resolution level J, (10.8) can be expressed as

$$y_i = g_J(x_i) + \eta_i + \varepsilon_i, \tag{10.23}$$

where $\eta_i = R_J(x_i)$. Because the amount of information available in the likelihood function to estimate the infinitely many parameters $\beta_{jk}, j > J, |k| \leq K_j$ (arising from the higher levels of resolution and appearing in η_i) is very limited, it is best to treat these η_i as nuisance parameters and eliminate them by integrating out with respect to the prior given in (10.24) while estimating g_J. Otherwise, one will need to elicit some very informative prior on these parameters, thus attracting prior robustness issues as well. One other important issue is how large J should be. Note that the number of unknown parameters in the model grows exponentially with J, so it cannot be large for practical reasons. Also, there is no need for large J because its purpose is to check for local details only.

10.2.2 Hierarchical Prior Structure and Posterior Computations

In the first-stage prior specification, α_k and β_{jl} are all assumed to be independent normal random variables with mean 0. A common prior variance of τ^2 is assigned for α_k, whereas to accommodate the decreasing effect of the 'detail' coefficients β_{jl}, their variance is assumed to be $2^{-2js}\tau^2$. Now a joint prior distribution on σ^2 and τ^2 completes the prior specification. Even though conditionally, given τ^2, α_k and β_{jl} are normally distributed, unconditionally they do have heavy tailed prior distributions possessing robustness properties.

Let us now introduce some notations to facilitate the derivation of posterior quantities. Let $\gamma = (\alpha', \beta')'$, where $\alpha = (\alpha_k)_{|k| \leq K_0}$, and $\beta = (\beta_{jk})_{0 \leq j \leq J, |k| \leq K_j}$. Then the first stage prior specified above is

$$\gamma | \tau^2 \sim N_{2K_0+1+M_\beta}(\mathbf{0}, \tau^2 \Gamma), \text{ where } \Gamma = \begin{pmatrix} I_{2K_0+1} & 0 \\ 0 & \Delta_{M_\beta} \end{pmatrix},$$

with $M_\beta = \sum_{j=0}^{J}(2K_j + 1)$ and the diagonal matrix Δ being the variance-covariance matrix of β. Also,

$$\eta = (\eta_1, \ldots \eta_n)' | \tau^2 \sim N_n(\mathbf{0}, \tau^2 Q_n), \tag{10.24}$$

where, to keep the covariance structure of η_i simple, we choose

$$(Q_n)_{ij} = \tau^2 2^{-2Js} \exp(-c|x_i - x_j|),$$

for some moderate value of c. Further, let $X = (\Phi', S')$ with the ith row of Φ' being $\{\phi_k(x_i)\}'_{|k| \leq K_0}$ and the ith row of S' being $\{\psi_{jk}(x_i)\}'_{0 \leq j \leq J, |k| \leq K_j}$. Then, given $\boldsymbol{\gamma}$, σ^2 and τ^2, we have the following linear model for the observation vector $\mathbf{Y} = (y_1, \ldots, y_n)'$:

$$\mathbf{Y} = X\boldsymbol{\gamma} + \mathbf{u}, \tag{10.25}$$

where $\mathbf{u} = \boldsymbol{\eta} + \boldsymbol{\varepsilon} \sim N_n(\mathbf{0}, \Sigma)$ with $\Sigma = \sigma^2 I_n + \tau^2 Q_n$. This follows from the fact that

$$\mathbf{Y}|\boldsymbol{\gamma}, \boldsymbol{\eta}, \sigma^2, \tau^2 \sim N_n(X\boldsymbol{\gamma} + \boldsymbol{\eta}, \sigma^2 I_n), \tag{10.26}$$
$$\boldsymbol{\eta}|\tau^2 \sim N_n(\mathbf{0}, \tau^2 Q_n).$$

From (10.25) and using standard hierarchical Bayes techniques (*cf.* Lindley and Smith (1972)) and matrix identities (*cf.* Searle (1982)), it follows that

$$\mathbf{Y}|\sigma^2, \tau^2 \sim N_n(\mathbf{0}, \sigma^2 I_n + \tau^2 (X\Gamma X' + Q_n)), \tag{10.27}$$
$$\boldsymbol{\gamma}|\mathbf{Y}, \sigma^2, \tau^2 \sim N(A\mathbf{Y}, B), \tag{10.28}$$

where

$$A = \tau^2 \Gamma X' \left(\sigma^2 I_n + \tau^2 (X\Gamma X' + Q_n)\right)^{-1},$$
$$B = \tau^2 \Gamma - \tau^4 \Gamma X' \left(\sigma^2 I_n + \tau^2 (X\Gamma X' + Q_n)\right)^{-1} X\Gamma.$$

To proceed to the second-stage calculations, some algebraic simplifications are needed (see Angers and Delampady (1992)). Spectral decomposition yields $X\Gamma X' + Q_n = HDH'$, where $D = \text{diag}(d_1, d_2, \ldots, d_n)$ is the matrix of eigenvalues and H is the orthogonal matrix of eigenvectors. Thus,

$$\sigma^2 I_n + \tau^2 (X\Gamma X' + Q_n) = H\left(\sigma^2 I_n + \tau^2 D\right) H'$$
$$= \tau^2 H\left(vI_n + D\right) H', \tag{10.29}$$

where $v = \sigma^2/\tau^2$. Using this spectral decomposition, the marginal density of $\mathbf{Y}$ given τ^2 and v can be written as

$$m(\mathbf{Y} \mid \tau^2, v) = \frac{1}{(2\pi\tau^2)^{n/2}} \frac{1}{\det(vI_n + D)^{1/2}}$$
$$\times \exp\left\{-\frac{1}{2\tau^2} \mathbf{Y}' H(vI_n + D)^{-1} H'Y\right\}$$
$$= \frac{1}{(2\pi\tau^2)^{n/2}} \frac{1}{\prod_{i=1}^{n}(v + d_i)^{1/2}} \exp\left\{-\frac{1}{2\tau^2} \sum_{i=1}^{n} \frac{t_i^2}{v + d_i}\right\}, \tag{10.30}$$

where $\mathbf{t} = (t_1, \ldots, t_n)' = H'\mathbf{Y}$.

To derive the wavelet smoother, all that we need to do now is to eliminate the hyper- and nuisance parameters from the first-stage posterior distribution by integrating out these variables with respect to the second-stage prior on them. This is what we will do now. Alternatively, one could employ an empirical Bayes approach and estimate σ^2 and τ^2 from equation (10.27) and replace σ^2 and τ^2 by their estimates in equation (10.28) to approximate $\widehat{\gamma}$. However, this will underestimate the variance of the wavelet estimator, $\widehat{\mathbf{Y}} = X\widehat{\gamma}$. Suppose, then, $\pi_2(\tau^2, v)$ is the second stage prior. It is well known in the context of hierarchical Bayesian analysis (see Chapter 9, specially equation (9.7) and Berger, 1985a) that the sensitivity of the second and higher stage hyper-priors on the final Bayes estimator is somewhat limited. Therefore, for computational ease, we choose $\pi_2(\tau^2, v) = \pi_{22}(v)(\tau^2)^{-a}$ for some suitable choice of $a > 0$; π_{22} is the prior specified for v.

Once a and π_{22} are specified, using equation (10.28) along with (10.29) and taking the expectation with respect to τ^2, we have that

$$E\left(\gamma \mid \mathbf{Y}\right) = \widehat{\gamma} = \Gamma X' H E\left[(vI_n + D)^{-1} \mid \mathbf{Y}\right] \mathbf{t}, \qquad (10.31)$$

where the expectation is taken with respect to $\pi_{22}(v \mid \mathbf{Y})$. Again using equations (10.28) and (10.29), the posterior covariance matrix of γ can be written as

$$
\begin{aligned}
Var(\gamma \mid \mathbf{Y}) = {} & \frac{1}{n + 2a} E\left[\sum_{i=1}^{n} \frac{t_i^2}{v + d_i} \mid \mathbf{Y}\right] \Gamma \\
& - \frac{1}{n + 2a} \Gamma X' H E\left[\left(\sum_{i=1}^{n} \frac{t_i^2}{v + d_i}\right)(vI_n + D)^{-1} \mid \mathbf{Y}\right] H' X \Gamma \\
& + E\left[\widehat{\gamma}(v)\widehat{\gamma}(v)' \mid \mathbf{Y}\right],
\end{aligned}
\qquad (10.32)
$$

where $\widehat{\gamma}(v) = \Gamma X' H(vI_n + D)^{-1}\mathbf{t}$.

To compute these expectations, one can use several techniques. Because they involve only single dimensional integrals, standard numerical integration methods will work quite well. Several versions of the standard Monte Carlo approach can be employed quite satisfactorily and efficiently also. An example illustrating the methodology follows.

Example 10.2. This is based on data provided by Prof. Abraham Verghese (F.R.E.S.) of the Indian Institute of Horticultural Research, Bangalore, India (personal communication), which have already been analyzed in Angers and Delampady (2001). The variable of interest y that we have chosen from the data set is the weekly average humidity level. The observations were made from June 1, 1995, to December 13, 1998. (For some reason, the observations were not recorded on the same day of the week every time.) We have chosen time (day of recording the observation) as the covariate x. (Any other available covariate can be used also because wavelet-based smoothing with respect to any arbitrary covariate (measured in some general way) can be handled with

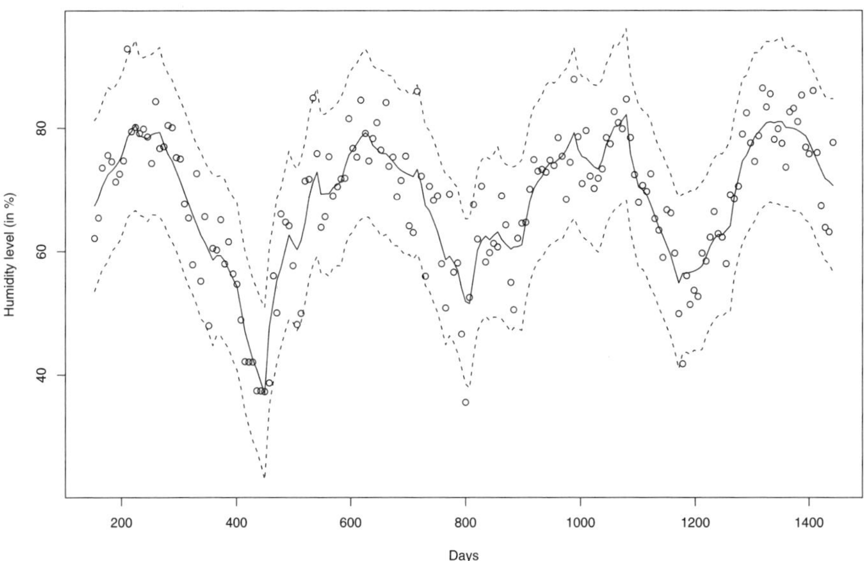

Fig. 10.1. Wavelet smoother and its error bands for the Humidity data.

our methodology.) For illustration purposes, we have chosen the model with $J = 6$; the hyperparameter a is 0.5 and the prior π_{22} corresponds with an F distribution with degrees of freedom 24 and 4. We have used compactly supported Daubechies wavelets for this analysis. As explained earlier, these cannot be expressed in closed form, but computations with these wavelets are possible using any of the several statistical and mathematical software packages. In Figure 10.1, we have plotted $\hat{g}_J$ (solid line) along with its error bands (dotted lines), $\pm 2\sqrt{Var(y \mid \mathbf{Y})}$, where

$$Var(y \mid \mathbf{Y}) = Var(g_J(x) + \eta + \varepsilon \mid \mathbf{Y}).$$

More details on this example as well as other studies can be found in Angers and Delampady (2001).

10.3 Estimation of Regression Function Using Dirichlet Multinomial Allocation

In Section 10.2, wavelets are used to represent the nonparametric regression function in (10.8) and a prior is put on the wavelet coefficients. Here we present an alternative approach based on the observation that the unknown regression function is locally linear and hence one may use a high-dimensional parametric family for modeling locally linear regression. Suppose we have a regression problem with a response variable Y and a regressor variable X. Let $(X_1, Y_1), \ldots (X_n, Y_n)$ be independent paired observations on (X, Y). Consider first the usual normal linear regression model where given values of the regressor variables x_i's, the Y_i's are independently normally distributed with common variance σ_Y^2 and mean $E(Y_i|x_i) = \beta_1 + \beta_2 x_i$, a linear function of x_i.

Let $Z_i = (X_i, Y_i)$ be independent, Z_i having the density

$$f(z|\phi_i) = f(x, y|\phi_i) = f_X(x|\mu_i, \sigma_i^2) f_Y(y|x, \beta_{1i}, \beta_{2i}, \sigma_Y^2)$$

where $f_X(x|\mu_i, \sigma_i^2)$ and $f_Y(y|x, \beta_{1i}, \beta_{2i}, \sigma_Y^2)$ denote respectively $N(\mu_i, \sigma_i^2)$ density for X_i and $N(\beta_{1i} + \beta_{2i}x, \sigma_Y^2)$ density for Y_i given x, $\phi_i = (\mu_i, \sigma_i^2, \beta_{1i}, \beta_{2i})$, $i = 1, \ldots, n$.

For simplicity we assume σ_Y^2 is known, say, equal to 1.

For the remaining parameters $\phi_i, i = 1, \ldots, n$, we have the Dirichlet multinomial allocation (DMA) prior, defined in the next paragraph.

(1) Let $k \sim p(k)$, a distribution on $\{1, 2, \ldots, n\}$.

(2) Given k, ϕ_i, $i = 1, \ldots, n$ have at most k distinct values $\theta_1, \ldots, \theta_k$, where θ_i's are i.i.d. $\sim G_0$ and G_0 is a distribution on the space of $(\mu, \sigma^2, \beta_1, \beta_2)$ (our choice of G_0 is mentioned below).

(3) Given k, the vector of weights $(w_1, \ldots, w_k) \sim$ Dirichlet $(\delta_1, \ldots, \delta_k)$.

(4) Allocation variables $a_1, \ldots, a_n$ are independent with

$$P(a_i = j) = w_j, \; j = 1, \ldots, k.$$

(5) Finally $\phi_i = \theta_{a_i}$, $i = 1, \ldots, n$.

For simplicity, we illustrate with a known k (which will be taken appropriately large). We refer to Richardson and Green (1997) for the treatment of the case with unknown k; see also the discussion of this paper by Gruet and Robert, and Green and Richardson (2001). Under this prior $\phi_i = (\mu_i, \sigma_i^2, \beta_{1i}, \beta_{2i})$, $i = 1, \ldots, n$ are exchangeable. This allows borrowing of strength, as in Chapter 9, from clusters of (x_i, y_i)'s with similar values. To see how this works, one has to calculate the Bayes estimate through MCMC.

We take G_0 to be the product of a normal distribution for μ, an inverse Gamma distribution for σ^2 and normal distributions for β_1, and β_2. The full conditionals needed for sampling from the posterior using Gibbs sampler can be easily obtained, see Robert and Casella (1999) in this context. For example, the conditional posterior distribution of $a_1, \ldots, a_n$ given other parameters are as follows:

$$a_i = j \text{ with probability } w_j f(Z_i|\theta_j) / \sum_{r=1}^{k} w_r f(Z_i|\theta_r).$$

$j = 1, \ldots, k$, $i = 1, \ldots, n$ and $a_1, \ldots, a_n$ are independent.

Due to conjugacy, the other full conditional distributions can be easily obtained. You are invited to calculate the conditional posteriors in Problem 4.

Note that given k, $\theta_1, \ldots, \theta_k$ and $w_1, \ldots, w_k$, we have a mixture with k components. Each mixture models a locally linear regression. Because θ_i and w_i are random, we have a rich family of locally linear regression models from which the posterior chooses different members and assigns to each member model a weight equal to its posterior probability density. The weight is a measure of how close is this member model to data. The Bayes estimate of the regression function is a weighted average of the (conditional) expectations of locally linear regressions.

We illustrate the use of this method with a set of data simulated form a model for which

$$E(Y|x) = \sin(2x) + \epsilon.$$

We generate 100 pairs of observations (X_i, Y_i) with normal errors ϵ_i. A scatter plot of the data points and a plot of the estimated regression at each X_i (using the Bayes estimates of β_{1i}, β_{2i}) together with the graph of $\sin(2x)$

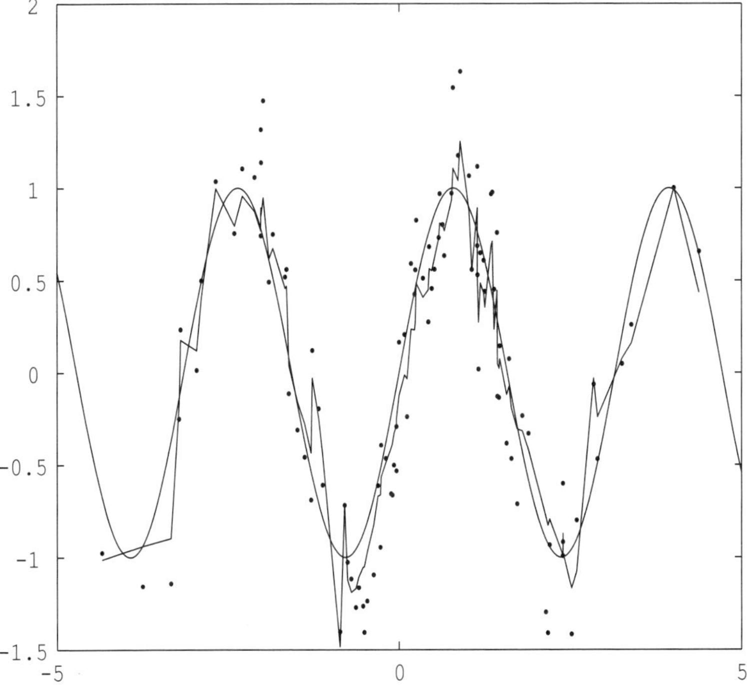

Fig. 10.2. Scatter plot, estimated regression, and true regression function.

are presented in Figure 10.2. In our calculation, we have chosen hyperparameters of the priors suitably to have priors with small information. Seo (2004) discusses the choice of hyperpriors and hyperparameters in examples of this kind.

Following Müller et al. (1996), Seo (2004) also uses a Dirichlet process prior instead of the DMA. The Dirichlet process prior is beyond the scope of our book. See Ghosh and Ramamoorthi (2003, Chapter 3) for details.

It is worth noting that the method developed works equally well if X is non-stochastic (as in Section 10.2) or has a known distribution. The trick is to ignore these facts and pretend that X is also random as above. See Müller et al. (1996) for further discussion of this point.

10.4 Exercises

1. Verify that Haar wavelets generate an MRA of $\mathcal{L}^2(\mathcal{R})$.
2. Indicate how Bayes factors can be used to obtain the optimal resolution level J in (10.21).
3. Derive an appropriate wavelet smoother for the data given in Table 5.1 and compare the results with those obtained using linear regression in Section 5.4.
4. For the problem in Section 10.3, explain how MCMC can be implemented, deriving explicitly all the full conditionals needed.
5. Choose any of the high-dimensional problems in Chapters 9 or 10 and suggest how hyperparameters may be chosen there. Discuss whether your findings will apply to all the higher levels of hierarchy.

A

Common Statistical Densities

For quick reference, listed below are some common statistical densities that are used in examples and exercise problems in the book. Only brief description including the name of the density, the notation (abbreviation) used in the book, the density itself, the range of the variable argument, and the parameter values and some useful moments are supplied.

A.1 Continuous Models

1. Univariate normal $(N(\mu, \sigma^2))$:

$$f(x|\mu, \sigma^2) = (2\pi\sigma^2)^{-1/2} \exp\left(-(x - \mu)^2/(2\sigma^2)\right),$$

   $-\infty < x < \infty,\ -\infty < \mu < \infty, \sigma^2 > 0.$
   Mean $= \mu$, variance $= \sigma^2$.
   Special case: $N(0, 1)$ is known as standard normal.
2. Multivariate normal $(N_p(\boldsymbol{\mu}, \Sigma))$:

$$f(\mathbf{x}|\boldsymbol{\mu}, \Sigma) = (2\pi)^{-p/2}|\Sigma|^{-1/2} \exp\left(-(\mathbf{x} - \boldsymbol{\mu})'\Sigma^{-1}(\mathbf{x} - \boldsymbol{\mu})\right),$$

   $\mathbf{x} \in \mathcal{R}^p, \boldsymbol{\mu} \in \mathcal{R}^p, \Sigma_{p \times p}$ positive definite.
   Mean vector $= \boldsymbol{\mu}$, covariance or dispersion matrix $= \Sigma$.
3. Exponential $(Exp(\lambda))$:

$$f(x|\lambda) = \lambda \exp(-\lambda x), x > 0, \lambda > 0.$$

   Mean $= 1/\lambda$, variance $= 1/\lambda^2$.
4. Double exponential or Laplace $(DExp(\mu, \sigma))$:

$$f(x|\mu, \sigma) = \frac{1}{2\sigma} \exp\left(-\frac{|x - \mu|}{\sigma}\right),$$

   $-\infty < x < \infty,\ -\infty < \mu < \infty,\ \sigma > 0.$
   Mean $= \mu$, variance $= 2\sigma^2$.

5. Gamma $(Gamma(\alpha, \lambda))$:

$$f(x|\alpha, \lambda) = \frac{\lambda^\alpha}{\Gamma(\alpha)} x^{\alpha-1} \exp(-\lambda x), x > 0, \alpha > 0, \lambda > 0.$$

Mean $= \alpha/\lambda$, variance $= \alpha/\lambda^2$.
Special cases:
(i) $Exp(\lambda)$ is $Gamma(1, \lambda)$.
(ii) Chi-square with n degrees of freedom (χ_n^2) is $Gamma(n/2, 1/2)$.
6. Uniform $(U(a, b))$:

$$f(x|a, b) = \frac{1}{b - a} I_{(a,b)}(x), -\infty < a < b < \infty.$$

Mean $= (a + b)/2$, variance $= (b - a)^2/12$.
7. Beta $(Beta(\alpha, \beta))$:

$$f(x|\alpha, \beta) = \frac{\Gamma(\alpha + \beta)}{\Gamma(\alpha)\Gamma(\beta)} x^{\alpha-1}(1 - x)^{\beta-1} I_{(0,1)}(x), \alpha > 0, \beta > 0.$$

Mean $= \alpha/(\alpha + \beta),$ variance $= \alpha\beta/\{(\alpha + \beta)^2(\alpha + \beta + 1)\}$.
Special case: $U(0, 1)$ is $Beta(1, 1)$.
8. Cauchy $(Cauchy(\mu, \sigma^2))$:

$$f(x|\mu, \sigma^2) = \frac{1}{\pi\sigma} \left(1 + \frac{(x - \mu)^2}{\sigma^2}\right)^{-1}, -\infty < x < \infty,$$

$-\infty < \mu < \infty, \sigma^2 > 0$. Mean and variance do not exist.
9. t distribution $(t(\alpha, \mu, \sigma^2))$:

$$f(x|\alpha, \mu, \sigma^2) = \frac{\Gamma((\alpha + 1)/2)}{\sigma\sqrt{\alpha\pi}\Gamma(\alpha/2)} \left(1 + \frac{(x - \mu)^2}{\alpha\sigma^2}\right)^{-(\alpha+1)/2},$$

$-\infty < x < \infty, \alpha > 0, -\infty < \mu < \infty, \sigma^2 > 0$.
Mean $= \mu$ if $\alpha > 1$,   variance $= \alpha\sigma^2/(\alpha - 2)$ if $\alpha > 2$.
Special cases:
(i) $Cauchy(\mu, \sigma^2)$ is $t(1, \mu, \sigma^2)$.
(ii) $t(k, 0, 1) = t_k$ is known as Student's t with k degrees of freedom.
10. Multivariate t $(t_p(\alpha, \boldsymbol{\mu}, \Sigma))$:

$$f(\mathbf{x}|\boldsymbol{\mu}, \Sigma) = \frac{\Gamma((\alpha + p)/2))}{(\alpha\pi)^{p/2}\Gamma(\alpha/2)} |\Sigma|^{-1/2} \left(1 + \frac{1}{\alpha}(\mathbf{x} - \boldsymbol{\mu})'\Sigma^{-1}(\mathbf{x} - \boldsymbol{\mu})\right)^{-(\alpha+p)/2},$$

$\mathbf{x} \in \mathcal{R}^p, \alpha > 0, \boldsymbol{\mu} \in \mathcal{R}^p, \Sigma_{p \times p}$ positive definite.
Mean vector $= \boldsymbol{\mu}$ if $\alpha > 1$,   covariance or dispersion matrix $=$
$\alpha\Sigma/(\alpha - 2)$ if $\alpha > 2$.

11. F distribution with degrees of freedom α and β $(F(\alpha, \beta))$:

$$f(x|\alpha, \beta) = \frac{\Gamma((\alpha + \beta)/2)}{\Gamma(\alpha/2)\Gamma(\beta/2)} \left(\frac{\alpha}{\beta}\right)^{\alpha/2} \frac{x^{\alpha/2-1}}{\left(1 + \frac{\alpha}{\beta}x\right)^{(\alpha+\beta)/2}}, x > 0, \alpha > 0, \beta > 0.$$

Mean $= \beta/(\beta-2)$ if $\beta > 2$,   variance $= 2\beta^2(\alpha+\beta-2)/\{\alpha(\beta-4)(\beta-2)^2\}$ if $\beta > 4$.

Special cases:

(i) If $X \sim t(\alpha, \mu, \sigma^2)$, $(X - \mu)^2/\sigma^2 \sim F(1, \alpha)$.

(ii) If $\mathbf{X} \sim t_p(\alpha, \boldsymbol{\mu}, \Sigma)$, $\frac{1}{p}(\mathbf{X} - \boldsymbol{\mu})'\Sigma^{-1}(\mathbf{X} - \boldsymbol{\mu}) \sim F(p, \alpha)$.

12. Inverse Gamma $(inverse\ Gamma(\alpha, \lambda))$:

$$f(x|\alpha, \lambda) = \frac{\lambda^\alpha}{\Gamma(\alpha)} x^{-(\alpha+1)} \exp(-\lambda/x), x > 0, \alpha > 0, \lambda > 0.$$

Mean $= \lambda/(\alpha - 1)$ if $\alpha > 1$,   variance $= \lambda^2/\{(\alpha - 1)^2(\alpha - 2)\}$ if $\alpha > 2$.

If $X \sim inverse\ Gamma(\alpha, \lambda)$, $1/X \sim Gamma(\alpha, \lambda)$.

13. Dirichlet (finite dimensional) $(D(\boldsymbol{\alpha}))$:

$$f(\mathbf{x}|\boldsymbol{\alpha}) = \frac{\Gamma(\sum_{i=1}^k \alpha_i)}{\prod_{i=1}^k \Gamma(\alpha_i)} \prod_{i=1}^k x_i^{\alpha_i-1},$$

$\mathbf{x} = (x_1, \ldots, x_k)'$ with $0 \leq x_i \leq 1$, for $1 \leq i \leq k$, $\sum_{i=1}^k x_i = 1$ and $\boldsymbol{\alpha} = (\alpha_1, \ldots, \alpha_k)'$ with $\alpha_i > 0$ for $1 \leq i \leq k$.

Mean vector $= \boldsymbol{\alpha}/(\sum_{i=1}^k \alpha_i)$,   covariance or dispersion matrix $= C_{k \times k}$ where

$$C_{ij} = \begin{cases} \dfrac{\alpha_i \sum_{l \neq i} \alpha_l}{\left(\sum_{l=1}^k \alpha_l\right)^2 \left(\sum_{l=1}^k \alpha_l+1\right)} & \text{if } i = j; \\ -\dfrac{\alpha_i \alpha_j}{\left(\sum_{l=1}^k \alpha_l\right)^2 \left(\sum_{l=1}^k \alpha_l+1\right)} & \text{if } i \neq j. \end{cases}$$

14. Wishart $(W_p(n, \Sigma))$:

$$f(A|\Sigma) = \frac{1}{2^{np/2}\Gamma_p(n/2)} |\Sigma|^{-n/2} \exp\left(-trace\{\Sigma^{-1}A\}/2\right) |A|^{(n-p-1)/2},$$

$A_{p \times p}$ positive definite, $\Sigma_{p \times p}$ positive definite, $n \geq p$, p positive integer,

$$\Gamma_p(a) = \int_{A \text{ positive definite}} \exp\left(-trace\{A\}\right) |A|^{a-(p+1)/2} \, dA,$$

for $a > (p - 1)/2$.

Mean $= n\Sigma$. For other moments, see Muirhead (1982).

Special case: χ_n^2 is $W_1(n, 1)$.

If $W^{-1} \sim W_p(n, \Sigma)$ then W is said to follow inverse-Wishart distribution.

15. Logistic $((Logistic(\mu, \sigma))$:

$$f(x|\mu, \sigma) = \frac{1}{\sigma} \frac{\exp(-\frac{x-\mu}{\sigma})}{\left(1 + \exp(-\frac{x-\mu}{\sigma})\right)^2},$$

$-\infty < x < \infty$, $-\infty < \mu < \infty$, $\sigma > 0$.
Mean $= \mu$, variance $= \pi^2 \sigma^2/3$.

A.2 Discrete Models

1. Binomial $(B(n, p))$:

$$f(x|n, p) = \binom{n}{x} p^x (1-p)^{n-x},$$

$x = 0, 1, \ldots, n$, $0 \le p \le 1$, $n \ge 1$ integer.
Mean $= np$,   variance $= np(1-p)$.
Special case: $Bernoulli(p)$ is $B(1, p)$.
2. Poisson $(\mathcal{P}(\lambda))$:

$$f(x|n, p) = \frac{\exp(-\lambda)\lambda^x}{x!},$$

$x = 0, 1, \ldots$, $\lambda > 0$.
Mean $= \lambda$,   variance $= \lambda$.
3. Geometric $(Geometric(p))$:

$$f(x|p) = (1-p)^x p,$$

$x = 0, 1, \ldots$, $0 < p \le 1$.
Mean $= (1-p)/p$,   variance $= (1-p)/p^2$.
4. Negative binomial $(Negative\ binomial(k, p))$:

$$f(x|k, p) = \binom{x + k - 1}{k - 1} (1-p)^x p^k,$$

$x = 0, 1, \ldots$, $0 < p \le 1$, $k \ge 1$ integer.
Mean $= k(1-p)/p$, variance $= k(1-p)/p^2$.
Special case: $Geometric(p)$ is $Negative\ binomial(1, p)$.
5. Multinomial $(Multinomial(n, \mathbf{p}))$:

$$f(\mathbf{x}|n, \mathbf{p}) = \frac{n!}{\prod_{i=1}^{k} x_i!} \prod_{i=1}^{k} p_i^{x_i},$$

$\mathbf{x} = (x_1, \ldots, x_k)'$ with x_i an integer between 0 and n, for $1 \le i \le k$, $\sum_{i=1}^{k} x_i = n$ and $\mathbf{p} = (p_1, \ldots, p_k)'$ with $0 \le p_i \le 1$ for $1 \le i \le k$, $\sum_{i=1}^{k} p_i = 1$.
Mean vector $= n\mathbf{p}$,   covariance or dispersion matrix $= C_{k \times k}$ where

$$C_{ij} = \begin{cases} np_i(1-p_i) & \text{if } i = j; \\ -np_i p_j & \text{if } i \ne j. \end{cases}$$

Birnbaum's Theorem on Likelihood Principle

The object of this appendix is to rewrite the usual proof of Birnbaum's theorem (e.g., as given in Basu (1988)) using only mathematical statements and carefully defining all symbols and the domain of discourse.

Let $\theta \in \Theta$ be the parameter of interest. A statistical experiment $\mathcal{E}$ is performed to generate a sample x. An experiment $\mathcal{E}$ is given by the triplet $(\mathcal{X}, \mathcal{A}, p)$ where $\mathcal{X}$ is the sample space, $\mathcal{A}$ is the class of all subsets of $\mathcal{X}$, and $p = \{p(\cdot|\theta), \theta \in \Theta\}$ is a family of probability functions on $(\mathcal{X}, \mathcal{A})$, indexed by the parameter space Θ. Below we consider experiments with a fixed parameter space Θ.

A (finite) mixture of experiments $\mathcal{E}_1, \ldots, \mathcal{E}_k$ with mixture probabilities $\pi_1, \ldots, \pi_k$ (non-negative numbers free of θ, summing to unity), which may be written as $\sum_{i=1}^{k} \pi_i \mathcal{E}_i$, is defined as a two stage experiment where one first selects $\mathcal{E}_i$ with probability π_i and then observes $x_i \in \mathcal{X}_i$ by performing the experiment $\mathcal{E}_i$.

Consider now a class of experiments closed under the formation of (finite) mixtures. Let $\mathcal{E} = (\mathcal{X}, \mathcal{A}, p)$ and $\mathcal{E}' = (\mathcal{X}', \mathcal{A}', p')$ be two experiments and $x \in \mathcal{X}, x' \in \mathcal{X}'$. By equivalence of the two points $(\mathcal{E}, x)$ and $(\mathcal{E}', x')$, we mean one makes the same inference on θ if one performs $\mathcal{E}$ and observes x or performs $\mathcal{E}'$ and observes x', and we denote this as

$$(\mathcal{E}, x) \sim (\mathcal{E}', x').$$

We now consider the following principles.

The likelihood principle (LP): We say that the equivalence relation "$\sim$" obeys the likelihood principle if $(\mathcal{E}, x) \sim (\mathcal{E}', x')$ whenever

$$p(x|\theta) = c\, p'(x'|\theta) \text{ for all } \theta \in \Theta \tag{B.1}$$

for some constant $c > 0$.

The weak conditionality principle (WCP): An equivalence relation "$\sim$" satisfies WCP if for a mixture of experiments $\mathcal{E} = \sum_{i=1}^{k} \pi_i \mathcal{E}_i$,

$$(\mathcal{E}, (i, x_i)) \sim (\mathcal{E}_i, x_i)$$

for any $i \in \{1, \ldots, k\}$ and $x_i \in \mathcal{X}_i$.

The sufficiency principle (SP): An equivalence relation "$\sim$" satisfies SP if $(\mathcal{E}, x) \sim (\mathcal{E}, x')$ whenever $S(x) = S(x')$ for some sufficient statistic S for θ (or equivalently, $S(x) = S(x')$ for a minimal sufficient statistic S).

It is shown in Basu and Ghosh (1967) (see also Basu (1969)) that for discrete models a minimal sufficient statistic exists and is given by the likelihood partition, i.e., the partition induced by the equivalence relation (B.1) for two points x, x' from the same experiment. The difference between the likelihood principle and sufficiency principle is that in the former, x, x' may belong to possibly different experiments while in the sufficiency principle they belong to the same experiment.

The weak sufficiency principle (WSP): An equivalence relation "$\sim$" satisfies WSP if $(\mathcal{E}, x) \sim (\mathcal{E}, x')$ whenever $p(x|\theta) = p(x'|\theta)$ for all θ.

If follows that SP implies WSP, which can be seen by noting that

$$S(x) = \left\{ \frac{p(x|\theta)}{\sum\limits_{\theta' \in \Theta} p(x|\theta')}, \ \theta \in \Theta \right\}$$

is a (minimal) sufficient statistic. We assume without loss of generality that

$$\sum\limits_{\theta \in \Theta} p(x|\theta) > 0 \text{ for all } x \in \mathcal{X}.$$

We now state and prove Birnbaum's theorem on likelihood principle (Birnbaum (1962)).

Theorem B.1. *WCP and WSP together imply LP, i.e., if an equivalence relation satisfies WCP and WSP then it also satisfies LP.*

Proof. Suppose an equivalence relation "$\sim$" satisfies WCP and WSP. Consider two experiments $\mathcal{E}_1 = (\mathcal{X}_1, \mathcal{A}_1, p_1)$ and $\mathcal{E}_2 = (\mathcal{X}_2, \mathcal{A}_2, p_2)$ with same Θ and samples $x_i \in \mathcal{X}_i$, $i = 1, 2$, such that

$$p_1(x_1|\theta) = cp_2(x_2|\theta) \text{ for all } \theta \in \Theta \tag{B.2}$$

for some $c > 0$.

We are to show that $(\mathcal{E}_1, x_1) \sim (\mathcal{E}_2, x_2)$. Consider the mixture experiment $\mathcal{E}$ of $\mathcal{E}_1$ and $\mathcal{E}_2$ with mixture probabilities $1/(1+c)$ and $c/(1+c)$ respectively, i.e.,

$$\mathcal{E} = \frac{1}{1+c} \mathcal{E}_1 + \frac{c}{1+c} \mathcal{E}_2.$$

The points $(1, x_1)$ and $(2, x_2)$ in the sample space of $\mathcal{E}$ have probabilities $p_1(x_1|\theta)/(1 + c)$ and $p_2(x_2|\theta)c/(1 + c)$, respectively, which are the same by (B.2). WSP then implies that

$$(\mathcal{E}, (1, x_1)) \sim (\mathcal{E}, (2, x_2)). \tag{B.3}$$

Also, by WCP

$$(\mathcal{E}, (1, x_1)) \sim (\mathcal{E}_1, x_1) \text{ and } (\mathcal{E}, (2, x_2)) \sim (\mathcal{E}_2, x_2). \tag{B.4}$$

From (B.3) and (B.4), we have $(\mathcal{E}_1, x_1) \sim (\mathcal{E}_2, x_2)$. $\square$

C

Coherence

Coherence was originally introduced by de Finetti to show any quantification of uncertainty that does not satisfy the axioms of a (finitely additive) probability distribution would lead to sure loss in a suitably chosen gample. This is formally stated in Theorem C.1 below. This section is based on Schervish (1995, pp. 654, 655) except that we use finite additivity instead of countable additivity.

Definition 1. For a bounded random variable X, the fair price or prevision $P(X)$ is a number p such that a gambler is willing to accept all gambles of the form $c(X - p)$ for all c in some sufficiently small symmetric interval around 0. Here $c(X - p)$ represents the gain to the gambler. That the values of c are sufficiently small ensures all losses are within the means of the gambler to pay, at least for bounded X.

Definition 2. Let $\{X_\alpha, \alpha \in A\}$ be a collection of bounded random variables. Suppose that for each $X_\alpha, P(X_\alpha)$ is the prevision of a gambler who is willing to accept all gambles of the form $c(X_\alpha - P(X_\alpha))$ for $-d_\alpha \leq c \leq d_\alpha$. These previsions are defined to be coherent if there do not exist a finite set $A_0 \subset A$ and $\{c_\alpha : -d \leq c_\alpha \leq d, \alpha \in A_0\}$, $d \leq \min\{d_\alpha, \alpha \in A_0\}$, such that $\sum_{\alpha \in A_0} c_\alpha(X_\alpha - P(X_\alpha)) < 0$ for all values of the random variables. It is assumed that a gambler willing to accept each of a finite number of gambles $c_\alpha(X_\alpha - P(X_\alpha)), \alpha \in A_0$ is also willing to take the gamble $\sum_{\alpha \in A_0} c(X_\alpha - P(X_\alpha))$, c sufficiently small, for finite sets A_0. If each X_α takes only a finite number of distinct values (as in Theorem C.1 below), then $\sum_{\alpha \in A_0} c_\alpha(X_\alpha - P(X_\alpha)) < 0$ for all values of X_α's if and only if $\sum_{\alpha \in A_0} c_\alpha(X_\alpha - P(X_\alpha)) < -\epsilon$ for all values of X_α's, for some $\epsilon > 0$. The second condition is what de Finetti requires.

If the previsions of a gambler are not coherent (incoherent), then he can be forced to lose money always in a suitably chosen gamble.

Theorem C.1. *Let $(S, \mathcal{A})$ be a "measurable" space. Suppose that for each $A \in \mathcal{A}$, the prevision is $P(I_A)$, where I_A denotes the indicator of A. Then the* *previsions are coherent if and only if the set function μ, defined as $\mu(A) = P(I_A)$, is a finitely additive probability on $\mathcal{A}$.*

Proof. Suppose μ is a finitely additive probability on $(S, \mathcal{A})$. Let $\{A_1, \ldots, A_m\}$ be any finite collection of elements in $\mathcal{A}$ and suppose the gambler is ready to accept gambles of the form $c_i(I_{A_i} - P(I_{A_i}))$. Then

$$Z = \sum_{i=1}^{m} c_i(I_{A_i} - P(I_{A_i}))$$

has μ-expectation equal to 0, and therefore it is not possible that Z is always less than 0. This implies incoherence cannot happen.

Conversely, assume coherence. We show that μ is a finitely additive probability by showing any violation of the probability axiom leads to a non-zero non-random gamble that can be made negative.

(i) $\mu(\phi) = 0$: Because $I(\phi) = 0$, $-c\mu(\phi) = c(I(\phi) - \mu(\phi)) \geq 0$ for some positive and negative values of c, implying $\mu(\phi) = 0$. Similarly $\mu(S) = 1$.

(ii) $\mu(A) \geq 0 \ \forall A \in \mathcal{A}$: If $\mu(A) < 0$, then for any $c < 0$, $c(I_A - \mu(A)) \leq -c\mu(A) < 0$. This means there is incoherence.

(iii) μ is finitely additive : Let $A_1, \cdots, A_m$ be disjoint sets in $\mathcal{A}$ and $U_{i=1}^{m} A_i = A$. Let

$$Z = \sum_{i=1}^{m} c(I(A_i) - \mu(A_i)) - c(I_A - \mu(A)) = c\left(\mu(A) - \sum_{i=1}^{m} \mu(A_i)\right).$$

If $\mu(A) < \sum_{i=1}^{m} \mu(A_i)$, then Z is always negative for any $c > 0$, whereas $\mu(A) > \sum_{i=1}^{m} \mu(A_i)$ implies Z is always negative for any $c < 0$. Thus $\mu(A) \neq \sum_{i=1}^{m} \mu(A_i)$ leads to incoherence. $\square$

D

Microarray

Proteins are essential for sustaining life of a living organism. Every cell in an individual has the same information for production of a large number of proteins. This information is encoded in the DNA. The information is transcribed and translated by the cell machinery to produce proteins. Different proteins are produced by different segments of the DNA that are called genes. Although every cell has the same information for production of the same set of proteins, all cells do not produce all proteins. Within an individual, cells are organized into groups that are specialized to perform specific tasks. Such groups of specialized cells are called tissues. A number of tissues makes up an organ, such as pancreas. Two tissues may produce completely disjoint sets of proteins; or, may produce the same protein in different quantities.

The molecule that transfers information from the genes for the production of proteins is called the messenger RNA (mRNA). Genes that produce a lot of mRNA are said to be upregulated and genes that produce little or no mRNA are said to be downregulated. For example, in certain cells of the pancreas, the gene that produces insulin will be upregulated (that is, large amounts of insulin mRNA will be produced), whereas it will be downregulated in the liver (because insulin is produced only by certain cells of the pancreas and by no other organ in the human body). In certain disease states, such as diabetes, there will be alteration in the amount of insulin mRNA.

A microarray is a tool for measuring the amount of mRNA that is circulating in a cell. Microarrays simultaneously measure the amount of circulating mRNA corresponding with thousands of different genes. Among various applications, such data are helpful in understanding the nature and extent of involvement of different genes in various diseases, such as cancer, diabetes, etc.

A generic microarray consists of multiple spots of DNA and is used to determine the quantities of mRNA in a collection of cells. The DNA in each spot is from a gene of interest and serves as a probe for the mRNA encoded by that gene. In general, one can think of a microarray as a grid (or a matrix) of several thousand DNA spots in a very small area (glass or polymer surface).

Each spot has a unique DNA sequence, different from the DNA sequence of the other spots around it.

mRNA from a clump of cells (that is, all the mRNAs produced by the different genes that are expressed in these cells) is extracted and experimentally converted (using a biochemical molecule called reverse transcriptase) to their complementary DNA strands (cDNA). A molecular tag that glows is attached to each piece of cDNA. This mixture is then "poured" over a microarray. Each DNA spot in the microarray will hybridize (that is, attach itself) only to its complementary DNA strand. The amount of fluorescence (usually measured using a laser beam) at a particular spot on the microarray gives an indication as to how much mRNA of a particular type was present in the original sample.

There are many sources of variability in the observations from a microarray experiment. Aside from the intrinsic biological variability across individuals or across tissues within the same individual, among the more important sources of variability are (a) method of mRNA extraction from the cells; (b) nature of fluorescent tags used; (b) temperature and time under which the experiment (that is, hybridization) was performed; (c) sensitivity of the laser detector in relation to the chemistry of the fluorescent tags used; and (d) the sensitivity and robustness of the image analysis system that is used to identify and quantify the fluorescence at each spot in the microarray. All of these experimental factors come in the way of comparing results across microarray experiments. Even within one experiment, the brightness of two spots can vary even when the same number of complementary DNA strands have hybridized to the spots, necessitating normalization of each image using statistical methods.

The amount of mRNA is quantified by a fluorescent signal. Some spots on a microarray, after the chemical reaction, show high levels of fluorescence and some show low or no fluorescence. The genes that show high level of fluorescence are likely to be expressed, whereas the genes corresponding with a low level of fluorescence are likely to be under-expressed or not expressed. Even genes that are not expressed may show low levels of fluorescence, which is treated as noise. The software package of the experimenter identifies background noise and calculates its mean, which is subtracted from all the measurements of fluorescence. This is the final observation X_i that we model with $\mu = 0$ indicating no expression, $\mu > 0$ indicating expression, and $\mu < 0$ indicating negative expression, i.e, under-expression. If the genes turn out in further studies to regulate growth of tumors, the expressed genes might help growth while the under-expressed genes could inhibit it.

E

Bayes Sufficiency

If T is a sufficient statistic, then, at least for the discrete and continuous case with p.d.f., an application of the factorization theorem implies posterior distribution of θ given $\boldsymbol{X}$ is the posterior given T. Thus in a Bayesian sense also, all information about θ contained in $\boldsymbol{X}$ is carried by T. In many cases, e.g., for multivariate normal, the calculation of posterior can be simplified by an application of this fact.

More importantly, these considerations suggest an alternative definition of sufficiency appropriate in Bayesian analysis.

Definition. *A statistic T is sufficient in a Bayesian sense, if for all priors $\pi(\theta)$, the posterior $\pi(\theta|\boldsymbol{X}) = \pi(\theta|T(\boldsymbol{X}))$*

Classical sufficiency always implies sufficiency in the Bayesian sense. It can be shown that if the family of probability measures in the model is dominated, i.e., the probability measures possess densities with respect to a σ-finite measure, then the factorization theorem holds, vide Lehmann (1986). In this case, it can be shown that the converse is true, i.e., T is sufficient in the classical sense if it is sufficient in the Bayesian sense.

A famous counter-example due to Blackwell and Ramamoorthi (1982) shows this is not true in the undominated case even under nice set theoretic conditions.

References

Abramowitz, M. and Stegun, I. (1970). *Handbook of Mathematical Functions*, **55**. National Bureau of Standards Applied Mathematics.

Agresti, A. and Caffo, B. (2000). Simple and effective confidence intervals for proportions and differences of proportions result from adding two successes and two failures. *Amer. Statist.* **54**, 280-288.

Akaike, H. (1983). *Information measure and model selection. Bull. Int. Statist. Inst.* **50**, 277-290.

Albert, J.H. (1990). A Bayesian test for a two-way contingency table using independence priors. *Canad. J. Statist.* **14**, 1583-1590.

Albert, J.H. and Chib, S. (1993). Bayesian analysis of binary and polychotomous response data. *J. Amer. Statist. Assoc.* **88**, 669-679.

Albert, J.H., Delampady, M., Polasek, W. (1991). A class of distributions for robustness studies. *J. Statist. Plann. Inference* **28**, 291-304.

Andersson, S. (1982). Distributions of maximal invariants using quotient measures. *Ann. Statist.* **10**, 955-961.

Andersson, S., Brons, H. and Jensen, S. (1983). Distribution of eigenvalues in multivariate statistical analysis. *Ann. Statist.* **11**, 392-415.

Angers, J-F. (2000). P-credence and outliers. *Metron* **58**, 81-108.

Angers, J-F. and Berger, J.O. (1991). Robust hierarchical Bayes estimation of exchangeable means. *Canad. J. Statist.* **19**, 39-56.

Angers, J-F. and Delampady, M. (1992). Hierarchical Bayesian estimation and curve fitting. *Canad. J. Statist.* **20**, 35-49.

Angers, J-F. and Delampady, M. (1997). Hierarchical Bayesian curve fitting and model choice for spatial data. *Sankhyā* (Ser. B) **59**, 28-43.

Angers, J-F. and Delampady, M. (2001). Bayesian nonparametric regression using wavelets. *Sankhyā* (Ser. A) **63**, 287-308.

Arnold, S.F. (1993). Gibbs sampling. In: Rao, C.R. (ed). *Handbook of Statistics* **9**, 599-625. Elsevier Science.

Athreya, K.B., Doss, H. and Sethuraman, J. (1996). On the convergence of the Markov chain simulation method. *Ann. Statist.* **24**, 69-100.

Athreya, K.B., Delampady, M. and Krishnan, T. (2003). Markov chain Monte Carlo methods. *Resonance* **8**, Part I, No. 4, 17-26, Part II, No. 7, 63-75, Part III, No. 10, 8-19, Part IV, No. 12, 18-32.

Bahadur, R.R. (1971). *Some Limit Theorems in Statistics*. CBMS Regional Conference Series in Applied Mathematics, **4**. SIAM, Philadelphia, PA.

Banerjee, S., Carlin, B.P. and Gelfand, A.E. (2004). *Hierarchical Modeling and Analysis for Spatial Data*. Chapman & Hall, London.

Barbieri, M.M. and Berger, J.O. (2004). Optimal predictive model selection. *Ann. Statist.* **32**, 870-897.

Basu, D. (1969). Role of the sufficiency and likelihood principles in sample survey theory. *Sankhyā* (Ser. A) **31**, 441-454.

Basu, D. (1988). *Statistical Information and Likelihood : A Collection of Critical Essays by Dr. D. Basu* (Ghosh, J.K. ed), Lecture Notes in Statistics, Springer-Verlag, New york.

Basu, D. and Ghosh, J. K. (1967). Sufficient statistics in sampling from a finite universe. *Bull. Int. Statist. Inst.* **42**, 850-858.

Basu, S. (1994). Variations of posterior expectations for symmetric unimodal priors in a distribution band. *Sankhyā* (Ser. A) **56**, 320-334.

Basu, S. (2000). Bayesian robustness and Bayesian nonparametrics. In: Ríos Insua, D. and Ruggeri, F. (eds) *Robust Bayesian Analysis*, 223-240, Springer-Verlag, New York.

Basu, S. and Chib, S. (2003). Marginal likelihood and Bayes factors from Dirichlet process mixture models. *J. Amer. Statist. Assoc.* **98**, 224-235.

Basu, R., Ghosh, J.K. and Mukerjee, R. (2003). Empirical Bayes prediction intervals in a normal regression model: higher order asymptotics. *Statist. Probab. Lett.* **63**, 197-203.

Bayarri, M.J. and Berger, J.O. (1998a). Quantifying surprise in the data and model verification. In: Bernardo, J.M. et al. (eds) *Bayesian Statistics* **6**, 53-82. Oxford Univ. Press, Oxford.

Bayarri, M.J. and Berger, J.O. (1998b). Robust Bayesian analysis of selection models. *Ann. Statist.* **26**, 645-659.

Benjamini, Y. and Hochberg, Y. (1995) Controlling the false discovery rate: a practical and powerful approach to multiple testing. *J. Roy. Statist. Soc.* (Ser. B) **57**, 289-300.

Benjamini, Y. and Liu, W. (1999). A step-down multiple hypotheses testing procedure that controls the false discovery rate under independence. Multiple comparisons (Tel Aviv, 1996). *J. Statist. Plann. Inference* **82**, 163-170.

Benjamini, Y. and Yekutieli, D. (2001). The control of the false discovery rate in multiple testing under dependency. *Ann. Statist.* **29**, 1165-1188.

Berger, J. (1982). Bayesian robustness and the Stein effect. *J. Amer. Statist. Assoc.* **77**, 358-368.

Berger, J.O. (1984). The robust Bayesian viewpoint(with discussion). In: *Robustness of Bayesian Analyses*. Studies in Bayesian Econometrics, **4**, 63-144. North-Holland Publishing Co., Amsterdam.

Berger, J.O. (1985a). *Statistical Decision Theory and Bayesian Analysis*, 2nd Ed. Springer-Verlag, New York.

Berger, J.O. (1985b). The frequentist viewpoint and conditioning. In: Le Cam, L. and Olshen, R.A. (eds) *Proc. Berkeley Conference in Honor of Jerzy Neyman and Jack Kiefer*, **I**, 15-44. Wadsworth, Inc., Monterey, California.

Berger, J. (1986). Are P-values reasonable measures of accuracy? In: Francis, I.S. et al. (eds) *Pacific Statistical Congress*. North-Holland, Amsterdam.

Berger, J.O. (1990). Robust Bayesian analysis: sensitivity to the prior. *J. Statist. Plann. Inference* **25**, 303-328.

Berger, J.O. (1994). An overview of robust Bayesian analysis (with discussion). *Test* **3**, 5-124.

Berger, J.O. (1997). Bayes factors. In: Kotz, S. et al. (eds) *Encyclopedia of Statistical Sciences* (Update) **3**, 20-29. Wiley, New York.

Berger, J. O. (2005). Generalization of BIC. Unpublished manuscript.

Berger, J. and Berliner, M.L. (1986). Robust Bayes and empirical Bayes analysis with ϵ-contaminated priors. *Ann. Statist.* **14**, 461-486.

Berger, J.O. and Bernardo, J.M. (1989). Estimating a product of means: Bayesian analysis with reference priors. *J. Amer. Statist. Assoc.* **84**, 200-207.

Berger, J.O. and Bernardo, J.M. (1992a). On the development of the reference priors. In: Bernardo, J.M. et al. (eds) *Bayesian Statistics* **4**, 35-60. Oxford Univ. Press, Oxford.

Berger, J. and Bernardo, J.M. (1992b). Ordered group reference priors with application to the multinomial problem. *Biometika* **79**, 25-37.

Berger, J., Bernardo, J.M. and Mendoza, M. (1989). On priors that maximize expected information. In: Klein, J. and Lee, J.C. (eds) *Recent Developments in Statistics and Their Applications*, 1-20. Freedom Academy Publishing, Seoul.

Berger, J., Bernardo, J.M. and Sun, D. (2006). A monograph on *Reference Analysis* (under preparation).

Berger, J.O., Betrò, B., Moreno, E., Pericchi, L.R., Ruggeri, F., Salinetti, G. and Wasserman, L. (eds) *Bayesian Robustness*. IMS, Hayward.

Berger, J.O. and Delampady, M. (1987). Testing precise hypotheses (with discussion). *Statist. Sci.* **2**, 317-352.

Berger, J. and Dey, D.K. (1983). Combining coordinates in simultaneous estimation of normal means. *J. Statist. Plann. Inference* **8**, 143-160.

Berger, J.O., Ghosh, J.K. and Mukhopadhyay, N. (2003). Approximations and consistency of Bayes factors as model dimension grows. *J. Statist. Plann. Inference* **112**, 241-258.

Berger, J.O., Liseo, B. and Wolpert, R.L. (1999). Integrated likelihood methods for eliminating nuisance parameters. *Statist. Sci.* **14**, 1-28.

Berger, J.O. and Moreno, E. (1994). Bayesian robustness in bidimensional models: prior independence (with discussion). *J. Statist. Plann. Inference* **40**, 161-176.

Berger, J.O. and Pericchi, L.R. (1996a). The intrinsic Bayes factor for model selection and prediction. *J. Amer. Statist. Assoc.* **91**, 109-122.

Berger, J.O. and Pericchi, L.R. (1996b). The intrinsic Bayes factor for linear models (with discussion). In: Bernardo, J.M. et al. (eds) *Bayesian Statistics* **5**, 25-44. Oxford Univ. Press, London.

Berger, J.O., Pericchi, L.R. and Varshavsky, J.A. (1998). Bayes factors and marginal distributions in invariant situations. *Sankhyā* (Ser. A) **60**, 307-321.

Berger, J.O. and Robert, C.P. (1990). Subjective hierarchical Bayes estimation of a multivariate normal mean: on the frequentist interface. *Ann. Statist.* **18**, 617-651.

Berger, J.O., Ríos Insua, D. and Ruggeri, F. (2000). Bayesian robustness. In: Ríos Insua, D. and Ruggeri, F. (eds) *Robust Bayesian Analysis*, 1-32, Springer-Verlag, New York.

Berger, J.O. and Sellke, T. (1987). Testing a point null hypothesis: the irreconcilability of p-values and evidence. *J. Amer. Statist. Assoc.* **82**, 112-122.

Berger, J.O. and Wolpert, R. (1988). *The Likelihood Principle*, 2nd Ed. IMS Lecture Notes - Monograph Ser. **9**. Hayward, California.

Bernardo, J.M. (1979). Reference posterior distribution for Bayesian inference (with discussion). *J. Roy. Statist. Soc.* (Ser. B) **41**, 113-147.

Bernardo, J.M. (1980). A Bayesian analysis of classical hypothesis testing. In: Bernardo, J.M. et al. (eds) *Bayesian Statistics*, 605-618. University Press, Valencia.

Bernardo, J.M. and Smith, A.F.M. (1994). *Bayesian Theory.* Wiley, Chichester, England.

Bernstein, S. (1917). *Theory of Probability* (in Russian).

Berti, P., Regazzini, E. and Rigo, P. (1991). Coherent statistical inference and Bayes theorem. *Ann. Statist.* **19**, 366-381.

Besag, J. (1974). Spatial interaction and the statistical analysis of lattice systems (with discussion). *J. Roy. Statist. Soc.* (Ser. B) **36**, 192-326.

Besag, J. (1986). On the statistical analysis of dirty pictures. *J. Roy. Statist. Soc.* (Ser. B) **48**, 259-279.

Betrò, B. and Guglielmi, A. (2000). Methods for global prior robustness under generalized moment conditions. In: Ríos Insua, D. and Ruggeri, F. (eds) *Robust Bayesian Analysis*, 273-293. Springer-Verlag, New York.

Bhattacharya, S. (2005). Model assessment using inverse reference distribution approach. Tech. Report.

Bickel, P.J. (1981). Minimax estimation of the mean of a normal distribution when the parameter space is restricted. *Ann. Statist.* **9**, 1301-1309.

Bickel, P.J. and Doksum, K.A. (2001). *Mathematical Statistics: Basic Ideas and Selected Topics.* Prentice Hall, Upper Saddle River, N.J.

Bickel, P.J. and Ghosh, J.K. (1990). A decomposition for the likelihood ratio statistic and the Bartlett correction – a Bayesian argument. *Ann. Statist.* **18**, 1070-1090.

Bickel, P.J. and Yahav, J. (1969) Some contributions to the asymptotic theory of Bayes solutions. *Z. Wahrsch. Verw. Gebiete* **11**, 257-275.

Birnbaum, A. (1962). On the foundations of statistical inference (with discussion). *J. Amer. Statist. Assoc.* **57**, 269-326.

Bishop, Y.M.M., Fienberg, S.E. and Holland, P.W. (1975). *Discrete Multivariate Analysis: Theory and Practice.* The MIT Press, Cambridge.

Blackwell, D. and Ramamoorthi, R.V. (1982). A Bayes but not classically sufficient statistic. *Ann. Statist.* **10**, 1025-1026.

Bondar, J.V. and Milnes, P. (1981). Amenability: A survey for statistical applications of Hunt-Stein and related conditions on groups. *Z. Wahrsch. Verw. Gebiete* **57**, 103-128.

Borwanker, J.D., Kallianpur, G. and Prakasa Rao, B.L.S. (1971). The Bernstein-von Mises theorem for stochastic processes. *Ann. Math. Statist.* **42**, 1241-1253.

Bose, S. (1994a). Bayesian robustness with more than one class of contaminations (with discussion). *J. Statist. Plann. Inference* **40**, 177-187.

Bose, S. (1994b). Bayesian robustness with mixture classes of priors. *Ann. Statist.* **22**, 652-667.

Box, G.E.P. (1980). Sampling and Bayes inference in scientific modeling and robustness. *J. Roy. Statist. Soc.* (Ser. A) **143**, 383-430.

Box, G.E.P. and Tiao, G. (1962). A further look at robustness via Bayes theorem. *Biometrika* **62**, 169-173.

Box, G.E.P. and Tiao, G. (1973). *Bayesian Inference in Statistical Analysis.* Addison-Wesley, Reading.

Brooks, S.P., Giudici, P. and Roberts, G.O. (2003). Efficient construction of reversible jump Markov chain Monte Carlo proposal distributions. *J. Roy. Statist. Soc.* (Ser. B) **65**, 3-55.

Brown, L.D. (1971). Admissible estimators, recurrent diffusions, and insoluble boundary-value problems. *Ann. Math. Statist.* **42**, 855-903.

Brown, L.D. (1986). *Foundations of Exponential Families.* IMS, Hayward.

Brown, L.D., Cai, T.T. and DasGupta, A. (2001). Interval estimation for a binomial proportion. *Statist. Sci.* **16**, 101-133.

Brown, L.D., Cai, T.T. and DasGupta, A. (2002). Confidence intervals for a binomial proportion and asymptotic expansions. *Ann. Statist.* **30**, 160-201.

Brown, L.D., Cai, T.T. and DasGupta, A. (2003). Interval estimation in exponential families. *Statist. Sinica* **13**, 19-49.

Burnham, K.P. and Anderson, D.R. (2002). *Model Selection and Multimodel Inference: A Practical Information Theoretic Approach.* Springer, New York.

Cai, T.T., Low, M. and Zhao, L. (2000). Sharp adaptive estimation by a blockwise method. Tech. Report, Dept. of Statistics, Univ. of Pennsylvania.

Carlin, B.P. and Chib, S. (1995). Bayesian model choice via Markov chain Monte Carlo methods. *J. Roy. Statist. Soc.* (Ser. B) **57**, 473-484.

Carlin, B.P. and Pérez, M.E. (2000). Robust Bayesian analysis in medical and epidemiological settings. In: Ríos Insua, D. and Ruggeri, F. (eds) *Robust Bayesian Analysis,* 351-372, Springer-Verlag, New York.

Carlin, B.P. and Louis, T.A. (1996). *Bayes and Empirical Bayes Methods for Data Analysis.* Chapman & Hall, London.

Casella, G. and Berger, R. (1987). Reconciling Bayesian and frequentist evidence in the one-sided testing problem (with discussion). *J. Amer. Statist. Assoc.* **82**, 106-111.

Casella, G. and Berger, R. (1990). *Statistical Inference.* Wadsworth, Belmont, California.

Cencov, N.N. (1982). *Statistical Decision Rules and Optimal Inference.* AMS, Providence, R.I., Translation from Russian edited by Lev L. Leifman.

Chakrabarti, A. (2004) Model selection for high dimensional problems with application to function estimation. Ph.D. thesis, Purdue Univ.

Chakrabarti, A. and Ghosh, J.K. (2005a). A generalization of BIC for the general exponential family. (In press)

Chakrabarti, A. and Ghosh, J.K. (2005b). Optimality of AIC in inference about Brownian Motion. (In press)

Chao, M.T. (1970). The asymptotic behaviour of Bayes estimators. *Ann. Math. Statist.* **41**, 601-609.

Chatterjee, S.K. and Chattopadhyay, G. (1994). On the nonexistence of certain optimal confidence sets for the rectangular problem. *Statist. Probab. Lett.* **21**, 263-269.

Chen, C.F. (1985). On asymptotic normality of limiting density functions with Bayesian implications. *J. Roy. Statist. Soc.* (Ser. B) **47**, 540-546.

Chib, S. (1995). Marginal likelihood from the Gibbs output. *J. Amer. Statist. Assoc.* **90**, 1313-1321.

Chib, S. and Greenberg, E. (1998). Analysis of multivariate probit models. *Biometrika* **85**, 347-361.

Clarke, B. and Barron, A. (1990). Information-theoretic asymptotics of Bayes methods. *IEEE Trans. Inform. Theory* **36**, 453-471.

Clayton, D.G. and Kaldor, J.M. (1987). Empirical Bayes estimates of age-standardized relative risks for use in disease mapping. *Biometrics* **43**, 671-681.

Clyde, M. (1999). Bayesian model averaging and model search strategies. In: Bernardo, J.M. et al. (eds) *Bayesian Statistics* **6**, 157-185. Oxford Univ. Press, Oxford.

Clyde, M. and Parmigiani, G. (1996). Orthogonalizations and prior distributions for orthogonalized model mixing. In: Lee, J.C. et al. (eds) *Modelling and Prediction*, 206-227. Springer, New York.

Congdon, P. (2001). *Bayesian Statistical Modelling.* Wiley, Chichester, England.

Congdon, P. (2003). *Applied Bayesian Modelling.* Wiley, Chichester, England.

Cox, D.R. (1958). Some problems connected with statistical inference. *Ann. Math. Statist.* **29**, 357-372.

Cox, D.R. and Reid, N. (1987). Orthogonal parameters and approximate conditional inference (with discussion). *J. Roy. Statist. Soc.* (Ser. B) **49**, 1-18.

Csiszár, I. (1978). Information measures: a critical survey. In: *Trans. 7th Prague Conf. on Information Theory, Statistical Decision Functions and the Eighth European Meeting of Statisticians* (Tech. Univ. Prague, Prague, 1974) **B**, 73-86. Academia, Prague.

Cuevas, A. and Sanz, P. (1988). On differentiability properties of Bayes operators. In: Bernardo, J.M. et al. (eds) *Bayesian Statistics* **3**, 569-577. Oxford Univ. Press, Oxford.

Dalal, S.R. and Hall, G.J. (1980). On approximating parametric Bayes models by nonparametric Bayes models. *Ann. Statist.* **8**, 664-672.

DasGupta, A. and Delampady, M. (1990). Bayesian hypothesis testing with symmetric and unimodal priors. Tech. Report, 90-43, Dept. Statistics, Purdue Univ.

DasGupta, A., Casella, G., Delampady, M., Genest, C., Rubin, H. and Strawderman, W.E. (2000). Correlation in a formal Bayes framework. *Can. J. Statist.* **28**, 675-687

Datta, G.S. (1996). On priors providing frequentist validity for Bayesian inference for multiple parametric functions. *Biometrika* **83**, 287-298.

Datta, G.S. and Ghosh, M. (1996). On the invariance of noninformative priors. *Ann. Statist.* **24**, 141-159.

Datta, G.S., Ghosh, M. and Mukerjee, R. (2000). Some new results on probability matching priors. *Calcutta Statist. Assoc. Bull.* **50**, 179-192.

Datta, G.S. and Mukerjee, R. (2004). *Probability Matching Priors: Higher Order Asymptotics.* Lecture Notes in Statistics. Springer, New York.

Daubechies, I. (1992). *Ten Lectures on Wavelets.* SIAM, Philadelphia.

Dawid, A.P. (1973). Posterior expectations for large observations. *Biometrika* **60**, 664-667.

Dawid, A.P., Stone, M. and Zidek, J.V. (1973). Marginalization paradoxes in Bayesian and structural inference (with discussion). *J. Roy. Statist. Soc.* (Ser. B) **35**, 189-233.

de Finetti, B. (1972). *Probability, Induction, and Statistics.* Wiley, New York.

de Finetti, B. (1974, 1975). *Theory of Probability*, Vols. **1, 2**. Wiley, New York.

DeGroot, M.H. (1970). *Optimal Statistical Decisions.* McGraw-Hill, New York.

DeGroot, M.H. (1973). Doing what comes naturally: Interpreting a tail area as a posterior probability or as a likelihood ratio. *J. Amer. Statist. Assoc.* **68**, 966-969.

Delampady, M. (1986). Testing a precise hypothesis: Interpreting P-values from a robust Bayesian viewpoint. Ph.D. thesis, Purdue Univ.

Delampady, M. (1989a). Lower bounds on Bayes factors for invariant testing situations. *J. Multivariate Anal.* **28**, 227-246.

Delampady, M. (1989b). Lower bounds on Bayes factors for interval null hypotheses. *J. Amer. Statist. Assoc.* **84**, 120-124.

Delampady, M. (1992). Bayesian robustness for elliptical distributions. *Rebrape: Brazilian J. Probab. Statist.* **6**, 97-119.

Delampady, M. (1999). Robust Bayesian outlier detection. *Brazilian J. Probab. Statist.* **13**, 149-179.

Delampady, M. and Berger, J.O. (1990). Lower bounds on Bayes factors for multinomial distributions, with application to chi-squared tests of fit. *Ann. Statist.* **18**, 1295-1316.

Delampady, M. and Dey, D.K. (1994). Bayesian robustness for multiparameter problems. *J. Statist. Plann. Inference* **40**, 375-382.

Delampady, M., Yee, I. and Zidek, J.V. (1993). Hierarchical Bayesian analysis of a discrete time series of Poisson counts. *Statist. Comput.* **3**, 7-15.

Delampady, M., DasGupta, A., Casella, G., Rubin, H., and Strawderman, W.E. (2001). A new approach to default priors and robust Bayes methodology. *Can. J. Statist.* **29**, 437-450.

Dempster, A.P. (1967). Upper and lower probabilities induced from a multivalued mapping. *Ann. Math. Statist.* **38**, 325-339.

Dempster, A.P. (1968). A generalization of Bayesian inference (with discussion). *J. Roy. Statist. Soc.* (Ser. B) **30**, 205-247.

Dempster, A.P. (1973). The direct use of likelihood for significance testing. In: Godambe, V.P. and Sprott,D.A. (eds) *Proceedings of the Conference on Foundational Questions in Statistical Inference.* Holt, Rinehart, and Winston, Toronto.

Dempster, A.P., Laird, N.M. and Rubin, D.B. (1977). Maximum likelihood from incomplete data via the EM algorithm (with discussion). *J. Roy. Statist. Soc.* (Ser. B) **39**, 1-38.

DeRobertis, L. and Hartigan, J.A. (1981). Bayesian inference using intervals of measures. *Ann. Statist.* **9**, 235-244.

Dey, D.K. and Birmiwal, L. (1994). Robust Bayesian analysis using divergence measures. *Statist. Probab. Lett.* **20**, 287-294.

Dey, D.K. and Berger, J.O. (1983). On truncation of shrinkage estimators in simultaneous estimation of normal means. *J. Amer. Statist. Assoc.* **78**, 865-869.

Dey, D.K., Ghosh, S.K. and Lou, K. (1996). On local sensitivity measures in Bayesian analysis. In: Berger, J.O. et al. (eds) *Bayesian Robustness*, IMS Lecture Notes, **29**, 21-39.

Dey, D.K., Lou, K. and Bose, S. (1998). A Bayesian approach to loss robustness. *Statist. Decisions* **16**, 65-87.

Dey, D.K. and Micheas, A.C. (2000). Ranges of posterior expected losses and ϵ-robust actions. In: Ríos Insua, D. and Ruggeri, F. (eds) *Robust Bayesian Analysis*, 145-159, Springer-Verlag, New York.

Dey, D.K. and Peng, F. (1996). Bayesian analysis of outlier problems using divergence measures. *Canad. J. Statist.* **23**, 194-213.

Dharmadhikari, S. and Joag-Dev, K. (1988). *Unimodality, Convexity, and Applications*. Academic Press, San Diego.

Diaconis, P. and Freedman, J. (1986). On the consistency of Bayes estimates. *Ann. Statist.* **14**, 1-26.

Diaconis, P. and Ylvisaker, D. (1979). Conjugate priors for exponential families. *Ann. Statist.* **7**, 269-281.

Diamond, G.A. and Forrester, J.S. (1983). Clinical trials and statistical verdicts: Probable grounds for appeal. *Ann. Intern. Med.* **98**, 385-394.

Dickey, J.M. (1971). The weighted likelihood ratio, linear hypotheses on normal location parameters. *Ann. Math. Statist.* **42**, 204-223.

Dickey, J.M. (1973). Scientific reporting. *J. Roy. Statist. Soc.* (Ser. B) **35**, 285-305.

Dickey, J.M. (1974). Bayesian alternatives to the F-test and least squares estimate in the linear model. In: Fienberg, S.E. and Zellner, A. (eds) *Studies in Bayesian Econometrics and Statistics*, 515-554. North-Holland, Amsterdam.

Dickey, J.M. (1976). Approximate posterior distributions. *J. Amer. Statist. Assoc.* **71**, 680-689.

Dickey, J.M. (1977). Is the tail area useful as an approximate Bayes factor? *J. Amer. Statist. Assoc.* **72**, 138-142.

Dickey, J.M. (1980). Approximate coherence for regression model inference – with a new analysis of Fisher's Broadback Wheatfield example. In: Zellner, A. (ed). *Bayesian Analysis in Econometrics and Statistics: Essays in Honour of Harold Jeffreys*, 333-354. North-Holland, Amsterdam.

Dmochowski, J. (1994). Intrinsic priors via Kullback-Leibler geometry. Tech. Report 94-15, Dept. Statistics, Purdue Univ.

Donoho, D. and Jin, J. (2004). Higher criticism for detecting sparse heterogeneous mixtures. *Ann. Staist.* **32**, 962-94.

Eaton, M.L. (1983). *Multivariate Statistics - A Vector Space Approach*. Wiley, New York.

Eaton, M.L. (1989). *Group Invariance Applications in Statistics*. Regional Conference Series in Probability and Statistics, **1**. IMS, Hayward, California.

Eaton, M.L. (1992). A statistical diptych: admissible inferences – recurrence of symmetric Markov chains. *Ann. Statist.* **20**, 1147-1179.

Eaton, M.L. (1997). Admissibility in quadratically regular problems and recurrence of symmetric Markov chains: Why the connection? *J. Statist. Plann. Inference* **64**, 231-247.

Eaton, M.L. (2004). Evaluating improper priors and the recurrence of symmetric Markov chains: an overview. A Festschrift for Herman Rubin, 5-20, IMS Lecture Notes Monogr. Ser. **45**. IMS, Beachwood, OH.

Eaton, M.L. and and Sudderth, W.D. (1998). A new predictive distribution for normal multivariate linear models. *Sankhyā* (Ser. A) **60**, 363-382.

Eaton, M.L. and and Sudderth, W.D. (2004). Properties of right Haar predictive inference. *Sankhyā* **66**, 487-512.

Edwards, W., Lindman, H. and Savage, L.J. (1963). Bayesian statistical inference for psychological research. *Psychol. Rev.* **70**, 193-242.

Efron, B. (1982). *The Jackknife, the Bootstrap and Other Resampling Plans*. SIAM, Philadelphia.

Efron, B. (2003). Robbins, empirical Bayes and microarrays. Dedicated to the memory of Herbert E. Robbins. *Ann. Statist.* **31**, 366-378.

Efron, B. (2004). Large-scale simultaneous hypothesis testing: The choice of a null hypothesis. *J. Amer. Statist. Assoc.* **99**, 96-104.

Efron, B. and Morris, C. (1971). Limiting the risk of Bayes and empirical Bayes estimators–Part I: The Bayes case. *J. Amer. Statist. Assoc.* **66**, 807-815.

Efron, B. and Morris, C. (1972). Limiting the risk of Bayes and empirical Bayes estimators–Part II: The empirical Bayes case. *J. Amer. Statist. Assoc.* **67**, 130-139.

Efron, B. and Morris, C. (1973). Stein's estimation rule and its competitors – an empirical Bayes approach. *J. Amer. Statist. Assoc.* **68**, 117-130.

Efron, B. and Morris, C. (1973). Combining possibly related estimation problems (with discussion). *J. Roy. Statist. Soc.* (Ser. B) **35**, 379-421.

Efron, B. and Morris, C. (1975). Data analysis using Stein's estimator and its generalizations. *J. Amer. Statist. Assoc.* **70**, 311-319.

Efron, B. and Morris, C. (1976). Multivariate empirical Bayes and estimation of covariance matrices. *Ann. Statist.* **4**, 22-32.

Efron, B., Tibshirani, R., Storey, J.D. and Tusher, V. (2001a). Empirical Bayes analysis of a microarray experiment. *J. Amer. Statist. Assoc.* **96**, 1151-1160.

Efron, B., Storey, J. and Tibshirani, R. (2001b). Microarrays, empirical Bayes methods, and false discovery rates. Tech. Report, Stanford Univ.

Ezekiel, M. and Fox, F.A. (1959). *Methods of Correlation and Regression Analysis.* Wiley, New York.

Fan, T-H. and Berger, J.O. (1992). Behavior of the posterior distribution and inferences for a normal mean with t prior convolutions. *Statist. Decisions* **10**, 99-120.

Fang, K.T., Kotz, S. and Ng, K.W. (1990). *Symmetric Multivariate and Related Distributions.* Chapman & Hall, London.

Farrell, R.H. (1985). *Multivariate Calculation - Use of the Continuous Groups.* Springer-Verlag, New York.

Feller, W. (1973). *Introduction to Probability Theory and Its Applications.* Vol. **1**, 3rd Ed. Wiley, New York.

Ferguson, T.S. (1967). *Mathematical Statistics: A Decision-Theoretic Approach.* Academic Press, New York.

Fernández, C., Osiewalski, J. and Steel, M.F.J. (2001). Robust Bayesian inference on scale parameters. *J. Multivariate Anal.* **77**, 54-72.

Finney, D.J. (1971). *Probit Analysis*, 3rd Ed. Cambridge Univ. Press, Cambridge, U.K.

Fishburn, P.C. (1981). Subjective expected utility: a review of normative theory. *Theory and Decision* **13**, 139-199.

Fisher, R.A. (1973). *Statistical Methods for Research Workers*, 14th Ed. Hafner, New York. Reprinted by Oxford Univ. Press, Oxford, 1990.

Flury, B. and Zoppè, A. (2000). Exercises in EM. *Amer. Statist.* **54**, 207-209.

Fortini, S. and Ruggeri, F. (1994). Concentration functions and Bayesian robustness (with discussion). *J. Statist. Plann. Inference* **40**, 205-220.

Fortini, S. and Ruggeri, F. (2000). On the use of the concentration function in Bayesian robustness. In: Ríos Insua, D. and Ruggeri, F. (eds) *Robust Bayesian Analysis*, 109-126. Springer-Verlag, New York.

Fortini, S., Ladelli, L. and Regazzini, E. (2000). Exchangeability, predictive distributions and parametric models. *Sankhyā* (Ser. A) **62**, 86-109.

Fraser, D.A.S., Monette, G. and Ng, K.W. (1995). Marginalization, likelihood and structured models. In: Krishnaiah, P. (ed) *Multivariate Analysis*, **6**, 209-217. North-Holland, Amsterdam.

Freedman, D.A. (1963). On the asymptotic behavior of Bayes estimates in the discrete case. *Ann. Math. Statist.* **34**, 1386-1403.

Freedman, D.A. (1965). On the asymptotic behavior of Bayes estimates in the discrete case. II. *Ann. Math. Statist.* **36**, 454-456.

Freedman, D.A. and Purves, R.A. (1969). Bayes methods for bookies. *Ann. Math. Statist.* **40**, 1177-1186.

French, S. (1986). *Decision Theory*. Ellis Horwood Limited, Chichester, England.

French, S. and Ríos Insua, D. (2000). *Statistical Decision Theory*. Oxford Univ. Press, New York.

Gardner, M. (1997). *Relativity Simply Explained*. Dover, Mineola, New York.

Garthwaite, P.H. and Dickey, J.M. (1988). Quantifying expert opinion in linear regression problems. *J. Roy. Statist. Soc.* (Ser. B) **50**, 462-474.

Garthwaite, P.H. and Dickey, J.M. (1992). Elicitation of prior distributions for variable-selection problems in regression. *Ann. Statist.* **20**, 1697-1719.

Garthwaite, P.H., Kadane, J.B., and O'Hagan, A. (2005). Statistical methods for eliciting probability distributions. *J. Amer. Statist. Assoc.* **100**, 680-701.

Gelfand, A.E. and Dey, D.K. (1991). On Bayesian robustness of contaminated classes of priors. *Statist. Decisions* **9**, 63-80.

Gelfand, A.E. and Dey, D.K. (1994). Bayesian model choice: Asymptotics and exact calculations. *J. Roy. Statist. Soc.* (Ser. B) **56**, 501-514.

Gelfand, A.E. and Ghosh, S.K. (1998). Model choice: a minimum posterior predictive loss approach. *Biometrika* **85**, 1-11.

Gelfand, A.E. and Smith, A.F.M. (1990). Sampling based approaches to calculating marginal densities. *J. Amer. Statist. Assoc.* **85**, 398-409.

Gelman, A., Carlin, J.B., Stern, H.S., and Rubin, D.B. (1995). *Bayesian Data Analysis*. Chapman & Hall, London.

Gelman, A., Meng, X., and Stern, H. (1996). Posterior predictive assessment of model fitness via realized discrepancies (with discussion). *Statist. Sinica* **6**, 733-807.

Genovese, C. and Wasserman, L. (2001). Operating characteristics and extensions of the FDR procedure. Tech. Report, Carnegie Mellon Univ.

Genovese, C. and Wasserman, L. (2002). Operating characteristics and extensions of the false discovery rate procedure. *J. Roy. Statist. Soc.* (Ser. B) **64**, 499-517.

George, E.I. and Foster, D.P. (2000). Calibration and empirical Bayes variable selection. *Biometrika* **87**, 731-747.

Geweke, J. (1999). Simulation methods for model criticism and robustness analysis. In: Bernardo, J.M. et al. (eds) *Bayesian Statistics* **6**, 275-299. Oxford Univ. Press, New York.

Ghosal, S. (1997). Normal approximation to the posterior distribution for generalized linear models with many covariates. *Math. Methods Statist.* **6**, 332-348.

Ghosal, S. (1999). Asymptotic normality of posterior distributions in high-dimensional linear models. *Bernoulli* **5**, 315-331.

Ghosal, S. (2000). Asymptotic normality of posterior distributions for exponential families when the number of parameters tends to infinity. *J. Multivariate Anal.* **74**, 49-68.

Ghosal, S., Ghosh, J.K. and Ramamoorthi, R.V. (1997). Noninformative priors via sieves and packing numbers. In: Panchapakesan, S. and Balakrishnan, N. (eds) *Advances in Statistical Decision Theory and Applications*, 119-132. Birkhäuser, Boston.

Ghosal, S., Ghosh, J.K. and Samanta, T. (1995). On convergence of posterior distributions. *Ann. Statist.* **23**, 2145-2152.

Ghosh, J.K. (1983). Review of *"Approximation Theorems of Mathematical Statistics"* by R.J. Serfling. *J. Amer. Statist. Assoc.* **78**.

Ghosh, J.K. (1994). *Higher Order Asymptotics*. NSF-CBMS Regional Conference Series in Probability and Statistics. IMS, Hayward.

Ghosh, J.K. (1997). Discussion of "Noninformative priors do not exist: a dialogue with J.M. Bernardo". *Jour. Statist. Plann. Inference* **65**, 159-189.

Ghosh, J.K. (2002). Review of *"Statistical Inference in Science"* by D.A. Sprott. *Sankhyā*. (Ser. B) **64**, 234-235.

Ghosh, J.K., Bhanja, J., Purkayastha, S., Samanta, T. and Sengupta, S. (2002). A statistical approach to geological mapping. *Mathematical Geology* **34**, 505-528.

Ghosh, J.K. and Mukerjee, R. (1992). Non-informative priors (with discussion). In: Bernardo, J. M. et al. (eds). *Bayesian Statistics* **4**, 195-210. Oxford Univ. Press, London.

Ghosh, J.K. and Mukerjee, R. (1993). On priors that match posterior and frequentist distribution functions. *Can. J. Statist.* **21**, 89-96.

Ghosh, J.K. and Ramamoorthi, R.V. (2003). *Bayesian Nonparametrics*. Springer, New York.

Ghosh, J.K. and Samanta, T. (2001). Model selection - an overview. *Current Science* **80**, 1135-1144.

Ghosh, J.K. and Samanta, T. (2002a). Nonsubjective Bayes testing – an overview. *J. Statist. Plann. Inference* **103**, 205-223.

Ghosh, J.K. and Samanta, T. (2002b). Towards a nonsubjective Bayesian paradigm. In: Misra, J.C. (ed) *Uncertainty and Optimality*, 1-69. World Scientific, Singapore.

Ghosh, J.K., Ghosal, S. and Samanta, T. (1994). Stability and convergence of posterior in non-regular problems. In: Gupta, S.S. and Berger, J.O. (eds). *Statistical Decision Theory and Related Topics*, V, 183-199.

Ghosh, J.K., Purkayastha, S. and Samanta, T. (2005). Role of P-values and other measures of evidence in Bayesian analysis. In: Dey, D.K. and Rao, C.R. (eds) *Handbook of Statistics* **25**, Bayesian Thinking: Modeling and Computation, 151-170.

Ghosh, J.K., Sinha, B.K. and Joshi, S.N. (1982). Expansion for posterior probability and integrated Bayes risk. In: Gupta, S.S. and Berger, J.O. (eds) *Statistical Decision Theory and Related Topics*, III, **1**, 403-456.

Ghosh, M. and Meeden, G. (1997). *Bayesian Methods for Finite Population Sampling*. Chapman & Hall, London.

Goel, P. (1983). Information measures and Bayesian hierarchical models. *J. Amer. Statist. Assoc.* **78**, 408-410.

Goel, P. (1986). Comparison of experiments and information in censored data. In: Gupta, S.S. and Berger, J.O. (eds) *Statistical Decision Theory and Related Topics*, IV. **2**, 335-349.

Good, I.J. (1950). *Probability and the Weighing of Evidence*. Charles Griffin, London.

Good, I.J. (1958). Significance tests in parallel and in series. *J. Amer. Statist. Assoc.* **53**, 799-813.

Good, I.J. (1965). *The Estimation of Probabilities: An Essay on Modern Bayesian Methods.* M.I.T. Press, Cambridge, Massachusetts.

Good, I.J. (1967). A Bayesian significance test for the multinomial distribution. *J. Roy. Statist. Soc.* (Ser. B) **29**, 399-431.

Good, I.J. (1975). The Bayes factor against equiprobability of a multinomial population assuming a symmetric Dirichlet prior. *Ann. Statist.* **3**, 246-250.

Good, I.J. (1983). *Good Thinking: The Foundations of Probability and its Applications.* Univ. Minnesota Press, Minneapolis.

Good, I.J. (1985). Weight of evidence: A brief survey. In: Bernardo, J.M. et al. (eds) *Bayesian Statistics* **2**, 249-270. North-Holland, Amsterdam.

Good, I.J. (1986). A flexible Bayesian model for comparing two treatments. *J. Statist. Comput. Simulation* **26**, 301-305.

Good, I.J. and Crook, J.F. (1974). The Bayes/non-Bayes compromise and the multinomial distribution. *J. Amer. Statist. Assoc.* **69**, 711-720.

Green, P.J. (1995). Reversible jump MCMC computation and Bayesian model determination. *Biometrika* **82**, 711-732.

Green, P.J. and Richardson, S. (2001). Modelling heterogeneity with and without the Dirichlet process. *Scand. J. Statist.* **28**, 355-375.

Gustafson, P. (2000). Local robustness in Bayesian analysis. In: Ríos Insua, D. and Ruggeri, F. (eds) *Robust Bayesian Analysis*, 71-88, Springer-Verlag, New York.

Gustafson, P. and Wasserman, L. (1995). Local sensitivity diagnostics for Bayesian inference. *Ann. Statist.* **23**, 2153-2167.

Guttman, I. (1967). The use of the concept of a future observation in goodness-of-fit problems. *J. Roy. Statist. Soc.* (Ser. B) **29**, 104-109.

Hájek, J. and Sidák, Z.V.(1967). *Theory of Rank Tests.* Academic Press, New York.

Halmos, P.R. (1950). *Measure Theory.* van Nostrand, New York.

Halmos, P.R. (1974). *Measure Theory*, 2nd Ed. Springer-Verlag, New York.

Hartigan J.A. (1983). *Bayes Theory.* Springer-Verlag, New York.

Heath, D. and Sudderth, W. (1978). On finitely additive priors, coherence, and extended admissibility. *Ann. Statist.* **6**, 335-345.

Heath, D. and Sudderth, W. (1989). Coherent inference from improper priors and from finitely additive priors. *Ann. Statist.* **17**, 907-919.

Hernández, E. and Weiss, G. (1996). *A First Course on Wavelets.* CRC Press Inc., Boca Raton.

Hewitt, E. and Savage, L.J. (1955). Symmetric measures on Cartesian products. *Trans. Amer. Math. Soc.* **80**, 907-919.

Hildreth, C. (1963). Bayesian statisticians and remote clients. *Econometrica* **31**, 422-438.

Hill, B. (1982). Comment on "Lindley's paradox," by G. Shafer. *J. Amer. Statist. Assoc.* **77**, 344-347.

Hoeting, J.A., Madigan, D., Raftery, A.E. and Volinsky, C.T. (1999). Bayesian model averaging: a tutorial (with discussion). *Statist. Sci.* **14**, 382-417.

Huber. P.J. (1964). Robust estimation of a location parameter. *Ann. Math. Statist.* **35**, 73-101.

Huber. P.J. (1974). Fisher information and spline interpolation. *Ann. Statist.* **2**, 1029-1034.

Huber. P.J. (1981). *Robust Statistics.* John Wiley, New York.

Hwang, J.T., Casella, G., Robert, C., Wells, M.T. and Farrell, R.H. (1992). Estimation of accuracy in testing. *Ann. Statist.* **20**, 490-509

Ibragimov, I.A. and Has'minskii, R.Z. (1981). *Statistical Estimation - Asymptotic Theory.* Springer-Verlag, New York.

Ickstadt, K. (1992). Gamma-minimax estimators with respect to unimodal priors. In: Gritzmann, P. et al. (eds) *Operations Research '91.* Physica-Verlag, Heidelberg.

James, W. and Stein, C. (1960). Estimation with quadratic loss. *Proc. Fourth Berkeley Symp. Math. Statist. Probab.* **1**, 361-380. Univ. California Press, Berkeley.

Jeffreys, H. (1946). An invariant form for the prior probability in estimation problems. *Proc. Roy. Soc. London* (Ser. A) **186**, 453-461.

Jeffreys, H. (1957). *Scientific Inference.* Cambridge Univ. Press, Cambridge.

Jeffreys, H. (1961). *Theory of Probability,* 3rd Ed. Oxford Univ. Press, New York.

Johnson, R.A. (1970). Asymptotic expansions associated with posterior distribution. *Ann. Math. Statist.* **42**, 1899-1906.

Kadane, J.B. (ed) (1984). *Robustness of Bayesian Analyses.* Studies in Bayesian Econometrics, **4**. North-Holland Publishing Co., Amsterdam.

Kadane, J.B., Dickey, J.M., Winkler, R.L., Smith, W.S. and Peters, S.C. (1980). Interactive elicitation of opinion for a normal linear model. *J. Amer. Statist. Assoc.* **75**, 845-854.

Kadane, J.B., Salinetti, G. and Srinivasan, C . (2000). Stability of Bayes decisions and applications. In: Ríos Insua, D. and Ruggeri, F. (eds) *Robust Bayesian Analysis,* 187-196, Springer-Verlag, New York.

Kadane, J.B., Schervish, M.J. and Seidenfeld, T. (1999). *Cambridge Studies in Probability, Induction, and Decision Theory.* Cambridge Univ. Press, Cambridge.

Kagan, A.M., Linnik, Y.V., Rao, C.R. (1973). *Characterization Problems in Mathematical Statistics.* Wiley, New York.

Kahneman, D., Slovic, P. and Tversky, A. (1982). *Judgement Under Uncertainty: Heuristics and Biases.* Cambridge Univ. Press, New York.

Kariya, T. and Sinha, B.K. (1989). *Robustness of Statistical Tests.* Statistical Modeling and Decision Science. Academic Press, Boston, MA.

Kass, R. and Raftery, A. (1995). Bayes factors. *J. Amer. Statist. Assoc.* **90**, 773-795.

Kass, R. and Wasserman, L. (1996). The selection of prior distributions by formal rules (review paper). *J. Amer. Statist. Assoc.* **91**, 1343-1370.

Kass, R.E., Tierney, L. and Kadane, J.B. (1988). Asymptotics in Bayesian computations. In: Bernardo, J.M. et al. (eds) *Bayesian Statistics* **3**, 261-278. Oxford Univ. Press, Oxford.

Kiefer, J. (1957). Invariance, minimax sequential estimation, and continuous time processes. *Ann. Math. Statist.* **28**, 573-601.

Kiefer, J. (1966). Multivariate optimality results. In: Krishnaiah, P.R. (ed) *Multivariate Analysis.* Academic Press, New York.

Kiefer, J. (1977). Conditional confidence statements and confidence estimators (with discussion). *J. Amer. Statist. Assoc.* **72**, 789-827.

Kiefer, J. and Wolfowitz, J. (1956). Consistency of the maximum likelihood estimator in the presence of infinitely many incidental parameters. *Ann. Math. Statist.* **27**, 887-906.

Laplace, P.S. (1986). Memoir on the probability of the causes of events (English translation of the 1774 French original by S.M. Stigler). *Statist. Sci.* **1**, 364-378.

Lavine, M. (1991). Sensitivity in Bayesian statistics: the prior and the likelihood. *J. Amer. Statist. Assoc.* **86**, 396-399.

Lavine, M., Pacifico, M.P., Salinetti, G. and Tardella, L. (2000). Linearization techniques in Bayesian robustness. In: Ríos Insua, D. and Ruggeri, F. (eds) *Robust Bayesian Analysis*, 261-272, Springer-Verlag, New York.

Leamer, E.E. (1978). *Specification Searches*. Wiley, New York.

Leamer, E.E. (1982). Sets of posterior means with bounded variance prior. *Econometrica* **50**, 725-736.

Le Cam, L. (1953). *On Some Asymptotic Properties of Maximum Likelihood Estimates and Related Bayes Estimates*. Univ. California Publications in Statistics, **1**, 277-330.

Le Cam, L. (1958). Les properties asymptotiques des solutions de Bayes. *Publ. Inst. Statist. Univ. Paris*, **7**, 17-35.

Le Cam, L. (1986). *Asymptotic Methods in Statistical Decision Theory*. Springer-Verlag, New York.

Le Cam, L. and Yang, G.L. (2000). *Asymptotics in Statistics: Some Basic Concepts*, 2nd Ed. Springer-Verlag, New York.

Lee, P.M. (1989). *Bayesian Statistics: An Introduction*. Oxford Univ. Press, New York.

Lehmann, E.L. (1986). *Testing Statistical Hypotheses*, 2nd Ed. Wiley, New York.

Lehmann, E.L. and Casella, G. (1998). *Theory of Point Estimation*, 2nd Ed. Springer-Verlag, New York.

Lempers, F.B. (1971). *Posterior Probabilities of Alternative Models*. Univ. of Rotterdam Press, Rotterdam.

Leonard, T. and Hsu. J.S.J. (1999). *Bayesian Methods*. Cambridge Univ. Press, Cambridge.

Li, K.C. (1987). Asymptotic optimality for c_p, c_l, cross validation and generalized cross validation: discrete index set. *Ann. Statist.* **15**, 958-975.

Liang, F., Paulo, R., Molina, G., Clyde, M.A. and Berger, J.O. (2005) Mixtures of g-priors for Bayesian variable selection. Unpublished manuscript.

Lindley, D.V. (1956). On a measure of the information provided by an experiment. *Ann. Math. Statist.* **27**, 986-1005.

Lindley, D.V. (1957). A statistical paradox. *Biometrika* **44**, 187-192.

Lindley, D.V. (1961). The use of prior probability distributions in statistical inference and decisions. *Proc. Fourth Berkeley Symp. Math. Statist. Probab.* **1**, 453-468. Univ. California Press, Berkeley.

Lindley, D.V. (1965). *An Introduction to Probability and Statistics from a Bayesian Viewpoint*, **1, 2**. Cambridge Univ. Press, Cambridge.

Lindley, D.V. (1977). A problem in forensic science. *Biometrika* **64**, 207-213.

Lindley, D.V. and Phillips, L.D. (1976). Inference for a Bernoulli process (a Bayesian view). *Amer. Statist.* **30**, 112-119.

Lindley, D.V. and Smith, A.F.M. (1972). Bayes estimates for the linear model. *J. Roy. Statist. Soc.* (Ser. B) **34**, 1-41.

Liseo, B. (2000). Robustness issues in Bayesian model selection. In: Ríos Insua, D. and Ruggeri, F. (eds) *Robust Bayesian Analysis*, 197-222, Springer-Verlag, New York.

Liseo, B., Pettrella, L. and Salinetti, G. (1996). Robust Bayesian analysis: an interactive approach. In: Bernardo, J.M. et al. (eds) *Bayesian Statistics*, **5**, 661-666. Oxford Univ. Press, London.

Liu, R.C. and Brown, L.D. (1992). Nonexistence of informative unbiased estimators in singular problems. *Ann. Statist.* **21**, 1-13.

Lu, K. and Berger, J.O. (1989a). Estimated confidence procedures for multivariate normal means. *J. Statist. Plann. Inference* **23**, 1-19.

Lu, K. and Berger, J.O. (1989b). Estimation of normal means: frequentist estimators of loss. *Ann. Statist.* **17**, 890-907.

Madigan, D. and Raftery, A.E. (1994). Model selection and accounting for model uncertainty in graphical models using Occam's window. *J. Amer. Statist. Assoc.* **89**, 1535-1546.

Madigan, D. and York, J. (1995). Bayesian graphical models for discrete data. *Int. Statist. Rev.* **63**, 215-232.

Martin, J.T. (1942). The problem of the evaluation of rotenone-containing plants. VI. The toxicity of l-elliptone and of poisons applied jointly, with further observations on the rotenone equivalent method of assessing the toxicity of derris root. *Ann. Appl. Biol.* **29**, 69-81.

Martín, J. and Arias, J.P. (2000). Computing efficient sets in Bayesian decision problems. In: Ríos Insua, D. and Ruggeri, F. (eds) *Robust Bayesian Analysis*, 161-186. Springer-Verlag, New York.

Martín, J., Ríos Insua, D. and Ruggeri, F. (1998). Issues in Bayesian loss robustness. *Sankhyā* (Ser. A) **60**, 405-417.

Matsumoto, M. and Nishimura, T. (1998). Mersenne Twister: A 623-dimensionally equidistributed uniform pseudorandom number generator. *ACM Trans. on Modeling and Computer Simulation* **8**, 3-30.

McLachlan, G.J. and Krishnan, T. (1997). *The EM Algorithm and Extensions.* Wiley, New York.

Meng, X.L. (1994). Posterior predictive p-values. *Ann. Statist.* **22**, 1142-1160.

Meyn, S.P. and Tweedie, R.L. (1993). *Markov Chains and Stochastic Stability.* Springer-Verlag, New York.

Moreno, E. (2000). Global Bayesian robustness for some classes of prior distributions. In: Ríos Insua, D. and Ruggeri, F. (eds) *Robust Bayesian Analysis*, 45-70, Springer-Verlag, New York.

Moreno, E. and Cano, J.A. (1995). Classes of bidimensional priors specified on a collection of sets: Bayesian robustness. *J. Statist. Plann. Inference* **46**, 325-334.

Moreno, E. and Pericchi, L.R. (1993). Bayesian robustness for hierarchical ϵ-contamination models. *J. Statist. Plann. Inference* **37**, 159-167.

Morris, C.N. (1983). Parametric empirical Bayes inference: Theory and applications (with discussion). *J. Amer. Statist. Assoc.* **78**, 47-65.

Morris, C.N. and Christiansen, C.L. (1996). Hierarchical models for ranking and for identifying extremes, with application. In: Bernardo, J.M. et al. (eds) *Bayesian Statistics* **5**, 277-296. Oxford Univ. Press, Oxford.

Mosteller, F. and Tukey, J.W. (1977). *Data Analysis and Regression.* Addison-Wessley, Reading.

Muirhead, R.J. (1982). *Aspects of Multivariate Statistical Theory.* Wiley, New York.

Mukhopadhyay, N. (2000). Bayesian model selection for high dimensional models with prediction error loss and $0-1$ loss. Ph.D. thesis, Purdue Univ.

Mukhopadhyay, N. and Ghosh, J.K. (2004a). Parametric empirical Bayes model selection - some theory, methods and simulation. In: Bhattacharya, R.N. et al. (eds) *Probability, Statistics and Their Applications: Papers in Honor of Rabi Bhattacharya.* IMS Lecture Notes-Monograph Ser. **41**, 229-245.

Mukhopadhyay, N. and Ghosh, J.K. (2004b). Bayes rule for prediction and AIC, an asymptotic evaluation. Unpublished manuscript.

Müller, P. and Vidakovic, B. (eds) (1999). *Bayesian Inference in Wavelet-Based Models*. Lecture Notes in Statistics, **141**. Springer, New York.

Nachbin, L. (1965). *The Haar Integral*. van Nostrand, New York.

Neyman, J. and Scott, E. L. (1948). Consistent estimates based on partially consistent observations. *Econometrica* **16**, 1-32.

Newton, M.A., Yang, H., Gorman, P.A., Tomlinson, I. and Roylance, R.R. (2003). A statistical approach to modeling genomic aberrations in cancer cells (with discussion). In: Bernardo, J.M. et al. (eds) *Bayesian Statistics* **7**, 293-305. Oxford Univ. Press, New York.

Ogden, R.T. (1997). *Essential Wavelets for Statistical Applications and Data Analysis*. Birkhäuser, Boston.

O'Hagan, A. (1988). Modelling with heavy tails. In: Bernardo, J.M. et al. (eds) *Bayesian Statistics* **3**, 569-577. Oxford Univ. Press, Oxford.

O'Hagan, A. (1990). Outliers and credence for location parameter inference. *J. Amer. Statist. Assoc.* **85**, 172-176.

O'Hagan, A. (1994). *Bayesian Inference*. Kendall's Advanced Theory of Statistics, Vol. 2B. Halsted Press, New York.

O'Hagan, A. (1995). Fractional Bayes factors for model comparisons. *J. Roy. Statist. Soc.* (Ser. B) **57**, 99-138.

Pearson, Karl (1892). *The Grammar of Science*. Walter Scott, London. Latest Edition: (2004). Dover Publications.

Pericchi, L.R. and Pérez, M.E. (1994). Posterior robustness with more than one sampling model (with discussion). *J. Statist. Plann. Inference* **40**, 279-294.

Perone P.M., Salinetti, G. and Tardella, L. (1998). A note on the geometry of Bayesian global and local robustness. *J. Statist. Plann. Inference* **69**, 51-64.

Pettit, L.I. and Young, K.D.S. (1990). Measuring the effect of observations on Bayes factors. *Biometrika* **77**, 455-466.

Pitman, E.J.G. (1979). *Some Basic Theory for Statistical Inference*. Chapman & Hall, London; A Halsted Press Book, Wiley, New York.

Polasek, W. (1985). Sensitivity analysis for general and hierarchical linear regression models. In: Goel, P.K. and Zellner, A. (eds). *Bayesian Inference and Decision Techniques with Applications*. North-Holland, Amsterdam.

Pratt, J.W. (1961). Review of *"Testing Statistical Hypotheses"* by E.L. Lehmann. *J. Amer. Statist. Assoc.* **56**, 163-166.

Pratt, J.W. (1965). Bayesian interpretation of standard inference statements (with discussion). *J. Roy. Statist. Soc.* (Ser. B) **27**, 169-203.

Raftery, A.E., Madigan, D. and Hoeting, J.A. (1997). Bayesian model averaging for linear regression models. *J. Amer. Statist. Assoc.* **92**, 179-191.

Raiffa, H. and Schlaiffer, R. (1961). *Applied Statistical Decision Theory*. Division of Research, School of Business Administration, Harvard Univ.

Ramsey, F.P. (1926). *Truth and probability*. Reprinted in: Kyburg, H.E. and Smokler, H.E. (eds) *Studies in Subjective Probability*. Wiley, New York.

Rao, C.R. (1973). *Linear Statistical Inference and Its Applications*, 2nd Ed. Wiley, New York.

Rao, C.R. (1982). Diversity: its measurement, decomposition, apportionment and analysis. *Sankhyā* (Ser. A) **44**, 1-22.

Richardson, S, and Green, P.J. (1997). On Bayesian analysis of mixtures with an unknown number of components (with discussion). *J. Roy. Statist. Soc.* (Ser. B) **59**, 731-792.

Rissanen, J. (1987). Stochastic complexity. *J. Roy. Statist. Soc.* (Ser. B) **49**, 223-239.

Ríos Insua, D. and Criado, R. (2000). Topics on the foundations of robust Bayesian analysis. In: Ríos Insua, D. and Ruggeri, F. (eds) *Robust Bayesian Analysis*, 33-44. Springer-Verlag, New York.

Ríos Insua, D. and Martn, J. (1994). Robustness issues under imprecise beliefs and preferences. *J. Statist. Plann. Inference* **40**, 383-389.

Ríos Insua, D. and Ruggeri, F. (eds) (2000). *Robust Bayesian Analysis*. Lecture Notes in Statistics, **152**. Springer-Verlag, New York.

Robbins, H. (1951). Asymptotically subminimax solutions of compound statistical decision problems. *Proc. Second Berkeley Symp. Math. Statist. Probab.* 131-148. Univ. California Press, Berkeley.

Robbins, H. (1955). An empirical Bayes approach to statistics. *Proc. Third Berkeley Symp. Math. Statist. Probab.* **1**, 157-164. Univ. California Press, Berkeley.

Robbins, H. (1964). The empirical Bayes approach to statistical decision problems. *Ann. Math. Statist.* **35**, 1-20.

Robert, C.P. (1994). *The Bayesian Choice.* Springer-Verlag, New York.

Robert, C.P. (2001). *The Bayesian Choice*, 2nd Ed. Springer-Verlag, New York.

Robert, C.P. and Casella, G. (1999). *Monte Carlo Statistical Methods.* Springer-Verlag, New York.

Robinson, G.K. (1976). Conditional properties of Student's t and of the Behrens-Fisher solution to the two means problem. *Ann. Statist.* **4**, 963-971.

Robinson, G.K. (1979). Conditional properties of statistical procedures. *Ann. Statist.* **7**, 742-755. Conditional properties of statistical procedures for location and scale parameters. *Ann. Statist.* **7**, 756-771.

Roussas, G.G. (1972). *Contiguity of Probability Measures: Some Applications in Statistics.* Cambridge Tracts in Mathematics and Mathematical Physics, **63**. Cambridge Univ. Press, London-New York.

Rousseau, J. (2000). Coverage properties of one-sided intervals in the discrete case and application to matching priors. *Ann. Inst. Statist. Math.* **52**, 28-42.

Rubin, D.B. (1984). Bayesianly justifiable and relevant frequency calculations for the applied statistician. *Ann. Statist.* **12**, 1151-1172.

Rubin, H. (1971). A decision-theoretic approach to the problem of testing a null hypothesis. In: Gupta, S.S. and Yackel, J. (eds) *Statistical Decision Theory and Related Topics.* Academic Press, New York.

Rubin, H. and Sethuraman, J. (1965). Probabilities of moderate deviations. *Sankhyā* (Ser. A) **27**, 325-346.

Ruggeri, F. and Sivaganesan, S. (2000). On a global sensitivity measure for Bayesian inference. *Sankhyā* (Ser. A) **62**, 110-127.

Ruggeri, F. and Wasserman, L. (1995). Density based classes of priors: infinitesimal properties and approximations. *J. Statist. Plann. Inference* **46**, 311-324.

Sarkar, S. (2003). FDR-controlling stepwise procedures and their false negative rates. Tech. Report, Temple Univ.

Satterwaite, F.E. (1946). An approximate distribution of estimates of variance components. *Biometrics Bull.* (now called *Biometrics*), **2**, 110-114.

Savage, L.J. (1954). *The Foundations of Statistics.* Wiley, New York.

Savage, L.J. (1972). *The Foundations of Statistics*, 2nd revised Ed. Dover, New York.

Schervish, M.J. (1995). *Theory of Statistics.* Springer-Verlag, New York.

Schwarz, G. (1978). Estimating the dimension of a model. *Ann. Statist.* **6**, 461-464.

Scott, J.G. and Berger, J.O. (2005). An exploration of aspects of Bayesian multiple testing. To appear in *J. Statist. Plan. Inference.*

Searle, S.R. (1982). *Matrix Algebra Useful for Statistics.* Wiley, New York.

Seidenfeld, T., Schervish, M.J. and Kadane, J.B. (1995). A representation of partially ordered preferences. *Ann. Statist.* **23**, 2168-2217.

Seo, J. (2004). Some classical and Bayesian nonparametric regression methods in a longitudinal marginal model. Ph.D. thesis, Purdue Univ.

Serfling, R.J. (1980). *Approximation Theorems of Mathematical Statistics.* Wiley, New York.

Shafer, G. (1976). *A Mathematical Theory of Evidence.* Princeton Univ. Press, NJ.

Shafer, G. (1979). Allocations of Probability. *Ann. Probab.* **7**, 827-839.

Shafer, G. (1982). Lindley's paradox (with discussion). *J. Amer. Statist. Assoc.* **77**, 325-351.

Shafer, G. (1982). Belief functions and parametric models (with discussion). *J. Roy. Statist. Soc.* (Ser. B) **44**, 322-352.

Shafer, G. (1987). Probability judgment in artificial intelligence and expert systems (with discussion). *Statist. Sci.* **2**, 3-44.

Shannon, C.E. (1948). A mathematical theory of communication. *Bell System Tech. J.* **27**, 379-423 and 623-656. Reprinted in *The Mathematical Theory of Communication* (Shannon, C.E. and Weaver, W., 1949). Univ. Illinois Press, Urbana, IL.

Shao, J. (1997). An asymptotic theory for linear model selection. *Statist. Sinica* **7**, 221-264.

Shibata, R. (1981). An optimal selection of regression variables. *Biometrika* **68**, 45-54.

Shibata, R. (1983). Asymptotic mean efficiency of a selection of regression variables. *Ann. Inst. Statist. Math.* **35**, 415-423.

Shyamalkumar, N.D. (2000). Likelihood robustness. In: Ríos Insua, D. and Ruggeri, F. (eds) *Robust Bayesian Analysis*, 109-126, Springer-Verlag, New York.

Sivaganesan, S. (1989). Sensitivity of posterior mean to unimodality preserving contaminations. *Statist. and Decisions* **7**, 77-93.

Sivaganesan, S. (1993). Robust Bayesian diagnostics. *J. Statist. Plann. Inference* **35**, 171-188.

Sivaganesan, S. (2000). Global and local robustness approaches: uses and limitations. In: Ríos Insua, D. and Ruggeri, F. (eds) *Robust Bayesian Analysis*, 89-108, Springer-Verlag, New York.

Sivaganesan, S. and Berger, J.O. (1989). Ranges of posterior means for priors with unimodal contaminations. *Ann. Statist.* **17**, 868-889.

Smith, A.F.M. and Roberts, G.O. (1993). Bayesian computation via the Gibbs sampler and related Markov chain Monte Carlo methods (with discussion). *J. Roy. Statist. Soc.* (Ser. B) **55**, 3-24.

Smith, A.F.M. and Spiegelhalter, D.J. (1980). Bayes factors and choice criteria for linear models. *J. Roy. Statist. Soc.* (Ser. B) **42**, 213-220.

Smith, C.A.B. (1965). Personal probability and statistical analysis. *J. Roy. Statist. Soc.* (Ser. A) **128**, 469-499.

Sorensen, D. and Gianola, D. (2002). *Likelihood, Bayesian, and MCMC Methods in Quantitative Genetics.* Springer-Verlag, New York.

Spiegelhalter, D.J., Best, N.G., Carlin, B.P. and van der Linde, A. (2002). Bayesian measures of model complexity and fit. *J. Roy. Statist. Soc.* (Ser. B) **64**, 583-639.

Spiegelhalter, D.J. and Smith, A.F.M. (1982). Bayes factors for linear and log-linear models with vague prior information. *J. Roy. Statist. Soc.* (Ser. B) **44**, 377-387.

Sprott, D.A. (2000). *Statistical Inference in Science.* Springer-Verlag, New York.

Srinivasan, C. (1981). Admissible generalized Bayes estimators and exterior boundary value problems. *Sankhyā* (Ser. A) **43**, 1-25.

Stein, C. (1955). Inadmissibility of the usual estimator for the mean of a multivariate normal distribution. *Proc. Third Berkeley Symp. Math. Statist. Probab.* **1**, 197-206. Univ. California Press, Berkeley.

Stein, C. (1956). Some problems in multivariate analysis, Part I. Tech. Report, No. 6, Dept. Statistics, Stanford Univ.

Stein, C. (1981). Estimation of the mean of a multivariate normal distribution. *Ann. Statist.* **9**, 1135-1151.

Stigler, S.M. (1977). Do robust estimators work with real data? (with discussion). *Ann. Statist.* **5**, 1055-1098.

Stone, M. (1979). Comments on model selection criteria of Akaike and Scwarz. *J. Roy. Stat. Soc.* (Ser. B) **41**, 276-278.

Storey, J.D. (2002). A direct approach to false discovery rates. *J. Roy. Stat. Soc.* (Ser. B) **64**, 479-498.

Storey, J.D. (2003). The positive false discovery rate: a Bayesian interpretation and the q-value. *Ann. Statist.* **31**, 2013–2035.

Strawderman, W.E. (1971). Proper Bayes minimax estimators of the multivariate normal mean. *Ann. Math. Statist.* **42**, 385-388.

Strawderman, W.E. (1978). Minimax adaptive generalized ridge regression estimators. *J. Amer. Statist. Assoc.* **73**, 623-627.

Stromborg, K.L., Grue, C.E., Nichols, J.D., Hepp, G.R., Hines, J.E. and Bourne, H.C. (1988). Postfledging survival of European starlings exposed as nestlings to an organophosphorus insecticide. *Ecology* **69**, 590-601.

Sun, D. and Berger, J.O. (1998). Reference priors with partial information. *Biometrika* **85**, 55-71.

Tanner, M.A. (1991). *Tools for Statistical Inference.* Springer-Verlag, New York.

Tierney, L. (1994). Markov chains for exploring posterior distributions (with discussion). *Ann. Statist.* **22**, 1701-1762.

Tierney, L. and Kadane, J.B. (1986). Accurate approximations for posterior moments. *J. Amer. Statist. Assoc.* **81**, 82-86.

Tierney, L., Kass, R.E. and Kadane, J.B. (1989). Fully exponential Laplace approximations to expectations and variances of nonpositive functions. *J. Amer. Statist. Assoc.* **84**, 710-716.

Tversky, A. and Kahneman, D. (1981). The framing of decisions and the psychology of choice. *Science* **211**, 453-458.

Vidakovic, B. (1999). *Statistical Modeling by Wavelets.* Wiley, New York.

Vidakovic, B. (2000). Γ-Minimax: A paradigm for conservative robust Bayesian. In: Ríos Insua, D. and Ruggeri, F. (eds) *Robust Bayesian Analysis*, 241-259, Springer-Verlag, New York.

von Mises, R. (1931). *Wahrscheinlichkeitsrechnung.* Springer, Berlin.

von Mises, R. (1957). *Probability, Statistics and Truth,* 2nd revised English Ed. (prepared by Hilda Geiringer). The Macmillan Company, New York.

Waagepetersen, R. and Sorensen, D. (2001). A tutorial on reversible jump MCMC with a view towards applications in QTL-mapping. *Int. Statist. Rev.* **69**, 49-61.

Wald, A. (1950). *Statistical Decision Functions*. Wiley, New York and Chapman & Hall, London.

Walker, A.M. (1969). On the asymptotic behaviour of posterior distributions. *J. Roy. Statist. Soc.* (Ser. B) **31**, 80-88.

Walley, P. (1991). *Statistical Reasoning with Imprecise Probabilities*. Chapman & Hall, London.

Wasserman, L. (1990). Prior envelopes based on belief functions. *Ann. Statist.* **18**, 454-464.

Wasserman, L. (1992). Recent methodological advances in robust Bayesian inference. In: Bernardo, J.M. et al. (eds) *Bayesian Statistics* **4**, 35-60. Oxford Univ. Press, Oxford.

Wasserman, L. and Kadane, J. (1990). Bayes' theorem for Choquet capacities. *Ann. Statist.* **18**, 1328-1339.

Wasserman, L., Lavine, M. and Wolpert, R.L. (1993). Linearization of Bayesian robustness problems. *J. Statist. Plann. Inference* **37**, 307-316.

Weiss, R. (1996). An approach to Bayesian sensitivity analysis. *J. Roy. Statist. Soc.* (Ser. B) **58**, 739-750.

Welch, B.L. (1939). On confidence limits and sufficiency with particular reference to parameters of location. *Ann. Math. Statist.* **10**, 58-69.

Welch, B.L. (1949). Further notes on Mrs. Aspin's tables. *Bimometrika* **36**, 243-246.

Welch, B.L. and Peers, H.W. (1963). On formulae for confidence points based on integrals of weighted likelihoods. *J. Roy. Statist. Soc. B*, **25**, 318-329.

Wijsman, R.A. (1967). Cross-sections of orbits and their application to densities of maximal invariants. *Proc. Fifth Berkeley Symp. Math. Statist. Probab.* **1**, 389-400. Univ. California Press, Berkeley.

Wijsman, R.A. (1985). Proper action in steps, with application to density ratios of maximal invariants. *Ann. Statist.* **13**, 395-402.

Wijsman, R.A. (1986). Global cross sections as a tool for factorization of measures and distribution of maximal invariants. *Sankhyā* (Ser. A) **48**, 1-42.

Wijsman, R.A. (1990). *Invariant Measures on Groups and Their Use in Statistics*. IMS Lecture Notes–Monograph Ser. **14**. IMS, Hayward, CA.

Woods, H., Steinour, H.H. and Starke, H.R. (1932). Effect of composition of Portland cement on heat evolved during hardening. *Industrial and Engineering Chemistry*, **24**, 1207-1214.

Woodward, G., Lange, S.W., Nelson, K.W., and Calvert, H.O. (1941). The acute oral toxocity of acetic, chloracetic, dichloracetic and trichloracetic acids. *J. Industrial Hygiene and Toxicology*, **23**, 78-81.

Young, G.A. and Smith, R.L. (2005). *Essentials of Statistical Inference*. Cambridge Univ. Press, Cambridge, U.K.

Zellner, A. (1971). *An Introduction to Bayesian Inference in Economics*. Wiley, New York.

Zellner, A. (1984). Posterior odds ratios for regression hypotheses: General considerations and some specific results. In: Zellner, A. (ed) *Basic Issues in Econometrics*, 275-305. Univ. of Chicago Press, Chicago.

Zellner, A. (1986). On assessing prior distributions and Bayesian regression analysis with g-prior distributions. In: Goel, P.K. and Zellner, A. (eds) *Basic Bayesian Inference and Decision Techniques: Essays in Honor of Bruno de Finetti*, 233-243. North-Holland, Amsterdam.

Zellner, A. and Siow, A. (1980). Posterior odds ratios for selected regression hypotheses. In: Bernardo, J.M. et al. (eds) *Bayesian Statistics*, 585-603. University Press, Valencia.

Author Index

Subject Index

Springer
the language of science

springeronline.com

Semiparametric Theory and Missing Data

Anastasios A. Tsiatis

This book combines much of what is known in regard to the theory of estimation for semiparametric models with missing data in an organized and comprehensive manner. It starts with the study of semiparametric methods when there are no missing data. The description of the theory of estimation for semiparametric models is at a level that is both rigorous and intuitive, relying on geometric ideas to reinforce the intuition and understanding of the theory. These methods are then applied to problems with missing, censored, and coarsened data with the goal of deriving estimators that are as robust and efficient as possible.

2006. 395 p. (Springer Series in Statistics) Hardcover ISBN 0-387-32448-8

Inference in Hidden Markov Models

O. Cappé, E. Moulines, and T. Rydén

This book is a comprehensive treatment of inference for hidden Markov models, including both algorithms and statistical theory. Topics range from filtering and smoothing of the hidden Markov chain to parameter estimation, Bayesian methods and estimation of the number of states.In a unified way the book covers both models with finite state spaces, which allow for exact algorithms for filtering, estimation etc. and models with continuous state spaces (also called state-space models) requiring approximate simulation-based algorithms that are also described in detail.

2005. 664 p. (Springer Series in Statistics) Hardcover ISBN 0-387-40264-0

Nonparametric Functional Data Analysis

F. Ferraty and P. Vieu

This book links two fields of modern statistics by explaining how functional data can be studied through parameter-free statistical ideas. It starts from theoretical foundations including functional nonparametric modeling, description of the mathematical framework, construction of the statistical methods, and statements of their asymptotic behaviors. It proceeds to computational issues including R and S-PLUS routines.

2006. 296 p. (Springer Series in Statistics) Hardcover ISBN 0-387-30369-3